Lecture Notes in Computer Science 12615

More information about this subseries at http://www.springer.com/series/7409

Madhusudan Singh · Dae-Ki Kang ·
Jong-Ha Lee · Uma Shanker Tiwary ·
Dhananjay Singh · Wan-Young Chung (Eds.)

Intelligent Human Computer Interaction

12th International Conference, IHCI 2020
Daegu, South Korea, November 24–26, 2020
Proceedings, Part I

 Springer

Editors
Madhusudan Singh ⓘ
Woosong University
Daejeon, Korea (Republic of)

Dae-Ki Kang ⓘ
Dongseo University
Busan, Korea (Republic of)

Jong-Ha Lee ⓘ
Keimyung University
Daegu, Korea (Republic of)

Uma Shanker Tiwary ⓘ
Indian Institute of Information Technology
Allahabad, India

Dhananjay Singh ⓘ
Hankuk University of Foreign Studies
Yongin, Korea (Republic of)

Wan-Young Chung ⓘ
Pukyong National University
Busan, Korea (Republic of)

ISSN 0302-9743   ISSN 1611-3349 (electronic)
Lecture Notes in Computer Science
ISBN 978-3-030-68448-8   ISBN 978-3-030-68449-5 (eBook)
https://doi.org/10.1007/978-3-030-68449-5

LNCS Sublibrary: SL3 – Information Systems and Applications, incl. Internet/Web, and HCI

This Springer imprint is published by the registered company Springer Nature Switzerland AG
The registered company address is: Gewerbestrasse 11, 6330 Cham, Switzerland

Preface

The science and technology of Human Computer Interaction (HCI) has taken a giant leap forward in the last few years. This has given impetus to two opposing trends. One divergent trend is to organize separate conferences on focused topics such as Interaction Design, User-Centered Design, etc., which earlier would have been covered under HCI. The other convergent trend is to assimilate new areas into HCI conferences, such as Computing with Words, Prosocial Agents Development, Attention-based Applications, etc. IHCI is one of the rare conferences focusing on those issues of Intelligence and Human Computer Interaction which exist at the crossroads of the abovementioned trends. IHCI is an annual international conference in the Human Computer Interaction field, where we explore research challenges emerging in the complex interaction between machine intelligence and human intelligence. It is a privilege to present the proceedings of the 12th International Conference on Intelligent Human Computer Interaction (IHCI 2020), organized on site and online by the Korea Institute of Convergence Signal Processing (KICSP) during November 24–26, 2020 at EXCO Daegu, South Korea. The twelfth instance of the conference was on the theme of "Intelligent Interaction for Smart Community Services", having 9 special sessions related to the main theme of the conference.

Out of 175 submitted papers, 93 papers were accepted for oral presentation and publication by the Program Committee, in each case based on the recommendations of at least 3 expert reviewers. The proceedings are organized in 9 sections corresponding to the 9 tracks of the conference. The 12th IHCI conference included keynote and invited talks with powerful expert session chairs who have worked in both industry and academia. It attracted more than 200 participants from more than 27 countries.

IHCI has emerged as the foremost worldwide gathering of the field's academic researchers, graduate students, top research think tanks, and industry technology developers. Therefore, we believe that the biggest benefit to the participant is the actualization of their goals in the field of HCI. That will ultimately lead to greater success in business, which is ultimately beneficial to society. Moreover, our warm gratitude should be given to all the authors who submitted their work to IHCI 2020. During the submission, review, and editing stages, the EasyChair conference system proved very helpful. We are grateful to the technical program committee (TPC) and local organizing committee for their immeasurable efforts to ensure the success of this conference. Finally we would like to thank our speakers, authors, and participants for their contribution to making IHCI 2020 a stimulating and productive conference. This IHCI conference series cannot achieve yearly milestones without their continued support in future.

November 2020

Wan-Young Chung
Dhananjay Singh

Organization

General Chairs

Wan-Young Chung	Pukyong National University (PKNU), Busan, Korea
Dhananjay Singh	Hankuk University of Foreign Studies (HUFS), Seoul, Korea

Technical Program Chairs

Uma Shanker Tiwary	IIIT-Allahabad, Allahabad, India
Dae-Ki Kang	Dongseo University, Busan, Korea
Jong-Ha Lee	Keimyung University, Daegu, Korea
Madhusudan Singh	Woosong University, Daejeon, Korea

Steering Committee

Uma Shanker Tiwary	IIIT-Allahabad, India
Santanu Chaudhury	IIT Jodhpur, India
Tom D. Gedeon	Australian National University, Australia
Debasis Samanta	IIT Kharagpur, India
Atanendu Sekhar Mandal	CSIR-CEERI, Pilani, India
Tanveer Siddiqui	University of Allahabad, India
Jaroslav Pokorný	Charles University, Czech Republic
Sukhendu Das	IIT Madras, India
Samit Bhattacharya	IIT Guwahati, India

Special Session Chairs

Uma Shanker Tiwary	IIIT-Allahabad, Allahabad, India
Suzana Brown	The State University of New York, Korea
Mark D. Whitaker	The State University of New York, Korea
Arvind W. Kiwelekar	Dr. Babasaheb Ambedkar Technological University, India
Kenneth A. Yates	University of Southern California, USA
Mohd Helmy Abd Wahab	Universiti Tun Hussein Onn Malaysia, Malaysia
Masoud Mohammadian	University of Canberra, Australia
Eui-Chul Lee	Sangmyung University, Korea
Hakimjon Zaynidinov	Tashkent University of Information Technologies, Uzbekistan
Jan-Willem van 't Klooster	University of Twente, The Netherlands
Thierry Oscar Edoh	University of Bonn, Germany
Zia Uddin	Woosong University, Korea

Muhammad Sohaib	Lahore Garrison University, Pakistan
Jong-Ha Lee	Keimyung University, South Korea
Shyam Perugu	National Institute of Technology Warangal, India
Nagamani M.	University of Hyderabad, India
Irish Singh	Ajou University, Korea

Publicity Chairs

Mario José Diván	National University of La Pampa, Argentina
Amine Chellali	University of Évry Val d'Essonne, France
Nirmalya Thakur	University of Cincinnati, USA

Industry Chairs

Antonio Jara	HOPU, Spain
Sangsu Jung	VESTELLA, Korea
Gyanendra Kumar	infoTrust, Singapore
Prem Singh	COIKOSITY, India

Local Organizing Committee

Daejin Park	Kyungpook National University, Korea
Jonghun Lee	DGIST, Korea
Do-Un Jeong	Dongseo University, Korea
Hoon-Jae Lee	Dongseo University, Korea
Sang-Gon Lee	Dongseo University, Korea
Yeon Ho Chung	Pukyong National University, Korea
Andrew Min-Gyu Han	Hansung University, Korea
Paul Moon Sub Choi	Ewha Womans University, Korea
Jae Hee Park	Keimyung University, Korea
Sukho Lee	Dongseo University, Korea
Sang-Joong Jung	Dongseo University, Korea
Pamul Yadav	GREW Creative Lab, Korea
Hyo-Jin Jung	Daegu Convention and Visitors Bureau, Korea

Technical Program Committee

Jong-Hoon Kim	Kent State University, USA
N. S. Rajput	Indian Institute of Technology (BHU) Varanasi, India
Ho Jiacang	Dongseo University, Korea
Ahmed Abdulhakim Al-Absi	Kyungdong University, Korea
Rodrigo da Rosa Righi	Unisinos, Brazil
Nagesh Yadav	IBM Research, Ireland
Jan Willem van 't Klooster	University of Twente, The Netherlands
Hasan Tinmaz	Woosong University, Korea

Zhong Liang Xiang	Shandong Technology and Business University, China
Hanumant Singh Shekhawat	Indian Institute of Technology Guwahati, India
Md. Iftekhar Salam	Xiamen University, Malaysia
Alvin Poernomo	University of New Brunswick, Canada
Surender Reddy Salkuti	Woosong University, Korea
Suzana Brown	State University of New York, Korea
Dileep Kumar	MR Research, Siemens HealthCare, India
Gaurav Trivedi	Indian Institute of Technology Guwahati, India
Prima Sanjaya	University of Helsinki, Finland
Thierry Oscar Edoh	University of Bonn, Germany
Garima Agrawal	Vulcan AI, Singapore
David (Bong Jun) Choi	Soongsil University, Korea
Gyanendra Verma	NIT Kurukshetra, India
Jia Uddin	Woosong University, Korea
Arvind W. Kiwelekar	Dr. Babasaheb Ambedkar Technological University, India
Alex Wong Ming Hui	Osaka University, Japan
Bharat Rawal	Gannon University, USA
Wesley De Neve	Ghent University Global Campus, Korea
Satish Kumar L. Varma	Pillai College of Engineering, India
Alex Kuhn	State University of New York, Korea
Mark Whitaker	State University of New York, Korea
Satish Srirama	University of Hyderabad, India
Nagamani M.	University of Hyderabad, India
Shyam Perugu	National Institute of Technology Warangal, India
Neeraj Parolia	Towson University, USA
Stella Tomasi	Towson University, USA
Marcelo Marciszack	National Technological University, Argentina
Andrés Navarro Newball	Pontificia Universidad Javeriana Cali, Colombia
Marcelo Marciszack	National Technological University, Argentina
Indranath Chatterjee	Tongmyong University, Korea
Gaurav Tripathi	BEL, India
Bernardo Nugroho Yahya	HUFS, Korea
Carlene Campbell	University of Wales Trinity Saint David, UK

Keynote Speakers

P. Nagabhushan	IIIT-Allahabad, India
Maode Ma	Nanyang Technological University. Singapore
Dugan Um	Texas A&M University, USA
Ajay Gupta	Western Michigan University, USA

Invited Speakers

Yeon-Ho Chung	Pukyong National University, Korea
James R. Reagan	IdeaXplorer, Korea

Mario J. Diván	Universidad Nacional de La Pampa, Argentina
Jae-Hee Park	Keimyung University, Korea
Antonio M. Alberti	Inatel, Brazil
Rodrigo Righi	Unisinos, Brazil
Antonio Jara	HOP Ubiquitous, Spain
Boon Giin Lee	University of Nottingham Ningbo China, China
Gaurav Trivedi	Indian Institute of Technology Guwahati, India
Madhusudan Singh	Woosong University, Korea
Mohd Helmy Abd Wahab	Universiti Tun Hussein Onn Malaysia, Malaysia
Masoud Mohammadian	University of Canberra, Australia
Jan Willem van' t Klooster	University of Twente, The Netherlands
Thierry Oscar Edoh	University of Bonn, Germany

Organizing Chair

Dhananjay Singh	Hankuk University of Foreign Studies (HUFS), Korea

Contents – Part I

Cognitive Modeling and System

A Two-Systems Perspective for Computational Thinking 3
Arvind W. Kiwelekar, Swanand Navandar, and Dharmendra K. Yadav

Detection of Semantically Equivalent Question Pairs 12
*Reetu Kumari, Rohit Mishra, Shrikant Malviya,
and Uma Shanker Tiwary*

Concentration Level Prediction System for the Students Based
on Physiological Measures Using the EEG Device 24
Varsha T. Lokare and Laxman D. Netak

A Correlation Analysis Between Cognitive Process and Knowledge
Dimension in Software Engineering by Using the Revised
Bloom's Taxonomy... 34
Manjushree D. Laddha, Laxman D. Netak, and Hansaraj S. Wankhede

A Study on Comparative Analysis of the Effect of Applying DropOut
and DropConnect to Deep Neural Network...................... 42
Hyun-il Lim

HHAR-net: Hierarchical Human Activity Recognition using
Neural Networks... 48
*Mehrdad Fazli, Kamran Kowsari, Erfaneh Gharavi, Laura Barnes,
and Afsaneh Doryab*

Analysis of Streaming Information Using Subscriber-Publisher Architecture... 59
Aparajit Talukdar

Fraudulent Practices in Dispensing Units and Remedies 70
Undru Vimal Babu, M. Nagamani, Shalam Raju, and M. Rama Krishna

Biomedical Signal Processing and Complex Analysis

Digital Processing of Blood Image by Applying Two-Dimensional
Haar Wavelets 83
H. N. Zaynidinov, I. Yusupov, J. U. Juraev, and Dhananjay Singh

Development of the Method, Algorithm and Software of a Modern
Non-invasive Biopotential Meter System   95
   *J. X. Djumanov, F. F. Rajabov, K. T. Abdurashidova, D. A. Tadjibaeva,
and N. S. Atadjanova*

Deception Detection Using a Multichannel Custom-Design EEG System
and Multiple Variants of Neural Network.   104
   Ngoc-Dau Mai, Trung-Hau Nguyen, and Wan-Young Chung

A Study on the Possibility of Measuring the Non-contact Galvanic Skin
Response Based on Near-Infrared Imaging   110
   Geumbi Jo, Seunggeon Lee, and Eui Chul Lee

EEG Motor Classification Using Multi-band Signal and Common
Spatial Filter.   120
   *Tan Yu Xuan, Norashikin Yahya, Zia Khan, Nasreen Badruddin,
and Mohd Zuki Yusoff*

Segmentation of Prostate in MRI Images Using Depth Separable
Convolution Operations   132
   Zia Khan, Norashikin Yahya, Khaled Alsaih, and Fabrice Meriaudeau

Bone Age Assessment for Lower Age Groups Using Triplet Network
in Small Dataset of Hand X-Rays   142
   Shipra Madan, Tapan Gandhi, and Santanu Chaudhury

A Development of Enhanced Contactless Bio Signal Estimation Algorithm
and System for COVID19 Prevention   154
   Chan-il Kim and Jong-ha Lee

Stress Detection from Different Environments for VIP Using EEG Signals
and Machine Learning Algorithms   163
   *Mohammad Safkat Karim, Abdullah Al Rafsan,
Tahmina Rahman Surovi, Md. Hasibul Amin,
and Mohammad Zavid Parvez*

Natural Language, Speech, Voice and Study

Analysis of Emotional Content in Indian Political Speeches   177
   Sharu Goel, Sandeep Kumar Pandey, and Hanumant Singh Shekhawat

A Bengali Voice-Controlled AI Robot for the Physically Challenged.   186
   *Abul Bashar Bhuiyan, Anamika Ahmed, Sadid Rafsun Tulon,
Md. Rezwan Hassan Khan, and Jia Uddin*

How to Enhance the User Experience of Language Acquisition
in the Mobile Environment: A Case Study of Amkigorae(암기고래),
a Vocabulary Acquisition Mobile Application...................... 195
    Chiwon Lee, Donggyu Kim, Eunsuh Chin, and Jihyun Kim

Screening Trauma Through CNN-Based Voice Emotion Classification...... 208
    Na Hye Kim, So Eui Kim, Ji Won Mok, Su Gyeong Yu, Na Yeon Han,
    and Eui Chul Lee

Interchanging the Mode of Display Between Desktop and Immersive
Headset for Effective and Usable On-line Learning.................. 218
    Jiwon Ryu and Gerard Kim

Verification of Frequently Used Korean Handwritten Characters Through
Artificial Intelligence.. 223
    Kyung Won Jin, Mi Kyung Lee, Woohyuk Jang, and Eui Chul Lee

Study of Sign Language Recognition Using Wearable Sensors........... 229
    Boon Giin Lee and Wan Young Chung

Automated Grading of Essays: A Review............................ 238
    Jyoti G. Borade and Laxman D. Netak

Voice Attacks to AI Voice Assistant.............................. 250
    Seyitmammet Alchekov Saparmammedovich,
    Mohammed Abdulhakim Al-Absi, Yusuph J. Koni, and Hoon Jae Lee

Skills Gap is a Reflection of What We Value: A Reinforcement
Learning Interactive Conceptual Skill Development Framework
for Indian University... 262
    Pankaj Velavan, Billy Jacob, and Abhishek Kaushik

PRERONA: Mental Health Bengali Chatbot for Digital Counselling....... 274
    Asma Ul Hussna, Azmiri Newaz Khan Laz, Md. Shammyo Sikder,
    Jia Uddin, Hasan Tinmaz, and A. M. Esfar-E-Alam

Speech Based Access of Kisan Information System in Telugu Language 287
    Rambabu Banothu, S. Sadiq Basha, Nagamani Molakatala,
    Veerendra Kumar Gautam, and Suryakanth V. Gangashetty

Voice Assistant for Covid-19.................................... 299
    Shokhrukhbek Primkulov, Jamshidbek Urolov, and Madhusudan Singh

Combining Natural Language Processing and Blockchain for Smart
Contract Generation in the Accounting and Legal Field............... 307
    Emiliano Monteiro, Rodrigo Righi, Rafael Kunst, Cristiano da Costa,
    and Dhananjay Singh

Algorithm and Related Applications

Fault Identification of Multi-level Gear Defects Using Adaptive Noise
Control and a Genetic Algorithm   325
   Cong Dai Nguyen, Alexander Prosvirin, and Jong-Myon Kim

Applying Multiple Models to Improve the Accuracy of Prediction
Results in Neural Networks   336
   Hyun-il Lim

OST-HMD for Safety Training   342
   Christopher Koenig, Muhannad Ismael, and Roderick McCall

One-Dimensional Mathematical Model and a Numerical Solution
Accounting Sedimentation of Clay Particles in Process of Oil Filtering
in Porous Medium.   353
   Elmira Nazirova, Abdug'ani Nematov, Rustamjon Sadikov,
   and Inomjon Nabiyev

A Novel Diminish Smooth L1 Loss Model with Generative
Adversarial Network   361
   Arief Rachman Sutanto and Dae-Ki Kang

Interactive Machine Learning Approach for Staff Selection Using
Genetic Algorithm.   369
   Preethi Ananthachari and Nodirbek Makhtumov

Software of Linear and Geometrically Non-linear Problems Solution Under
Spatial Loading of Rods of Complex Configuration.   380
   Sh. A. Anarova, SH. M. Ismoilov, and O. Sh. Abdirozikov

Mathematical Modeling of Pascal Triangular Fractal Patterns
and Its Practical Application.   390
   Sh. A. Anarova, Z. E. Ibrohimova, O. M. Narzulloyev,
   and G. A. Qayumova

Crowd Sourcing and Information Analysis

An NLP and LSTM Based Stock Prediction and Recommender System
for KOSDAQ and KOSPI   403
   Indranath Chatterjee, Jeon Gwan, Yong Jin Kim, Min Seok Lee,
   and Migyung Cho

The Commodity Ecology Mobile (CEM) Platform Illustrates Ten Design
Points for Achieving a Deep Deliberation in Sustainable Development
Goal #12   414
   Mark D. Whitaker

The Design and Development of School Visitor Information Management
System: Malaysia Perspective   431
    Check-Yee Law, Yong-Wee Sek, Choo-Chuan Tay, Wei-Ann Lim,
    and Tze-Hui Liew

The Impact of the Measurement Process in Intelligent System of Data
Gathering Strategies   445
    Mario José Diván and Madhusudan Singh

Detecting Arson and Stone Pelting in Extreme Violence: A Deep Learning
Based Identification Approach   458
    Gaurav Tripathi, Kuldeep Singh, and Dinesh Kumar Vishwakarma

A Scientometric Review of Digital Economy for Intelligent
Human-Computer Interaction Research   469
    Han-Teng Liao, Chung-Lien Pan, and Jieqi Huang

eGovernance for Citizen Awareness and Corruption Mitigation   481
    A. B. Sagar, M. Nagamani, Rambabu Banothu, K. Ramesh Babu,
    Venkateswara Rao Juturi, and Preethi Kothari

Towards a Responsible Intelligent HCI for Journalism: A Systematic
Review of Digital Journalism   488
    Yujin Zhou and Zixian Zhou

A Systematic Review of Social Media for Intelligent Human-Computer
Interaction Research: Why Smart Social Media is Not Enough   499
    Han-Teng Liao, Zixian Zhou, and Yujin Zhou

Author Index   511

Contents – Part II

Intelligent Usability and Test System

IoT System for Monitoring a Large-Area Environment Sensors and Control
Actuators Using Real-Time Firebase Database   3
  Giang Truong Le, Nhat Minh Tran, and Thang Viet Tran

A Method for Localizing and Grasping Objects in a Picking Robot System
Using Kinect Camera   21
  Trong Hai Nguyen, Trung Trong Nguyen, and Thang Viet Tran

A Comparative Analysis on the Impact of Face Tracker and Skin
Segmentation onto Improving the Performance of Real-Time
Remote Photoplethysmography   27
  *Kunyoung Lee, Kyungwon Jin, Youngwon Kim, Jee Hang Lee,
  and Eui Chul Lee*

Fuzzy-PID-Based Improvement Controller for CNC Feed Servo System   38
  Nguyen Huu Cuong, Trung Trong Nguyen, and Tran Viet Thang

Plastic Optical Fiber Sensors Based on in-Line Micro-holes: A Review   47
  Hyejin Seo and Jaehee Park

Long-Distance Real-Time Rolling Shutter Optical Camera Communication
Using MFSK Modulation Technique   53
  *Md Habibur Rahman, Mohammad Abrar Shakil Sejan,
  and Wan-Young Chung*

A Systematic Review of Augmented Reality in Multimedia Learning
Outcomes in Education   63
  *Hafizul Fahri Hanafi, Mohd Helmy Abd Wahab, Abu Zarrin Selamat,
  Abdul Halim Masnan, and Miftachul Huda*

RFID Technology for UOH Health Care Center   73
  *Velugumetla Siddhi Chaithanya, M. Nagamani, Venkateswara Juturi,
  C. Satyanarayana Prasad, and B. Sitaram*

An IOT Based Smart Drain Monitoring System with Alert Messages.   84
  *Samiha Sultana, Ananya Rahaman, Anita Mahmud Jhara,
  Akash Chandra Paul, and Jia Uddin*

Comparison of SVM and Random Forest Methods for Online
Signature Verification   288
  Leetesh Meena, Vijay Kumar Chaurasiya, Neetesh Purohit,
  and Dhananjay Singh

Human-Centered AI Applications

Audio Augmented Reality Using Unity for Marine Tourism   303
  Uipil Chong and Shokhzod Alimardanov

A Prototype Wristwatch Device for Monitoring Vital Signs Using
Multi-wavelength Photoplethysmography Sensors   312
  Nguyen Mai Hoang Long, Jong-Jin Kim, and Wan-Young Chung

User Perception on an Artificial Intelligence Counseling App   319
  Hyunjong Joo, Chiwon Lee, Mingeon Kim, and Yeonsoo Choi

Authentication of Facial Images with Masks Using Periocular Biometrics . . .   326
  Na Yeon Han, Si Won Seong, Jihye Ryu, Hyeonsang Hwang,
  Jinoo Joung, Jeeghang Lee, and Eui Chul Lee

Analysis of User Preference of AR Head-Up Display Using Attrakdiff   335
  Young Jin Kim and Hoon Sik Yoo

IoT-Enabled Mobile Device for Electrogastrography Signal Processing   346
  Hakimjon Zaynidinov, Sarvar Makhmudjanov, Farkhad Rajabov,
  and Dhananjay Singh

A Study on the Usability Test Method of Collaborative Robot Based
on ECG Measurement   357
  Sangwoo Cho and Jong-Ha Lee

The Human Factor Assessment of Consumer Air Purifier Panel Using Eye
Tracking Device   363
  Shin-Gyun Kim and Jong-Ha Lee

AI-Based Voice Assistants Technology Comparison in Term
of Conversational and Response Time   370
  Yusuph J. Koni, Mohammed Abdulhakim Al-Absi,
  Seyitmammet Alchekov Saparmammedovich, and Hoon Jae Lee

HCI Based In-Cabin Monitoring System for Irregular Situations
with Occupants Facial Anonymization   380
  Ashutosh Mishra, Jaekwang Cha, and Shiho Kim

Achievement of Generic and Professional Competencies Through
Virtual Environments.   391
    Zhoe Comas-Gonzalez, Ronald Zamora-Musa, Orlando Rodelo Soto,
    Carlos Collazos-Morales, Carlos A. Sanchez, and Laura Hill-Pastor

AARON: Assistive Augmented Reality Operations and Navigation System
for NASA's Exploration Extravehicular Mobility Unit (xEMU).   406
    Irvin Steve Cardenas, Caitlyn Lenhoff, Michelle Park, Tina Yuqiao Xu,
    Xiangxu Lin, Pradeep Kumar Paladugula, and Jong-Hoon Kim

Socio-Cognitive Interaction Between Human and Computer/Robot
for HCI 3.0   423
    Sinae Lee, Dugan Um, and Jangwoon Park

Author Index   433

highlights the necessity of a rich framework grounded in Psychological theories to analyze the influences of these traits on CT.

Hence this position paper suggests adopting one such theory i.e. Kahneman's two-systems model of thinking, from the field of Psychology, to analyze the human aspects involved in computational thinking. The article defines Computational Thinking and elements of Kahneman's two systems model of thinking in Sects. 2, 3, and 4. Section 5 identifies CT activities and map them on Kahneman's two systems model of thinking. Section 6 relates our proposal with the existing applications of Kahneman's model of thinking in Computer Science. The paper concludes with directions for future work.

2   Computational Thinking (CT)

In the seminal paper on *Computational Thinking*, J M Wing [18] clearly explains the breadth of the CT as a thought process. This broad definition of CT includes a set of skills such as solving a problem using computers, designing and evaluating a complex system, and understanding human behaviour.

For computer programmers, CT is a way of representing a real-life problem and solving it with the help of computers. For example, writing a program to find an optimal path to travel from a source to a destination.

For software engineers, CT refers to designing and evaluating a complex information processing system such as an online railway or an airline reservation system.

For computer scientists, CT refers to getting insights about human behaviour by answering questions such as (i) *what are the limitations and power of computation?*, (ii) *what does it mean by intelligence?*, (iii) *what motivates us as a human being to perform or not to perform a specific action?*.

This broad coverage of topics included under computational thinking highlights that computational thinking is beyond mere computer programming or coding.

Further, Aho [1] brings out the differences between the terms *Computation* and *Computational Thinking*. He recommends that the term *Computation* shall be restricted to denote those tasks for which the semantics can be described through a formal mathematical model of computation (e.g., Finite Automata, Pi-Calculus, Turing Machine). The tasks for which no such appropriate models exist, it is necessary to invent such formal models.

Computational thinking being a complex thought process, the paper proposes to analyze it through the cognitive model of thinking propagated by Psychologist, Economist, and Nobel Laureate Prof. Daniel Kahneman. Though the Kahneman's model of thinking is not a formal model useful to describe exact semantics of CT activities, the cognitive model helps us to fix the errors in our reasoning and to sharpen our thought process.

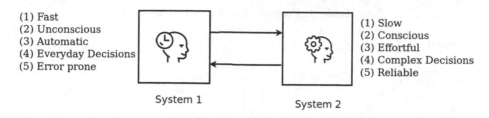

(1) Fast
(2) Unconscious
(3) Automatic
(4) Everyday Decisions
(5) Error prone

(1) Slow
(2) Conscious
(3) Effortful
(4) Complex Decisions
(5) Reliable

System 1   System 2

Fig. 1. Kahneman's two-systems model

3   Kahneman's Two-Systems Model of Thinking

The human cognitive processes, such as judgement and decision making, are complex and intricate. To understand these processes in a better and simplified way, many psychologists have proposed that human thinking operates at two different levels [7]. First one is a fast, intuitive, and effortless way of thinking requiring less or no attention. The second one is a slow, intentional, and effortful way of thinking, often requiring forceful attention or focus. The theories accepting this separation are also known as dual-system or dual-process theories of human cognition. In this paper, this model is referred as a Kahneman's two-systems model of thinking because non-psychologists [11,14,15] have started using it to understand the cognitive processes involved in their domains after the publication of the book titled *Thinking, Fast and Slow* [10].

Kahneman's model primarily consists of two systems, as shown in Fig. 1. These systems are labelled as *System 1* and *System 2*. These systems can be considered as mental constructs or fictitious agents driving the thought process. The *System 1* usually responds to routine operations quickly in an unconscious way while *System 2* is called in action in a novel situation, responds consciously and slowly. In comparison with *System 2*, the *System 1* is more error-prone and unreliable. The following examples illustrate the existence of such two different modes of thinking in the context of Basic Algebra.

Example 1. We answer the question *2 + 2 = ?* without any effort, quickly, and accurately. But, to answer the question *17 X 24 =?*, we require to put effort and do deliberate calculations.

Example 2. Consider the following example from Kahneman's book [10] page 44:

> A bat and ball cost $1.10. The bat costs one dollar more than the ball. How much does the ball cost?

Most of the people answer the question as 10¢ using their intuitive thinking (*System 1*) which is a wrong answer, while the correct answer is 5¢.

This example is purposefully designed to demonstrate that *System 1* is error-prone.

Table 1. Program 1 and Program 2

Program 1	Program 2

```int main ()```
 ```{```
 ```int x=10, y=20;```
 	
 ```int temp = x;```
 ```x = y;```
 ```y = temp;```
 	
 ```printf("x= %d,y= %d", x,y);```
 ```return 0;```
 ```}```
	```int main ()```
	 ```{```
	 ```int x = 10, y = 20;```
	
	 ```x = x + y;```
	 ```y = x - y;```
	 ```x = x - y;```
	
	 ```printf("x= %d,y= %d", x,y);```
	 ```return 0;```
	 ```}```

Example 3. Let us consider the third example. This is a question asked in the end-semester examinations of First course on Computer Programming (C Programming) offered at Dr. B. A. Tech. University India to its undergraduate students of Engineering.

Question: Which of the program(s) shown in Table 1 swaps the values of the variables x and y?

(A) Program 1
(B) Program 2
(C) None of the programs Program 1 or Program 2.
(D) Both Program 1 and Program 2.

While answering this question, majority of the students (77%) out of the eighty three students enrolled for the course have answered it as (A) i.e. Program 1 while the correct answer is (D) i.e. Both Program 1 and Program 2. This example demonstrates that, most of the students relied on their *System 1* while answering the question.

4   Characteristics of Dual System Thinking

The examples in the previous section demonstrate that thinking happens in two different modes. To elaborate the working further, we describe some of the characteristics crucial for understanding computational thinking. This section describes these characteristics in a general setting. However, these characteristics also apply in the context of computational thinking in its broader sense when we presume human as a computing agent.

1. **Both systems work concurrently.** The *System 1* and *System 2* work concurrently and cooperate most of the times while thinking and reasoning about the external world. When conflict arises, *System 2* attempts to regulate *System 1*. The *System 1* generates feelings, impressions, and inclinations. When asked by *System 1*, the *System 2* endorses and transforms them into beliefs and attitudes. Neither system can be turned off but *System 2* is lazy to respond as compared to *System 1* .

2. **The two-systems model represents a division of labour.** The *System 1* performs some tasks efficiently while others are performed by *System 2*. The *System 1* performs tasks such as: (i) To execute skilled responses after imparting proper training. For example, applying brakes while driving. (ii) Recognizing typical situations, recognizing norms and standards and complying to conventions. For instance, recognizing irritations in the voice of a known person. (iii) To identify causes and intentions. For example, identifying reasons behind delayed arrival of an aircraft.
   The *System 2* is called in action when a task requires more effort and attention, such as filling a form to apply for graduate studies or selecting a University for graduate studies.

3. **Biases and heuristics guide the *System 1* responses.** The responses of *System 1* are quick and error-prone. They are quick because heuristics drives them and they are error-prone because biases guide them. For example, we often quickly judge the level of confidence of a person through the external attributes such as being well-dressed and well-groomed, which is an instance of the use of a heuristic called *halo effect*. For example, the Example 3 from the previous section, students who responded with the wrong option (A), they relied on a bias called *availability bias*. The *availability bias* selects a familiar and widely exposed option over the least exposed and un-familiar one. The option (A) fulfills this criteria and majority of the students select it.

These characteristics play a significant role in understanding human thought process in general and computational thinking in our case.

5   Two-Systems Model and Computational Thinking

We often consider computational thinking is a deliberate thought process driven by the goals to be achieved. So it is a slow and effortful activity requiring a high level of focus and attention. Hence, we may conclude that computational thinking is a domain of *System 2*, and there is no or minimal role to play for *System 1*. In this section, we hypothesize that two systems govern computational thinking. First one is fast, automatic and intuitional. The second one is slow, effortful, and systematic.

To support our argument, we identify smaller and primitive computational thinking activities and map them on *System 1* and *System 2*. Table 2 shows some of the computational thinking activities mapped to *System 1* and *System 2*.

While defining this mapping, we assumed a minimum level of knowledge and skills that students acquire after the courses on Computer Programming, Data

7 Conclusion

The paper identifies the necessity of investigating the psychological dimension of Computational Thinking skill. Further, it proposes to adopt Kahneman's Two-systems model of thinking for this purpose because it is simple to utilise, and it is rich enough in terms of analytical tools. Primarily, it separates the human thought process in two broad categories: (i) Fast and intuitional activities, and (ii) Slow and deliberate one.

The paper illustrates the applicability of the approach by mapping CT activities on two systems requiring fast and slow thinking as a baseline for further empirical investigation. The identified mapping needs to be substantiated by carrying out either controlled experiments or through the detailed analyses of students' responses in an educational setting.

Kahneman's two-systems model is rich as a cognitive analysis framework providing a broad set of biases and heuristics, which can be used to study the human aspects of computational thinking. It will be interesting to explore the role of these biases and heuristics in the context of Computational Thinking to make it less error-prone and a faster reasoning activity.

Acknowledgement. The first author would like to acknowledge his younger brother Mr. Nagesh Kiwelekar for inspiring him to explore the connections between dual system theories and Computer Science.

References

1. Aho, A.V.: Computation and computational thinking. Comput. J. **55**(7), 832–835 (2012)
2. Behimehr, S., Jamali, H.R.: Cognitive biases and their effects on information behaviour of graduate students in their research projects. J. Inf. Sci. Theory Practice **8**(2), 18–31 (2020)
3. Çalıklı, G., Bener, A.B.: Influence of confirmation biases of developers on software quality: an empirical study. Softw. Qual. J. **21**(2), 377–416 (2013)
4. Chen, D., Bai, Y., Zhao, W., Ament, S., Gregoire, J.M., Gomes, C.P.: Deep reasoning networks: thinking fast and slow, for pattern de-mixing (2019)
5. Csernoch, M.: Thinking fast and slow in computer problem solving. J. Softw. Eng. Appl. **10**(1), 11–40 (2017)
6. Denning, P.J., Tedre, M.: Computational Thinking. MIT Press, Cambridge (2019)
7. Evans, J.S.B., Stanovich, K.E.: Dual-process theories of higher cognition: advancing the debate. Perspect. Psychol. Sci. **8**(3), 223–241 (2013)
8. Goel, G., Chen, N., Wierman, A.: Thinking fast and slow: optimization decomposition across timescales. In: 2017 IEEE 56th Annual Conference on Decision and Control (CDC), pp. 1291–1298. IEEE (2017)
9. Iyer, S.: Teaching-learning of computational thinking in k-12 schools in India. Comput. Thinking Educ. p. 363 (2019)
10. Kahneman, D.: Thinking, Fast and Slow. Macmillan, New York (2011)
11. Kannengiesser, U., Gero, J.S.: Design thinking, fast and slow: a framework for Kahneman's dual-system theory in design. Des. Sci. **5** (2019)

12. Mittal, S., Joshi, A., Finin, T.: Thinking, fast and slow: Combining vector spaces and knowledge graphs. arXiv preprint arXiv:1708.03310 (2017)
13. Mohanani, R., Salman, I., Turhan, B., Rodríguez, P., Ralph, P.: Cognitive biases in software engineering: a systematic mapping study. IEEE Trans. Softw. Eng. (2018)
14. Murdock, C.W., Sullivan, B.: What Kahneman means for lawyers: Some reflections on thinking, fast and slow. Loy. U. Chi. LJ **44**, 1377 (2012)
15. Preisz, A.: Fast and slow thinking; and the problem of conflating clinical reasoning and ethical deliberation in acute decision-making. J. Paediatrics Child Health **55**(6), 621–624 (2019)
16. Rossi, F., Loreggia, A.: Preferences and ethical priorities: thinking fast and slow in AI. In: Proceedings of the 18th International Conference on Autonomous Agents and MultiAgent Systems, pp. 3–4 (2019)
17. Rüping, A.: Taming the biases: a few patterns on successful decision-making. In: Proceedings of the 19th European Conference on Pattern Languages of Programs, pp. 1–5 (2014)
18. Wing, J.M.: Computational thinking. Commun. ACM **49**(3), 33–35 (2006)
19. Zalewski, A., Borowa, K., Kowalski, D.: On cognitive biases in requirements elicitation. In: Jarzabek, S., Poniszewska-Marańda, A., Madeyski, L. (eds.) Integrating Research and Practice in Software Engineering. SCI, vol. 851, pp. 111–123. Springer, Cham (2020). https://doi.org/10.1007/978-3-030-26574-8_9

Detection of Semantically Equivalent Question Pairs

Reetu Kumari, Rohit Mishra(✉) ⓘ, Shrikant Malviya ⓘ,
and Uma Shanker Tiwary ⓘ

Department of Information Technology, Indian Institute of Information Technology,
Allahabad, Prayagraj 211012, India
reetu.iiita2018@gmail.com, rohit129iiita@gmail.com,
s.kant.malviya@gmail.com, ustiwary@gmail.com

Abstract. Knowledge sharing platforms like Quora have millions and billions of questions. With such a vast number of questions, there will be a lot of duplicates in it. Duplicate questions in these sites are normal, especially with the increasing number of questions asked. These redundant queries reduce efficiency and create repetitive data on the data server. Because these questions have the same answers, the user has to write the same content for each of these questions, which is a waste of time. Dealing with this issue would be significant for helping community question answering websites to sort out this problem and deduplicate their database. In this paper, we augment the Siamese-LSTM in two ways to achieve better results than the previous works. First, we augment basic Siamese-LSTM with a dense-layer (Model-1) to observe the improvement. In the second part, Siamese-LSTM is augmented with the machine learning classifier (Model-2). In both scenarios, we observed the improved results when we include the Hand-Engineered features. The proposed model (Model-1) achieves the highest accuracy of 89.11%.

Keywords: LSTM · Word Embedding · Hand engineered features · Deep learning

1 Introduction

Estimating the likenesses between two sentences is an important task in natural language processing. In NLU, It is considered as a hard AI problem. Duplicate question detection is somehow similar to sentence similarity tasks. Detecting duplicate questions on any knowledge sharing platform is a challenging problem because one can write the same question in lots of different ways with different sentence structuring. Many questions have been daily asked on quora. If the user puts duplicate questions on quora and ask people to again answer it, as a new question, then it is just a waste of time and effort of those people who are providing answers. So it makes sense for quora to merge duplicate questions.

R. Kumari and R. Mishra—Equally contributed.

© Springer Nature Switzerland AG 2021
M. Singh et al. (Eds.): IHCI 2020, LNCS 12615, pp. 12–23, 2021.
https://doi.org/10.1007/978-3-030-68449-5_2

If two questions are differently worded but roughly or mostly the same, then as soon as somebody asks this question then it is necessary for quora to understand that these two questions are the same and connect both questions so that all answers of the previous question also become relevant to this one.

There is lots of research that has been done in this area. Hand feature engineering is the most popular. In this paper, a globally released quora question pair dataset [9] is used, which is downloaded from Kaggle[1]. The dataset contains a binary class label is_duplicate, which means this problem is solved by supervised algorithms. Finding duplicate questions will help these forums to suggest similar questions to the users instead of posing a new one.

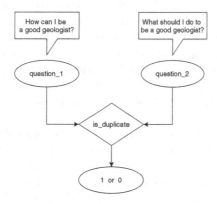

Fig. 1. Question Similarity as a binary classfication problem

1.1 Problem Definition

This is a binary classification problem, for a given pair of questions, the task is to determine if the given pair is duplicate or not, as shown in Fig. 1. In a more formal way:

$$f(q_1, q_2) \rightarrow 0/1$$

Here f is a model-learned function. 0 and 1 is the output of the model where 0 represents non-duplicate question pair and 1 represents a duplicate question pair.

2 Related Work

To solve the question duplication problem, various models have been proposed by the researchers. Feature engineering based models are very popular in machine learning [12, 18]. But nowadays deep learning models are trending in the NLP

---

[1] Quora Question Pairs: https://www.kaggle.com/c/quora-question-pairs.

Dataset contains the following fields:

1. *id*: Each question pair has a unique id.
2. *qid1*: Unique id of first question.
3. *qid2*: Unique id of second question.
4. *question1*: English text representation of first question.
5. *question2*: English text representation of second question.
6. *is_duplicate*: This is a class label that contains two values. If a pair of questions is the same then this field has a value of 1 otherwise 0.

4   Preprocessing

Preprocessing of the text data is the very beginning and important step for solving any NLP problem. In this step, different writing forms, which represent the same thing are unified (i.e., all the same things are written in one way). It is important so that the model should not learn differently for the same things written in different notations which results in underfitting. Preprocessing involves the following operations:

1. Special characters removal
2. Unified different interpretations
3. Converting sentences into lower case
4. Lemmatization

These all operations are mainly used for removing the noise from the data. Special characters such as !, #, %, &, @ etc., are removed from the question during preprocessing steps. Stop words are not removed from the training corpus. In the lemmatization process, a word is converted into its base form. There are few question fields that contain NaN (Not-a-Number) values. NaN is replaced with an empty string.

5   Experiments

5.1   Embedding

GLOVE(300D)[3] embedding [16] is used for vector representation of questions. An embedding matrix is created by using a glove embedding file[4]. In the embedding matrix, Index 0 and 1 are reserved for *STOP* and *UNK* words. Embedding vector of a word is placed in the $(i + 2)$ position in the matrix, where i is the corresponding word index in the dictionary. This embedding matrix is passed to LSTM, and trainable parameter is set to *false* in the embedding layer, because pre-trained embedding is used for question embedding, so there is no need for training.

---

[3] GLOVE: Global Vectors for Word Representation.
[4] GLOVE Repository: https://github.com/stanfordnlp/GloVe.

5.2   Hand-Engineered Features

The following Hand-Engineered features and Fuzzy-based features [17] are extracted with the help of the python *spaCy*[5] library.

- *Fuzzy ratio*: This function internally uses Levenshtein distance. It gives an integer value between 0 to 100 and can be considered as a similarity score.
- *Partial fuzzy ratio*: It partially checks the string match and gives us a score between 0 to 100.
- *Token sort ratio*: It captures the out of order string match. It sorts both strings and matches the similarity using fuzzy ratio. Its output is an integer value between 0 to 100.
- *Token set ratio*: It is similar but more adjustable where before proceeding to compare, the string is tokenize and splitted into two groups (i.e., sorted intersection and sorted remainder). Its output in the range 0–100.
- *Partial token sort ratio*: Partial token sort ratio sorts both strings and then check similarity using a fuzzy partial ratio. It gives the integer value between 0 to 100.
- *Partial token set ratio*: Partial token set ratio and token set ratio is almost the same. There is only one difference: it uses a partial set ratio.
- *Longest substring ratio*: It is the ratio of the longest common substring to the minimum of question-1 and question-2 length.
- *W-ratio*: It uses a different algorithm and returns us an integer value between 0 to 100.
- *Common word count (min)*: It is ratio of common word count to the minimum of question-1 words and question-2 words length.
- *Common word count (max)*: It is the ratio of the common word count to the maximum of question-1 words and question-2 words length.
- *Common stop word count (min)*: It is the ratio of common stop word count to the minimum of question-1 stop words and question-2 stop word length.
- *Common stop word count (max)*: It is the ratio of common stop word count to the maximum of question-1 stop words and question-2 stop words length.
- *Common token count (min)*: It is the ratio of common token count to the minimum of question-1 tokens and question-2 tokens length.
- *Common token count (max)*: It is ratio of common token count to the maximum of question-1 tokens and question-2 tokens length.
- *Absolute length difference*: Absolute length difference of question-1 and question-2 tokens.
- *Mean length*: It is average of question-1 and question-2 tokens.
- *Last word equal*: The last word of both questions are equal or not equal.
- *First word equal*: The first word of both questions are equal or not equal.

---

[5] spaCy: https://spacy.io/.

5.3   Siamese LSTM

Siamese LSTM network[6] is proposed for signature verification tasks [3]. It contains two subnetworks with the same parameter values. Each network receives different input and gives us an encoding vector representation for both inputs. These vectors are used for further processing to predict the final output. This network was first proposed by Bromley in 1994 [3].

5.4   Proposed Model-1 (SiameseLSTM + Dense Layer)

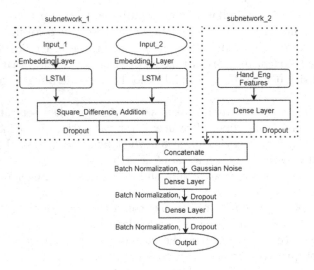

Fig. 2. SiameseLSTM + Dense Layer (Model-1)

This model is a combination of two subnetworks as shown in Fig. 2, the first network is based on the Siamese LSTM network, in which a shared LSTM is used to encode both questions. Word embedding features are used in the input layer to get final representation for questions. Two higher level vector representations $h_*^{(a)}$ and $h_*^{(b)}$ is obtained from the shared LSTM layer, then two commutative operation, addition and square difference is performed using these vectors, and the following vector is used for further processing.

$$v = \left[\left(h_*^{(a)} - h_*^{(b)} \right)^2 h_*^{(a)} + h_*^{(b)} \right] \tag{1}$$

This is passed through a dropout layer. Let's assume this as a subnetwork_1 output. And In subnetwork_2, Hand engineered features are passed through a

---

[6] "Siamese-LSTM" is a neural-network framework consists of two identical subnetworks (LSTMs) joined at their outputs.

dense layer. Concatenation of subnetwork_1 output and subnetwork_2 output is then passed through some dense layers for final class label predictions. Swish activation function is used in the dense layer and sigmoid activation for final output prediction.

5.5 Proposed Model-2 (SiameseLSTM + Machine Learning Classifier)

This architecture (Fig. 3) is almost the same as previous Siamese LSTM with dense-layer (model-1), instead of passing the concatenation layer output to the dense layers, we passed this layer output to a machine learning classifier, i.e. SVM, Logistic Regression or XGBoost[7].

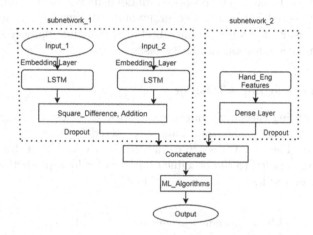

Fig. 3. SiameseLSTM + Machine Learning classifier (Model-2)

6   Results and Discussion

The proposed model is trained with Hand-Engineered features or without Hand-Engineered features. Also In another set of experiments, machine learning classifier are built on top of Siamese LSTM. 3,63,861 data is used for training and the remaining 40,429 is for testing the model. The experiment is performed on UBUNTU, which has 32 GB RAM and 16 cores. It takes about 30–40 min to run each experiment. All codes are written in python version 3.7.

---

[7] XGBoost: e**X**treme **G**radient **B**oosting.

6.1   Siamese LSTM + Dense Layer (Model-1)

Table 2. Siamese LSTM +Dense layer result (Model-1)

Model	Hand Engineered Features	Train_Loss	Test_Loss	Accuracy (%)
LSTM	Yes	0.1717	0.2647	**89.11%**
LSTM	No	0.2276	0.3063	86.91%

It can be clearly observed that the model with Hand engineered Features are giving better results as compared to the model without hand engineered features as shown in below Table 2. The proposed model achieves an accuracy of 89.11% with Hand-engineered features and accuracy of 86.91% without hand engineered features. This result shows that siameseLSTM with hand-engineered features is effective in semantic question similarity tasks.

6.2   Machine Learning Classifier

For machine learning classifier models, some basic features and fuzzy-based features [17] are extracted. The same hand-engineered features are used for deep learning models. These features are combined with GLOVE vector representation of questions and fed to the baseline classifiers for final predictions. Machine learning classifier results are shown in Table 3.

Table 3. Machine learning classifier model result

Model	Hand engineered features	Train_Loss	Test_Loss	Accuracy (%)
Logistic Regression	Yes	0.4451	0.4495	77.48
SVM	Yes	0.4883	0.4901	74.95
XGBoost	Yes	0.3497	0.3566	82.17

6.3   Siamese LSTM + Machine Learning Classifier (Model-2)

In this architecture commutative operation is performed on the high-level representation of question embedding, then this commutative vector result is concatenated with hand engineered features and fed to Machine Learning Models to predict the final output. The experiment shows that machine learning models built on top of siamese LSTM outperforms the machine learning models. Table 4 shows the experiment results of this architecture.

Table 4. Siamese LSTM + Machine learning classifier model Result (Model-2)

Model	Hand Engineered Features	Train_Loss	Test_Loss	Accuracy (%)
Logistic Regression	Yes	0.2102	0.3105	86.66
SVM	Yes	0.1992	0.3139	86.77
XGBoost	Yes	0.2549	0.3075	86.30

6.4   Some Previous Study Results

For comparing model performance, some previous study results are selected as baseline results, shown in Table 5. It can be observed that the Proposed model outperforms these results.

Table 5. Some previous study results

Paper	Model	Accuracy(%)
Quora Question Duplication [4]	Siamese with bag of words	77.3
	Siamese with LSTM	83.2
	seq2seq LSTM with attention	80.8
	Ensemble	83.8
Duplicate question pair Detection with DL [1]	LSTM (Twitter word embedding)	81.07
Determining Entailment of Questions in the quora dataset [22]	LSTM	78.4
	LSTM with Attention	81
	LSTM with two way attention	81.4
	Decomposable Attention Model	79.8
Quora's Text Corpus [11]	LSTM with concatenation	87
	LSTM with distance and angle	87
	Decomposable Attention	86
Bilateral Multi-perspective Matching for Natural Language Sentences [24]	BiMPM	88
Proposed Model-1 (SiameseLSTM + Dense layer)	Siamese LSTM with Square Difference and Addition(+Hand-Engineered features) + dense layer	**89.11**

7   Conclusion

In this work, we explored the problem of detecting semantically equivalent question-pairs from the quora-question corpus. We explore the basic Siamese-LSTM with several augmentations to the architecture. The proposed Siamese-LSTM + dense-layer (Model-1) with hand-engineered features outperformed the other state-of-the-art models. The Machine learning based classifier, i.e. SVM, Logistic-Regression, XGBoost are also evaluated as alternatives for the dense-layer under proposed Model-2 architecture. But, the Model-1 has shown higher accuracy than the Model-2 scenarios. The proposed Model-1 contains two sub-network one is based on the Siamese-LSTM, and the other is a simple dense-layer which takes hand-engineered features and gives its high-level representation. In the experiments, it is observed that passing some additional hand-engineered features to the model provides a boost to accuracy. It helps the model to achieve better results considering the basic Siamese-LSTM model. The main motive behind using it as a base model is to build a model which is lightweight and fast because it is important for the websites to take less time to suggest similar questions to the users on knowledge sharing platforms. The proposed model-1 achieves an accuracy of 89.11%. The result is compared with other existing works as shown in Table 5.

For future work, besides the hand-engineered features, some machine translation based statistical features [5] can be included to avoid the influence of lexeme-gaps between the question-pairs. It is also interesting to explore how the proposed augmented-DNN models could be extended to apply in automatic short-answer grading systems and textual entailment detection problems [8]. Higher-order fuzzy sets can be used to further increase the accuracy of the proposed model [13].

References

1. Addair, T.: Duplicate question pair detection with deep learning. Stanf. Univ. J. (2017)
2. Bogdanova, D., dos Santos, C., Barbosa, L., Zadrozny, B.: Detecting semantically equivalent questions in online user forums. In: Proceedings of the Nineteenth Conference on Computational Natural Language Learning, pp. 123–131 (2015)
3. Bromley, J., Guyon, I., LeCun, Y., Säckinger, E., Shah, R.: Signature verification using a "siamese" time delay neural network. In: Advances in Neural Information Processing Systems, pp. 737–744 (1994)
4. Dadashov, E., Sakshuwong, S., Yu, K.: Quora question duplication (2017)
5. Dhariya, O., Malviya, S., Tiwary, U.S.: A hybrid approach for Hindi-English machine translation. In: 2017 International Conference on Information Networking (ICOIN), pp. 389–394. IEEE (2017)
6. Gong, Y., Luo, H., Zhang, J.: Natural language inference over interaction space. arXiv preprint arXiv:1709.04348 (2017)
7. Homma, Y., Sy, S., Yeh, C.: Detecting duplicate questions with deep learning. In: Proceedings of the International Conference on Neural Information Processing Systems (NIPS) (2016)
8. Jain, S., Malviya, S., Mishra, R., Tiwary, U.S.: Sentiment analysis: an empirical comparative study of various machine learning approaches. In: Proceedings of the 14th International Conference on Natural Language Processing (ICON-2017), pp. 112–121 (2017)
9. Kaggle: Quora Question Pairs. https://www.kaggle.com/quora/question-pairs-dataset. Accessed 4 June 2020
10. Kim, S., Kang, I., Kwak, N.: Semantic sentence matching with densely-connected recurrent and co-attentive information. In: Proceedings of the AAAI Conference on Artificial Intelligence, vol. 33, pp. 6586–6593 (2019)
11. Lili Jiang, S.C., Dandekar, N.: Semantic Question Matching with Deep Learning, Blog. https://www.quora.com/q/quoraengineering/Semantic-Question-Matching-with-Deep-Learning. Accessed 7 July 2020

12. Mishra, R., Barnwal, S.K., Malviya, S., Mishra, P., Tiwary, U.S.: Prosodic feature selection of personality traits for job interview performance. In: Abraham, A., Cherukuri, A.K., Melin, P., Gandhi, N. (eds.) ISDA 2018 2018. AISC, vol. 940, pp. 673–682. Springer, Cham (2020). https://doi.org/10.1007/978-3-030-16657-1_63

13. Mishra, R., et al.: Computing with words through interval type-2 fuzzy sets for decision making environment. In: Tiwary, U.S., Chaudhury, S. (eds.) IHCI 2019. LNCS, vol. 11886, pp. 112–123. Springer, Cham (2020). https://doi.org/10.1007/978-3-030-44689-5_11

14. Mueller, J., Thyagarajan, A.: Siamese recurrent architectures for learning sentence similarity. In: Thirtieth AAAI Conference on Artificial Intelligence (2016)

15. Parikh, A.P., Täckström, O., Das, D., Uszkoreit, J.: A decomposable attention model for natural language inference. arXiv preprint arXiv:1606.01933 (2016)

16. Pennington, J., Socher, R., Manning, C.D.: Glove: Global vectors for word representation. In: Proceedings of the 2014 Conference on Empirical Methods in Natural Language Processing (EMNLP), pp. 1532–1543 (2014)

17. Revanuru, K.: Quora Questinn Pairs Report. https://karthikrevanuru.github.io/assets/documents/projects/Quora_Pairs.pdf. Accessed 5 June 2020

18. Pathak, S., Ayush Sharma, S.S.S.: Semantic string similarity for quora question pairs (2019)

19. Silva, J., Rodrigues, J., Maraev, V., Saedi, C., Branco, A.: A 20% jump in duplicate question detection accuracy? Replicating IBM teams experiment and finding problems in its data preparation. META 20(4k), 1k (2018)

20. Singh, S., Malviya, S., Mishra, R., Barnwal, S.K., Tiwary, U.S.: RNN based language generation models for a Hindi dialogue system. In: Tiwary, U.S., Chaudhury, S. (eds.) IHCI 2019. LNCS, vol. 11886, pp. 124–137. Springer, Cham (2020). https://doi.org/10.1007/978-3-030-44689-5_12

21. Tai, K.S., Socher, R., Manning, C.D.: Improved semantic representations from tree-structured long short-term memory networks. arXiv preprint arXiv:1503.00075 (2015)

22. Tung, A., Xu, E.: Determining entailment of questions in the quora dataset (2017)

23. Vila, M., Martí, M.A., Rodríguez, H., et al.: Is this a paraphrase? What kind? Paraphrase boundaries and typology. Open J. Modern Linguist. 4(01), 205 (2014)

24. Wang, Z., Hamza, W., Florian, R.: Bilateral multi-perspective matching for natural language sentences. arXiv preprint arXiv:1702.03814 (2017)

25. Yu, J., et al.: Modelling domain relationships for transfer learning on retrieval-based question answering systems in e-commerce. In: Proceedings of the Eleventh ACM International Conference on Web Search and Data Mining, pp. 682–690 (2018)

26. Zhou, C., Sun, C., Liu, Z., Lau, F.: A C-LSTM neural network for text classification. arXiv preprint arXiv:1511.08630 (2015)

27. Zhu, W., Yao, T., Ni, J., Wei, B., Lu, Z.: Dependency-based siamese long short-term memory network for learning sentence representations. PloS One 13(3), e0193919 (2018)

Concentration Level Prediction System for the Students Based on Physiological Measures Using the EEG Device

Varsha T. Lokare[✉] and Laxman D. Netak

Department of Computer Engineering,
Dr. Babasaheb Ambedkar Technological University, Lonere-Raigad 402103, India
varsha.lokare@ritindia.edu, ldentak@dbatu.ac.in

Abstract. Concentration level plays a significant role while performing cognitive actions. There are many ways to predict the concentration level, such as with the help of physical reflection, facial expressions, and body language. Self- evaluation on the scale of 0 to 1 can also be used to measure the concentration level. In this paper, a publicly available dataset is used for classifying the concentration level using students' brain signals recorded through Electroencephalogram (EEG) device while performing different tasks that require varied concentration level. The study aims to find the appropriate Machine Learning (ML) model that predicts the concentration level through brain signal analysis. For this purpose, five different ML classifiers are used for comparative analysis, namely: Adaboost, Navie Bays, Artificial Neural Networr (ANN), Support Vector Machine (SVM) and Decision Tree. The ANN model gives the highest accuracy, i.e. 71.46% as compared to other classifiers for the concentration level measurement.

Keywords: Brain signals · EEG · Concentration level · ML classifiers

1 Introduction

The performance during cognitive actions is greatly influenced by the level of concentration or degree of focus of a performer. The performer's behaviour while performing repetitive tasks such as mathematical calculations, web browsing can be monitored to measure the level of concentration. In this paper, a performer's behaviour is observed by capturing brain signals through Electroencephalography (EEG). As shown in Fig. 1, brain signals are captured during the performance of the five tasks, namely: (i) Performing Arithmetic Calculations (ii) Reading Technical Articles (iii) Listening to the Technical Podcasts, Reading Transcripts (iv) Browsing Internet (v) Relaxing with Open or closed Eyes.

V. T. Lokare—Presently working as an Assistant Professor, Rajarambapu Institute of Technology, Sakharale, Affiliated to Shivaji University, Kolhapur.

M. Singh et al. (Eds.): IHCI 2020, LNCS 12615, pp. 24–33, 2021.
https://doi.org/10.1007/978-3-030-68449-5_3

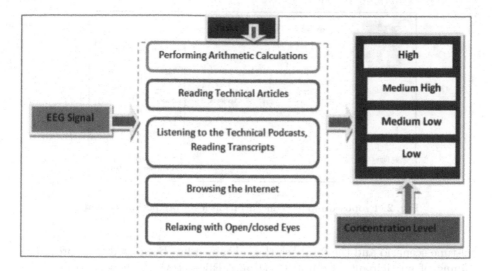

Fig. 1. Concentration level prediction system design

A muse headset is used to capture the brain waves. The different frequency bands *alpha, beta, gamma, theta* and *delta* are considered as features to build the classification models. The concentration levels are categorized into total four classes: *Low, Medium_Low, Medium_High* and *High*. Five machine learning classifiers namely: (i) Adaboost (ii) Navie Bays (iii)ANN (iv)SVM and (v)Decision Tree are built to classify the concentration levels. The primary motivation behind measuring the concentration level is to get feedback on why students fail to perform better while doing specific cognitive actions.

Rest of the paper is organized as follows: Sect. 2 reviews the existing literature. Section 3 explains the proposed methodologies for measuring concentration levels. Section 4 illustrates the working of four classifiers namely DT, ANN, SVM and ADB. Section 5 explains the experimental work and finally concluding remarks are given in the last section.

2 Earlier Approaches

Brain signal analysis is useful for many applications from healthcare and education. Researchers have analyzed brain waves to measure the concentration level in different contexts, such as game playing. In [11], the concentration level has monitored through EEG signals during reading practices. This paper focuses on capturing brain signals during the reading of paragraphs and answering objective questions based on that. ANN classifier is used to analyze EEG signals with an accuracy of 73.81%.

Friedman Nir and Fekete Tomer [6] have measured the impact of tricky questions on cognitive load through EEG signal analysis. Coyne, J. T., Baldwin [5] has adopted the analysis of EEG signals to cognitive load measurement in the

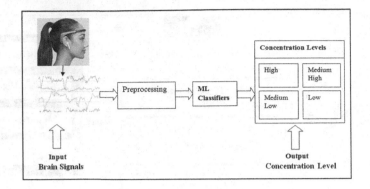

Fig. 2. Proposed model: concentration level prediction system

real-time scenario and the feedback thus obtained has been used to improve the training of participants. Conrad C. D. and Bliemel M. [4] have also used the same measures in the context of E-learning. Sports is another application area where concentration level plays a significant role. Few researchers [2,3,9] have used EEG signal-based measurement of concentration level for players during the game of golf putting.

In this paper, different tasks are considered that a student may repeatedly perform in his or her daily routine like Performing Arithmetic Calculations, Reading Technical Articles, Listening to the Technical Podcasts, Reading Transcripts, Browsing the Internet and Relaxing with Open/closed Eyes. This research aims to study and analyze brain signals during the above said tasks. This study will help to measure the concentration level of the students. Also, the classification is carried out with the help of five Machine Learning algorithms, as shown in the Fig. 2.

3   Proposed Approach

Brainwaves are generated when human participants are subjected to perform a task requiring a high level of concentration. These brainwaves are in the form of weak electric signals generated from hundreds of neurons communicating with each other. The variations in the electric signals depend on the tasks that an individual is performing. The study described in this paper captures brainwaves with the help of EEG device when an individual is performing five different functions, as mentioned in Fig. 1.

After the collection of the raw EEG data with the help of Muse headset, the preprocessing of the raw data has been carried out because recorded brain signals may contain some artifacts in the form of noise, eyes blinking, physical movement etc. Once pre-processed data is ready, it can provide as an input to any ML classifier. To measure the accuracy between the input brain signals and the predicted concentration level, total five Machine Learning classifiers namely: Adaboost, Navie Bays, Artificial Neural Network (ANN), Support Vector Machine (SVM) and Decision Tree have been applied.

Table 1. Concentration levels

Score (on the scale of 0 to1)	Class	Concentration level
0.8 to1	3	High
0.5 to 0.7	2	Medium high
0.3 to 0.4	1	Medium low
0 to 0.2	0	Low

3.1 Machine Learning Classifiers

To measure the accuracy of testing data, a total of five classifiers have been used, namely Adaboost, Navie Bays, ANN, SVM and Decision Tree. First, all the models are needed to train on the training dataset and then it is required to check its accuracy based on predictions made on the testing dataset. Here, 70% of the dataset has been used for training purpose, and 30% data are used for testing purpose.

1. **AdaBoost Adaptive Boosting (AdB)** [1]: It combines many weak classifiers like Decision Trees into a single strong classifier. It is used for text classification problem.
2. **Navie Bays (NB):** This classifier assumes that the features have no relation with each other and it is the extension of Bayes Theorem.
3. **Artificial Neural Network (ANN)** [8]: The functionality of this classifier is based on the human brain structure. Nodes are associated with the neurons in the brain. Each node in the network gets input from the other nodes while training the model and predicts the output.
4. **Support Vector Machine (SVM)** [7]: It finds the vectors on and nearest of the hyperplane to classify them among said concentration levels.
5. **Decision Tree (DT)** [10]: This classifier is mainly based on a supervised learning approach and is used for classification as well as regression purpose. This model helped in the classification of concentration level among low, *medium_low*, *medium_high* and high.

3.2 Concentration Levels

One can measure the performance of each task listed in Fig. 1 against the self-scored values for concentration levels as described in Table 1. Zero indicates low concentration and 1 indicates high concentration.

As per brain signal input, predicted concentration level based on various ML classifiers are compared and analyzed. The proposed model can predict the concentration level as per the input brain signals. The task with the highest concentration level can be used for assigning the most suitable task for learning purpose. Table 2 shows the expected Concentration Levels required for performing the various tasks.

4   Model Development

4.1   Dataset

The publicly available dataset `kaggle.com/dqmonn/personal-eeg-tasks` has been used for experimental purpose. The Muse headset has been used by the subject to record EEG signals during the performance of tasks as mentioned above. The subjects included male in the age-group of 21 at the time of recording. Total 7243 data entries have been collected with the concentration level self-classified by the subjects among different categories such as *high*, *medium_high*, *medium_low* and *low* concentration levels. These classification levels are encoded numerically on the scale of 0 to 1. From four electrodes of Muse headset, alpha, beta, gamma, delta and theta waves at different timestamp have been captured.

Table 2. Expected concentration levels required for performing the task

Task	Concentration level
Performing arithmetic calculations	High
Reading technical articles	Medium high
Listening to the technical podcasts, reading transcripts	Medium high
Browsing the Internet	Medium low
Relaxing	Low

4.2   Feature Selection

The raw data is preprocessed to remove noise in the collected data. Another critical step is to choose features which really affects the output. Here, we have applied the Pearson correlation filter method for the feature selection in which it has returned the rank of each feature based on various metrics like variance, correlation coefficient, etc. Total 20 parameters have been selected, namely *alPha0-3, beta0-3, theta0-3, gamma0-3, delta0-3* based on the filtering method out of 21 features recorded along with a timestamp.

As shown in Fig. 3, all frequency bands have a relative importance greater than 0.2. Hence, a total 20 features were considered, namely alha0-3, beta0-3, theta0-3, gamma0-3, delta0-3 and timestamp metric has importance 0.0. Hence, timestamp parameter is being removed from features list.

4.3   Performance Measurement Parameters

Following parameters have been considered to measure the accuracy of five classifiers.

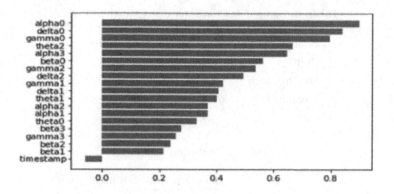

Fig. 3. Feature selection by pearson correlation filter method

1. **Accuracy.** It is the ratio of correct predictions by total predictions. Generally, to measure the percentage of correct predictions made over total predictions, accuracy metric is used.

2. **Precision.** The precision is defined as:

$$Precision = \frac{TP}{TP + FP}$$

3. **Recall.** It measures the number of positive class predictions made out of all positive examples in the dataset. The recall is defined as:

$$Recall = \frac{TP}{TP + FN}$$

4. **F1-Measure.** It gives the single value based on the precision and recall.

$$F1Score = \frac{2 * Recall \; X \; Precision}{Recall + Precision}$$

5. **Confusion Matrix.** This matrix is used to measure the performance of the model or classifier based on the Test data. As name given, it shows the matrix of values where the model gets confused to predict the correct output. In short, it summarizes the correct and wrong outputs in matrix form. Here, in this paper multi-class problem is discussed as the concentration level of the students is divided into low, medium_low, medium_high and high concentration.

As shown in Table 3, total testing dataset contains 2173 entries and out of that actual 158 belongs to class 0 or Low concentration level. Class 0: Low concentration level. As, shown in Table 4, 99 entries correctly predicted as Low concentration level called as True Positive (TP). 31 entries wrongly predicted as Low concentration level called as False Positive (FP).

Table 3. Sample test data per concentration level

Concentration level	Total test data inputs
High	589
Medium high	574
Medium low	852
Low	158

Table 4. Confusion matrix

	Low	Medium low	Medium high	High
Low	99	14	22	23
Medium low	6	695	109	42
Medium high	5	113	386	70
High	20	77	119	373

4.4   Observations and Result Analysis

Total 2173 data inputs are considered for testing purpose out of 7243 inputs as 70% data are used for Training and remaining 30% data is reserved for Testing. The class-wise data inputs are shown in Table 3.

1. **Accuracy.** As shown in Fig. 4, it is observed that the ANN classifier gives higher accuracy (71.46%) in predicting the correct concentration level based on the brain signal input. The second highest accuracy is observed in Decision Tree classifier. i.e. 66.35%. DT classifier works on the principle of tree-based structure in which splitting of data is performed at every level. In predicting the correct concentration level, DT shows better results than Navie Bays, Adaboost and SVM classifiers. The AdaBoost classifier is the ensemble of weak classifiers; also, it is a decision tree with only one level. Hence, the result of AdaBoost classifier is less than DT classifier. An SVM classifier works on the concept of finding appropriate hyper-plane that results in the classification of different output classes based on the distance between points and hyperplane. It is observed that the accuracy of the SVM classifier in predicting the concentration level is 53.17%. The lowest accuracy in predicting the correct concentration level is observed in Naive Bays classifier as it assumes that the features are independent with each other. But in this particular application, the frequency ranges for saying Beta0 to Beta3 waves are correlated with each other.

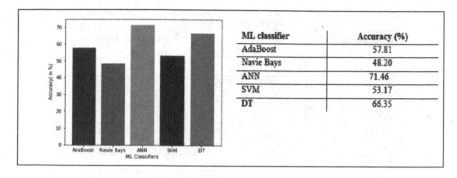

Fig. 4. Accuracy per classifier

2. **Precision, Recall and F1 Score.** From Fig. 5, it is observed that in terms of Precision and F1 score measuring parameters; ANN classifier outperforms other four classifiers. But in case of Recall parameter, DT classifier is better in predicting High and Low Concentration. At the same time, Navie Bays and ANN are better in predicting Medium High and Medium Low concentrations, respectively.

3. **Confusion Matrix.** From Table 5, it has been observed that in predicting the correct concentration level of the students' class-wise different models are better. i.e. for class0 or predicting low concentration level and class3 or predicting the high concentration level, DT classifier outperforms other classifiers. Similarly, for class1 or predicting medium low concentration and for class2 or predicting medium high concentration, ANN classifier outperforms other classifiers.

Table 5. Confusion matrix

Class 0					Class1				
AdB	NB	SVM	ANN	DT	AdB	NB	SVM	ANN	DT
71	13	43	98	100	153	659	616	697	597
Class 2					Class3				
AdB	NB	SVM	ANN	DT	AdB	NB	SVM	ANN	DT
219	90	153	382	330	226	111	151	376	388

Precision	Concentration Level			
	High (3)	Medium High (2)	Medium Low (1)	Low (0)
Adaboost	0.55	0.65	0.53	0.50
Navie Bays	0.46	0.49	0.32	0.69
ANN	0.76	0.77	0.61	0.73
SVM	0.63	0.54	0.43	0.61
DT	0.63	0.75	0.58	0.64
Recall				
Adaboost	0.54	0.75	0.46	0.46
Navie Bays	0.10	0.93	0.19	0.29
ANN	0.63	0.82	0.67	0.63
SVM	0.33	0.87	0.32	0.31
DT	0.65	0.73	0.59	0.65
F1 Score				
Adaboost	0.54	0.70	0.49	0.48
Navie Bays	0.16	0.64	0.24	0.34
ANN	0.69	0.79	0.64	0.68
SVM	0.43	0.67	0.37	0.41
DT	0.64	0.74	0.58	0.65

Fig. 5. Precision, Recall and F1 Score per classifier

5   Conclusion

The paper presents an approach to classify students based on their concentration levels when they are engaged in performing various kinds of cognitive tasks. Students' responses are monitored through EEG while performing tasks like Arithmetic Calculations, Reading Technical Articles, Listening to the Technical Podcasts, Reading Transcripts, Browsing the Internet and Relaxing with Open/closed Eyes. These signals have analyzed with the help of five machine learning algorithms. It is observed that ANN classifier gives the highest accuracy of around 71.46%. All frequency bands are considered as features because the tasks considered are related to all alpha, beta, gamma, theta and delta frequency ranges. Also, the low accuracy is achieved with Navie Bays Classifier as it works on the principle of no relation between the features. The proposed ANN-based model system helps predict the concentration level of the students during the performance of multiple tasks. This work can be extended with a deep learning algorithm to achieve more accuracy.

References

1. An, T.K., Kim, M.H.: A new diverse AdaBoost classifier. In: 2010 International Conference on Artificial Intelligence and Computational Intelligence, vol. 1, pp. 359–363. IEEE (2010)
2. Arns, M., Kleinnijenhuis, M., Fallahpour, K., Breteler, R.: Golf performance enhancement and real-life neurofeedback training using personalized event-locked EEG profiles. J. Neurother. **11**(4), 11–18 (2008)
3. Babiloni, C., et al.: Golf putt outcomes are predicted by sensorimotor cerebral EEG rhythms. J. Physiol. **586**(1), 131–139 (2008)
4. Conrad, C.D., Bliemel, M.: Psychophysiological measures of cognitive absorption and cognitive load in e-learning applications (2016)
5. Coyne, J.T., Baldwin, C., Cole, A., Sibley, C., Roberts, D.M.: Applying real time physiological measures of cognitive load to improve training. In: Schmorrow, D.D., Estabrooke, I.V., Grootjen, M. (eds.) FAC 2009. LNCS (LNAI), vol. 5638, pp. 469–478. Springer, Heidelberg (2009). https://doi.org/10.1007/978-3-642-02812-0_55
6. Friedman, N., Fekete, T., Gal, Y.K., Shriki, O.: EEG-based prediction of cognitive load in intelligence tests. Front. Hum. Neurosci. **13**, 191 (2019)
7. Hearst, M.A., Dumais, S.T., Osuna, E., Platt, J., Scholkopf, B.: Support vector machines. IEEE Intell. Syst. Appl. **13**(4), 18–28 (1998)
8. Mehta, H., Singla, S., Mahajan, A.: Optical character recognition (OCR) system for roman script & English language using artificial neural network (ANN) classifier. In: 2016 International Conference on Research Advances in Integrated Navigation Systems (RAINS), pp. 1–5. IEEE (2016)
9. Sakai, Y., Yagi, T., Ishii, W.: EEG analysis of mental concentration in golf putting. In: The 5th 2012 Biomedical Engineering International Conference, pp. 1–5. IEEE (2012)
10. Xiaochen, D., Xue, H.: Multi-decision-tree classifier in master data management system. In: 2011 International Conference on Business Management and Electronic Information, vol. 3, pp. 756–759. IEEE (2011)
11. Zaeni, I.A., Pujianto, U., Taufani, A.R., Jiono, M., Muhammad, P.S.: Concentration level detection using EEG signal on reading practice application. In: 2019 International Conference on Electrical, Electronics and Information Engineering (ICEEIE), vol. 6, pp. 354–357. IEEE (2019)

A Correlation Analysis Between Cognitive Process and Knowledge Dimension in Software Engineering by Using the Revised Bloom's Taxonomy

Manjushree D. Laddha[✉], Laxman D. Netak, and Hansaraj S. Wankhede

Department of Computer Engineering, Dr. Babasaheb Ambedkar Technological University Lonere, Raigad 402103, India
{mdladdha,ldnetak,hswankhede}@dbatu.ac.in

Abstract. In this competitive world, students are required to acquire hard power skills to sustain in the competition. It is also the responsibility of the university education system to develop students' hard power skills. The hard power skills include both technical abilities as well as professional skills. The necessary hard power skills are highly based on cognitive processes and knowledge dimensions. The purpose of this study was to investigate the correlation between the knowledge dimensions and cognitive processes as identified in the Revised Bloom's Taxonomy (RBT) for the course on Software Engineering. For this purpose, a correlation analysis of questions items collected from the various examinations conducted for the course on Software Engineering is performed. Each question item is first classified along the knowledge dimension and cognitive processes from RBT. The results of the Chi-Square test shows that both dimensions are significantly correlated.

Keywords: Revised Bloom's Taxonomy · Cognitive process · Knowledge dimension

1 Introduction

Education taxonomy plays a vital role for instructors and students. For instructors [9], it is useful to design the course with proper educational outcomes and to assess the student's attainment of it. For students, it is helpful to know how much skills and knowledge a student has acquired through learning. End semester examinations are the preferred mode of assessing student's knowledge and skills. The result of the assessment judges the student's performance, understanding and skills. Some researchers use the rule-based assessment [12], assessment framework [8,10], project-based assessment [6].

The paper presents a correlation analysis of the categories included in cognitive processes and knowledge dimensions in RBT. For this purpose, the question items from the examinations conducted for a course on Software Examinations

© Springer Nature Switzerland AG 2021
M. Singh et al. (Eds.): IHCI 2020, LNCS 12615, pp. 34–41, 2021.
https://doi.org/10.1007/978-3-030-68449-5_4

by Indian universities are used. The Software Engineering course is typically offered in the third year by a majority of the Indian universities. Question items from these examinations can be classified as per Revised Bloom's Taxonomy (RBT) [13]. Cognitive processes and Knowledge types are the two dimensions described in RBT along which the teaching, learning and assessment processes can be aligned [7].

2   Background

The Revised Bloom's Taxonomy (RBT) [1] includes six different cognitive processes involved in the learning activity. These are: *Remember*, *Understand*, *Apply*, *Analyze*, *Evaluate*, and *Create*. The second dimension in RBT specifies four kinds of knowledge required to answer a question item, namely, *Factual*, *Conceptual*, *Procedural*, and *Meta-cognitive* knowledge [1].

Prime applications of the RBT includes teaching, learning, assessment, designing questions, designing curriculum, mapping curriculum with learning outcomes and objectives. The combination of the frameworks of RBT and Efklides's metacognition framework [10] has been used to design question items to evaluate students' acquired skills. Teachers use these frameworks to design questions that make students think critically. Also, these frameworks help students to solve the questions that improve students' strategies to solve complex problems and to increase intelligence from lower order cognitive processes to higher order cognitive processes [5].

2.1   Revised Bloom's Taxonomy Applied to the Software Engineering Course

The Revised Bloom's Taxonomy [1] is mainly used for learning, teaching, and assessing. The uses of this taxonomy as a tool for alignment of curriculum, instructional materials and assessments are widely reported. This taxonomy is a two-dimensional matrix rows representing the knowledge dimensions and columns representing the cognitive processes. Generally, verbs refer to cognitive processes and objects in sentences that describe the knowledge dimension.

Some course assessment has been carried out for problem-based learning by using Bloom's taxonomy [3,4] and to compare the competencies with ordinary skills for passing the course. A set of guidelines [2] has been prepared to design and assess Software Engineering courses.

3   Method

3.1   Study Goals

This paper aims to examine the question items from Software Engineering examinations and identify both cognitive and knowledge skills associated to them. The main aim is to find the correlation between cognitive skills and knowledge dimensions. It aims to study how these two skills are related. Besides, it also examines whether any of these six categories of cognitive skills (remember, understand, apply, analyze, evaluate, create) are associated with any one of the knowledge dimension. (factual, conceptual and procedural knowledge.)

3.2   Study Design and Research Questions

Table 1. A random sample of data set constructed

QID	Question text	Cognitive process	Knowledge type
Q002	Compare waterfall model and spiral model	Analyze	Conceptual
Q006	Draw DFD (Level 0, 1 and 2) for a course management system, and explain	Apply	Procedural
Q009	Describe the activities of project scheduling and tracking	Understand	Conceptual
Q017	Justify any five guidelines of metrics with respect to your project of choice. Explain it with suitable diagrams. Of the above project of choice?	Evaluate	Procedural
Q044	What are the advantages of test driven development?	Remember	Factual

The following research questions are formulated:

1. **RQ1** . Is there any correlation between cognitive skills and knowledge dimension?
2. **RQ2** . Is there any relation between Factual knowledge and remember?
3. **RQ3** . Is there any relation between Conceptual knowledge and understand?
4. **RQ4** . Is there any relation between Procedural knowledge and apply, create and evaluate?

3.3   Data Exploration

In this paper, the main focus is to answer the question: *is there any relation between cognitive skills and knowledge dimension?* So it is needed to identify the question items from various universities for the course on Software Engineering. This course is typically included in the syllabus of the undergraduate programs of Computer Engineering and Information Technology. In some universities, the course is offered in the third year or few universities in the final year.

Questions asked in Software Engineering examinations are collected. The question papers collected from different universities are used. A total of 794 question items are collected. All the question items are manually classified for cognitive processes and knowledge dimensions from the RBT. A sample of the collected data set is shown in Table 1. The complete data set is available at https://github.com/akiwelekar/QIClassifier.

4   Knowledge Types in RBT

This section briefly explains various kinds of knowledge types included in the RBT.

4.1   Factual Knowledge

Factual Knowledge is basic knowledge that students must understand about Software Engineering and able to solve any problems. This category contains the sub-categories (i) Knowledge of terminology and such as definition of software, software engineering, process.(ii) Knowledge of specific details and elements such as format, attributes, factors.

In these categories the expectation from the students should understand the basic facts and findings of the course. It contains basic knowledge of the course. It is expected from learners to extract from already learned information and find what exactly the solution or answer to question items.

In this paper, question items were identified which fall under remember-factual, understand-factual, apply-factual, analysis-factual, evaluate-factual, create-factual. Then mapping of all the cognitive levels with factual knowledge dimension was carried out. Question items were collected from end semester examinations. As refer to Fig. 1 40.3% factual knowledge question items were there in end semester examinations.

4.2   Conceptual Knowledge

Conceptual Knowledge is the association between the basic elements within a major concept that allow them to work together.

(a) Knowledge of classifications and categories such as list down the activities carried out during scheduling and tracking.
(b) Knowledge of principles and generalizations such as explain the principles of Software engineering?
(c) Knowledge of theories, models, and structures such as explain the spiral process model.

For solving the problem under conceptual category implicit or explicit understanding of concepts is required. An implicit knowledge is the extraction of unknown procedures. An explicit knowledge is the definition of a concept; explaining why a procedure works, drawing a concept map [11] and finding the resemblance and disparity among the two concepts.

In this paper, question items were identified which fall under remember-conceptual, understand-conceptual, apply-conceptual, analysis-conceptual, evaluate-conceptual, create-conceptual. Then mapping of all the cognitive levels with conceptual knowledge dimension was carried out. Question items were collected from end semester examinations. As refer to Fig. 1 22.4% conceptual knowledge question items were there in end semester examinations.

4.3   Procedural Knowledge

Procedural knowledge is how to execute anything, procedure of analysis, basis for implementing skills, algorithms, strategies and methods. This category contains the sub-categories as:

(a) Knowledge of subject-specific skills and algorithms such as explain in detail algorithm of cost models in project planning,
(b) Knowledge of subject-specific techniques and methods such as explain the agile methods,
(c) Knowledge of determining when to use appropriate procedures such as what is the difference among formal, semi formal and informal methods of software development?

Out of 794 question, the distribution of question items as per knowledge dimensions is shown in Fig. 1.

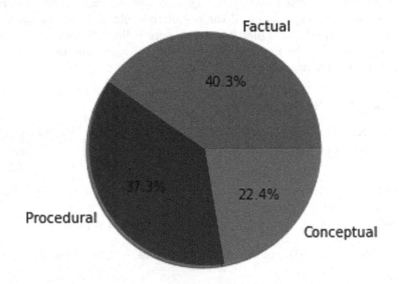

Fig. 1. Knowledge dimension wise question items

Table 2. Contingency table

Cognitive Process Knowledge Type	Analyze	Apply	Create	Evaluate	Remember	Understand
Observed Frequencies						
Conceptual	21	11	1	7	31	249
Factual	1 ·	3	0	2	99	73
Procedural	19	46	47	21	21	142
Expected Frequencies						
Conceptual	16	24	19	12	60	187
Factual	9	13	10	6	33	104
Procedural	15	22	17	11	56	172

Table 3. Results of Chi-square test

Probability	0.95	Critical Value	18.30
Statistic	336.77	p-value	$0.000 < 0.05$

5 Analysis and Results

The cognitive processes (X) and knowledge dimensions (Y) are both categorical variables. For finding the dependency between them, the Chi-square test is performed. This test is typically carried out to study dependency among categorical variables.

The following hypothesis statements are formulated to perform Chi-square test.

- **H0:** Cognitive skills and Knowledge dimensions are independent.
- **H1:** Cognitive skills and Knowledge dimensions are dependent.

For the purpose examining dependencies between cognitive processes and knowledge types, question items are paired along different categorical values and a contingency table as shown in Table 2 is prepared. The table shows observed frequencies in the dataset and expected frequencies calculated as per Chi-square test.

There are 794 question items in the dataset and they are presented in the three and six columns the contingency table. Hence, degrees of freedom is

$$degrees\ of\ freedom = (rows - 1) * (cols - 1) = 2 * 5 = 10$$

The Python's *chi2* module from *Scipy.stat* library is used to calculate Chi-square. The Python program implented for this purpose is also available at https://github.com/akiwelekar/QIClassifier. The calculated Chi-square statistic($x^2 = 336'77$), which is much greater than predetermined critical value of

18.30 hence the null hypothesis i.e. *(H0) Cognitive skills and Knowledge dimensions are independent* is rejected. Also the p-value is less than significance level of 0.05 indicating the results are statistically significant. In other words, there is statistically significant dependency between cognitive processes (X) and knowledge dimension (Y) is observed in the dataset.

6   Conclusion and Future Work

This paper presents the results of Chi-square test conducted to analyze the dependencies between cognitive processes and knowledge dimensions in the Revised Bloom's Taxonomy. This has been conducted for the dataset which includes 794 question items collected from various examinations on Software Engineering. It is observed that there is a statistically significant dependency is present between cognitive processes and knowledge dimension. Further, through comparison of expected and observed frequencies it is observed that

(i) the *factual* knowledge dimension and *remembering* process as a cognitive skill
(ii) the *Conceptual* knowledge dimension and *understanding* process as a cognitive skill
(iii) the *Procedural* knowledge dimension and *apply, evaluate, create* process as a cognitive skill, a significant level of dependencies are present.

Moreover, the research can be used to help students to upgrade their hard skills while learning the software engineering course. The research reported in this paper is concerned with only for the Software Engineering course. In future, few more courses can be considered for generalising the dependencies between the knowledge dimension and cognitive process.

References

1. Anderson, L.W., Krathwohl, D.R., Airasian, P.W., Cruikshank, K.A., Mayer, R.E., Pintrich, P.R., Raths, J., Wittrock, M.C.: p. 214 (2001)
2. Britto, R., Usman, M.: Bloom's taxonomy in software engineering education: a systematic mapping study. In: 2015 IEEE Frontiers in Education Conference (FIE), pp. 1–8. IEEE (2015)
3. Dolog, P., Thomsen, L.L., Thomsen, B.: Assessing problem-based learning in a software engineering curriculum using bloom's taxonomy and the IEEE software engineering body of knowledge. ACM Trans. Comput. Educ. (TOCE) 16(3), 1–41 (2016)
4. Dos Santos, S.C.: PBL-SEE: an authentic assessment model for PBL-based software engineering education. IEEE Trans. Educ. 60(2), 120–126 (2016)
5. Gomes, A., Correia, F.B.: Bloom's taxonomy based approach to learn basic programming loops. In: 2018 IEEE Frontiers in Education Conference (FIE), pp. 1–5. IEEE (2018)

6. Juárez-Ramírez, R., Jiménez, S., Huertas, C.: Developing software engineering competences in undergraduate students: a project-based learning approach in academy-industry collaboration. In: 2016 4th International Conference in Software Engineering Research and Innovation (CONISOFT), pp. 87–96. IEEE (2016)
7. Kiwelekar, A.W., Wankhede, H.S.: Learning objectives for a course on software architecture. In: Weyns, D., Mirandola, R., Crnkovic, I. (eds.) ECSA 2015. LNCS, vol. 9278, pp. 169–180. Springer, Cham (2015). https://doi.org/10.1007/978-3-319-23727-5_14
8. Lajis, A., Nasir, H.M., Aziz, N.A.: Proposed assessment framework based on bloom taxonomy cognitive competency: introduction to programming. In: Proceedings of the 2018 7th International Conference on Software and Computer Applications, pp. 97–101 (2018)
9. Masapanta-Carrión, S., Velázquez-Iturbide, J.Á.: Evaluating instructors' classification of programming exercises using the revised bloom's taxonomy. In: Proceedings of the 2019 ACM Conference on Innovation and Technology in Computer Science Education, pp. 541–547 (2019)
10. Radmehr, F., Drake, M.: An assessment-based model for exploring the solving of mathematical problems: utilizing revised bloom's taxonomy and facets of metacognition. Stud. Educ. Eval. **59**, 41–51 (2018)
11. Rittle-Johnson, B., Schneider, M., Star, J.R.: Not a one-way street: bidirectional relations between procedural and conceptual knowledge of mathematics. Educ. Psychol. Rev. **27**(4), 587–597 (2015)
12. Ullah, Z., Lajis, A., Jamjoom, M., Altalhi, A.H., Shah, J., Saleem, F.: A rule-based method for cognitive competency assessment in computer programming using bloom's taxonomy. IEEE Access **7**, 64663–64675 (2019)
13. Wankhede, H.S., Kiwelekar, A.W.: Qualitative assessment of software engineering examination questions with bloom's taxonomy. Indian J. Sci. Technol. **9**(6), 1–7 (2016)

A Study on Comparative Analysis of the Effect of Applying DropOut and DropConnect to Deep Neural Network

Hyun-il Lim[✉]

Department of Computer Engineering,
Kyungnam University, Changwon, Gyeongsangnam-do, South Korea
hilim@kyungnam.ac.kr

Abstract. Neural networks are increasingly applied to real-life problems by training models for solving prediction problems with known training data. Overfitting is one of the major problems that should be considered in designing neural network models. Among various approaches to treat the overfitting problem, DropOut and DropConnect are approaches to adjust the process of training by temporarily changing the structures of neural network models. In this study, we compare the effect of applying DropOut and DropConnect approaches to a deep neural network. The result of this study will help to understand the effect of applying DropOut and DropConnect to neural networks in terms of loss and the accuracy of the trained model. It is also expected to help to design the structure of deep neural networks by effectively applying DropOut and DropConnect to control overfitting in the training of machine learning models.

Keywords: Neural network · DropOut · DropConnect · Machine learning · Overfitting

1 Introduction

As neural networks are widely used to solve real-life problems, methods for improving neural networks are being studied in various studies [1, 2]. Learning models through vast amounts of training data show applicability for solving real-life problems that are difficult to solve with traditional algorithmic approaches. However, the practical application of neural networks requires the ability to design effective models that describe problems, and continued study on designing methods to improve the accuracy of model. When applying a neural network to real-life problems, its structure is needed to be designed so that the model can efficiently accept the features of many training data while suppressing the occurrence of overfitting.

Overfitting [3–6] is a problem that must be considered to improve the accuracy of neural networks. Overfitting may occur when training data contain a lot of noise, so various approaches are being studied to suppress the occurrence of overfitting. The DropOut [7–9] and DropConnect [10, 11] are approaches that can be applied in the process of training neural network models with training data to control such overfitting

© Springer Nature Switzerland AG 2021
M. Singh et al. (Eds.): IHCI 2020, LNCS 12615, pp. 42–47, 2021.
https://doi.org/10.1007/978-3-030-68449-5_5

problem by temporarily changing the structure of neural network models. The main point of this study is to understand the effects of applying DropOut and DropConnect to neural network models from comparative experimental analysis, and to help to apply such approaches to control overfitting problems in designing neural network models. The results of this study are expected to be considered in the design of neural network models to improve the accuracy of prediction results.

2 Applying DropOut and DropConnect to a Neural Network

It is important to improve the accuracy of prediction results on real-world data in generating neural network models. However, data used in training such a model may include noise that interferes with producing accurate prediction results. Such noise may make the trained model too finely overfitted for noise data. Effective control of such overfitting should be considered important in neural network design to increase the accuracy of prediction results [3, 4].

The concept of applying DropOut [7, 9] or DropConnect [10, 11] is the temporary removal of nodes or links of deep neural networks in training a prediction model with training data. Such temporary removal of nodes or links in the training phase is believed to reduce overfitting by simplifying the structure of nodes that could be excessively customized to noise data. However, inconsiderate application of DropOut or DropConnect can be disadvantageous in improving the accuracy of neural network models, so the structure of neural network needs to be designed considerately by understanding the effect of applying such techniques.

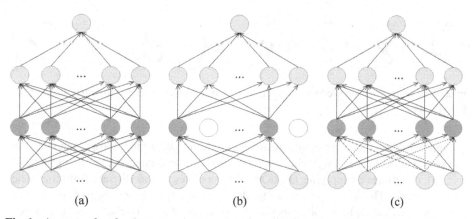

Fig. 1. An example of a deep neural network and application of the DropOut or DropConnect technique to the network. (a) The structure of a fully connected neural network. (b) After applying DropOut to the neural network. (c) After applying DropConnect to the neural network.

Figure 1 shows an example of a fully connected deep neural network and the changes of structures after applying DropOut or DropConnect to the hidden layers, respectively. The approaches affect the training of models by temporarily removing nodes or links of the neural network in each training phase. Such temporary removal of nodes or links in the training phase simplifies the structure of the neural network model, so the change of structure affects adjusting weights of nodes or links in the training phase. By temporarily removing some nodes or links, the excessive customization of the weights to noise contained in training data can be suppressed at the training phase. So, it is required to understand the effect of applying DropOut or DropConnect to a neural network to effectively design the structure of neural network models.

3   The Effect of Applying DropOut and DropConnect in a Deep Neural Network

In this section, an experiment is conducted to comparatively analyze the effect of applying DropOut and DropConnect in a deep neural network. The experiment was performed in a deep neural network model for recognizing the image of the handwritten digits of MNIST data [12]. Figure 2 shows the sample dataset of the MNIST data. To analyze the effect of applying DropOut or DropConnect, the basic neural network model is applied in this experiment. The structure of the neural network was designed with one input layer and two hidden layers of 128 and 64 nodes with ReLU activation functions, respectively. The output layer was designed with the softmax function to classify the input images of digits into their corresponding numbers. The neural network model was implemented in Python [13] and Keras [14] with TensorFlow backend. The second hidden layer of the model was experimented with by applying DropOut or DropConnect to compare the effect on the losses and accuracies of the trained model as the training epoch progressed. In this experiment, DropOut and DropConnect were applied with rates of 20% and 40%.

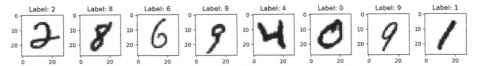

Fig. 2. The sample dataset used for the experiment for recognizing the handwritten digits of the MNIST data.

Figure 3 shows the experimental results of applying DropOut or DropConnect to the deep neural network for recognizing handwritten digits. In comparing the loss graphs of Fig. 3(a), the effects of applying DropOut or DropConnect was distinct from the basic neural network. The loss of the basic model did not improve any more after the 6th epoch that was the earliest epoch as compared to the other models. It was confirmed that the losses of models that were applied with DropOut or DropConnect were showing more improvement as the training epochs progressed. It showed that the overfitting that might occur in the basic model was effectively controlled and the loss could be reduced by applying DropOut or DropConnect. The applications of a higher rate of DropOut or DropConnect were more effective in reducing the losses of training. As the losses decreased, the accuracies increased as the training epochs progressed. DropOut had a greater effect on reducing losses than DropConnect, but accuracies were found to converge at a similar level as training epochs progressed.

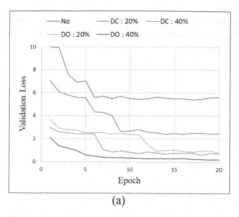
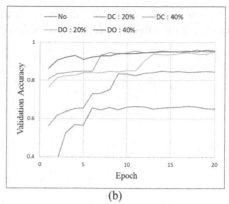

(a)   (b)

Fig. 3. The experimental results of the effect of applying DropOut or DropConnect to a deep neural network model. (a) The graph of validation losses of neural network with DropOut and DropConnect. (b) The graph of validation accuracies of neural network with DropOut and DropConnect.

It is confirmed that applying a higher rate of DropOut or DropConnect improves loss or accuracy more quickly. However, such an application has limitations in improving loss or accuracy resulting from the structure of a neural network model. While the converging speed of loss or accuracy varies, it can be seen that loss and accuracy converge at similar levels as the training epochs progresses.

4 The Effect on a Deep Neural Network with Unbalanced Numbers of Nodes

In this section, the effect of applying DropOut or DropConnect is analyzed in a neural network with unbalanced numbers of nodes. In this experiment, we evaluate whether the application of DropOut or DropConnect is effective in improving loss and accuracy even if the structure of a neural network is roughly designed with unbalanced numbers of nodes. The structure of the neural network was designed with two hidden layers of 128 and 1024 nodes, respectively. The DropOut or DropConnect were applied to the second hidden layer that has 1024 nodes.

Figure 4 shows the experimental results of applying DropOut or DropConnect to the deep neural network with unbalanced numbers of nodes. The losses and accuracies of this experiment showed different aspects from those of the previous experiment. The result showed that the losses were stagnated in a certain range without a gradual decrease as the training epochs progressed. The accuracy also did not improve further beyond certain values. It was confirmed that the structural imbalance of neural network nodes interfered with optimizing models in the training phase, and also confirmed that there were limits in improving loss and accuracy for such an unbalanced model. The losses and accuracies of the models with DropOut were rather worse than those of the basic model. Besides, even if the application of DropOut or DropConnect was effective in controlling overfitting in the previous experiment, it was confirmed that the limit of accuracy due to structural imbalances in the design of neural networks could not be controlled effectively by only applying DropOut or DropConnect.

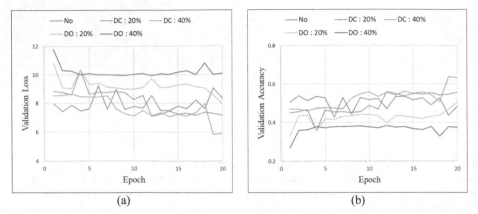

(a)   (b)

Fig. 4. The experimental results of the effect of applying DropOut or DropConnect to a deep neural network model with unbalanced numbers of nodes. (a) The graph of validation losses of neural network with DropOut and DropConnect. (b) The graph of validation accuracies of neural network with DropOut and DropConnect.

From the results of the experiments, the application of DropOut or DropConnect was helpful to control overfitting in a well-designed neural network model. However, it was more important to design the structure of the neural network well so that the model was suitable for learning from training data. Inconsiderate application of DropOut to a neural network with an unbalanced number of nodes has rather deteriorated the accuracy and the loss. So, it is necessary to understand the structure of neural network models by carefully applying a suitable rate of DropOut or DropConnect according to the features of neural network model to improve the accuracy of neural networks.

5 Conclusion

With the increasing popularity of neural networks in solving real-world problems, various studies on improving the accuracy of neural networks are in progress. Overfitting is one of the problems that should be considered in designing neural network models. In this study, the two approaches of DropOut and DropConnect were experimented to analyze the effect of applying the methods to a neural network model to control overfitting. In the experimental results, it is confirmed that applying DropOut or DropConnect to a well-designed neural network model is effective in improving losses and accuracies by controlling overfitting that may occur at the training phase. On the other hand, the effect of applying DropOut or DropConnect is not so advantageous in improving accuracies in a neural network model which have layers of unbalanced numbers of nodes. To make matters worse, the application of DropOut and DropConnect to an unsuitable neural network model may deteriorate the accuracy and the loss. So, it is necessary to understand the structure of neural network models and the effect of applying DropOut and DropConnect to the models. Considerate application of DropOut or DropConnect in designing neural network models is expected to improve the accuracy of the neural network by effectively suppressing the occurrence of overfitting to training data.

Acknowledgments. This work was supported by the National Research Foundation of Korea (NRF) grant funded by the Korea government (Ministry of Education) (No. NRF-2017R1D1A1B03034769).

References

1. Shai, S.-S., Shai, B.-D.: Understanding Machine Learning: From Theory to Algorithms. Cambridge University Press, Cambridge (2014)
2. Domingos, P.: A few useful things to know about machine learning. Commun. ACM **55**(10), 78–87 (2012)
3. Salman, S., Liu, X.: Overfitting Mechanism and Avoidance in Deep Neural Networks (2019)
4. Ying, X.: An overview of overfitting and its solutions. J. Phys.: Conf. Ser. **1168**(022022) (2019)
5. Cohen, P., Jensen, D.: Overfitting explained. In: Proceedings of the Sixth International Workshop on Artificial Intelligence and Statistics (2000)
6. Srivastava, N., Hinton, G., Krizhevsky, A., Sutskever, I., Salakhutdinov, R.: Dropout: a simple way to prevent neural networks from overfitting. J. Mach. Learn. Res. **15**, 1929–1958 (2014)
7. Baldi, P., Sadowski, P.: Understanding dropout. In: Proceedings of the 26th International Conference on Neural Information Processing Systems, pp. 2814–2822 (2013)
8. Duyck, J., Lee, M.H., Lei, E.: Modified dropout for training neural network (2014)
9. Labach, A., Salehinejad, H., Valaee, S.: Survey of Dropout Methods for Deep Neural Networks (2019)
10. Wan, L., Zeiler, M., Zhang, S., Cun, Y.L., Fergus, R.: Regularization of neural networks using dropconnect. In: Proceedings of the 30th International Conference on International Conference on Machine Learning, vol. 28, pp. 1058–1066 (2013)
11. Ravichandran, J., Saralajew, S., Villmann, T.: DropConnect for evaluation of classification stability in learning vector quantization. In: Proceedings of European Symposium on Artificial Neural Networks (ESANN 2019), pp. 19–24. Belgium (2019)
12. THE MNIST DATABASE of handwritten digits. https://yann.lecun.com/exdb/mnist/
13. Python programming language. https://www.python.org/
14. Keras: the Python deep learning API. https://keras.io/

HHAR-net: Hierarchical Human Activity Recognition using Neural Networks

Mehrdad Fazli[1(✉)], Kamran Kowsari[1,2], Erfaneh Gharavi[1], Laura Barnes[1], and Afsaneh Doryab[1]

[1] University of Virginia, Charlottesville, VA 22903, USA
{mf4yc,kk7nc,eg8qe,lb3dp,ad4ks}@virginia.edu
[2] University of California, Los Angeles, CA 90095, USA

Abstract. Activity recognition using built-in sensors in smart and wearable devices provides great opportunities to understand and detect human behavior in the wild and gives a more holistic view of individuals' health and well being. Numerous computational methods have been applied to sensor streams to recognize different daily activities. However, most methods are unable to capture different layers of activities concealed in human behavior. Also, the performance of the models starts to decrease with increasing the number of activities. This research aims at building a hierarchical classification with Neural Networks to recognize human activities based on different levels of abstraction. We evaluate our model on the Extrasensory dataset; a dataset collected in the wild and containing data from smartphones and smartwatches. We use a two-level hierarchy with a total of six mutually exclusive labels namely, "lying down", "sitting", "standing in place", "walking", "running", and "bicycling" divided into "stationary" and "non-stationary". The results show that our model can recognize low-level activities (stationary/non-stationary) with 95.8% accuracy and overall accuracy of 92.8% over six labels. This is 3% above our best performing baseline (HHAR-net is shared as an open source tool at https://github.com/mehrdadfazli/HHAR-Net).

Keywords: Deep learning · Human Activity Recognition · Hierarchical classification

1 Introduction

Human Activity Recognition (HAR) applied to data streams collected from mobile and embedded sensors [4,5,22] has numerous real-world applications in understanding human behavior in the wild for health monitoring [21] (mental and physical), smart environments [2,9], elderly care [14,15] and sports applications [1,8].

The advancements in mobile and wearable devices have made it possible to collect streams of data from built-in sensors in smartphones and fitness trackers

© Springer Nature Switzerland AG 2021
M. Singh et al. (Eds.): IHCI 2020, LNCS 12615, pp. 48–58, 2021.
https://doi.org/10.1007/978-3-030-68449-5_6

including accelerometer, gyroscope, magnetometer, GPS, and microphone. These sensor streams have been analyzed and modeled individually or in combination to recognize basic human activities such as running, walking, sitting, climbing stairs as well as daily activities such as cooking, shopping, and watching TV [3,17,33,36,37]. Detecting human activities and discovering behavioral patterns sets the ground for real-time monitoring of individuals' physical and mental health. This has urged researchers in the field of connected health to strive for building more accurate HAR systems.

Different computational methods ranging from classic machine learning (e.g., decision trees, naive Bayes [16,23], and Logistic Regression [35]) to graphical models such as HMM [11,28] and Conditional Random Fields [27,34] to Neural Networks [25,30,38] and Pattern Mining [6,12,13] have been applied to recognize human activities. These methods, however, are unable to recognize different levels of activities and their abstractions in one model.

In this paper, we present a tree-based hierarchical classifier using Neural Networks to both improve accuracy and capture different levels of abstraction. We use the Extrasensory dataset [35] to evaluate our proposed model. We test our approach on 6 activity labels, "Lying down", "sitting" and "standing" which are stationary activities and "walking", "running" and "bicycling" which are non-stationary activities. The results obtained on these basic activities are promising and indicate the capability of our method to be applied to more complex HAR systems.

2 Methodology

2.1 Data Processing

To clean the data, we discard all the samples other than those that have at least one relevant label from the six target labels. We do not remove samples with one or more missing labels for those six target labels as long as one of the six labels is relevant. For example, if the user was sitting at a certain time, it means no other activities could be performed. We also apply mean imputation for the missing features to avoid losing data.

2.2 Hierarchical Classification

Hierarchical Classification algorithms employ stacks of machine learning architectures to provide specialized understanding at each level of the data hierarchy [18] which have been used in many domains such as text and document classification, medical image classification [20], web content [10] and sensor data [24]. Human activity recognition is a multi-class (and in some cases multi-label) classification task. Examples of activities include sitting, walking, eating, watching TV, and bathing. The following contrasts the two types of classifiers we use in this paper for multi-class classification, namely *flat* and *hierarchical*. Given n classes as $C = \{c_1, c_2, ..., c_n\}$ in which C is the set of all classes, a flat classifier directly outputs one of these classes, whereas a hierarchical classifier first

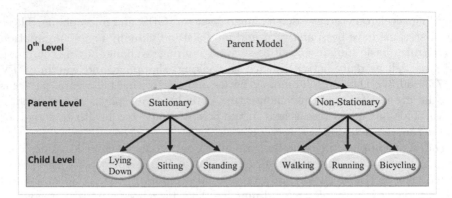

Fig. 1. Schematic of the hierarchical model employed for the classification task

segments the output in $\{c_p, c_p + 1, ..., c_q\}$ and then provides the final label in the next level c_i where $p \leq i \leq q$. C. N. Silla Jr., A. A. Freitas [31] conducted a comprehensive study on the hierarchical classifications across different fields whose terminology we use in this paper.

Our proposed method is to manually apply the inherent structure that exists in the context (activity) labels to our classifier. For example, activities done by individuals can very well be categorized as stationary and non-stationary activities. Lying down, standing, sleeping, and sitting are all considered stationary activities during which your net body displacement is negligible. On the other hand, activities such as running, walking, swimming, driving, bicycling and working out, are all examples of non-stationary activities where your net body displacement is non-negligible.

Activities can also be divided into indoor and outdoor activities. Indoor activities can be washing dishes, doing laundry and watching TV as opposed to biking and driving that are examples of outdoor activities. However, this classification can be more challenging as many activities can occur both indoor and outdoor. Additionally, many exclusively indoor activities are rare activities that Extrasensory dataset does not have sufficient data for training a model. Hence, we only focused on stationary and non-stationary activities in this paper.

In this experiment, we use the stationary vs non-stationary grouping of six mutually exclusive activities including "lying down", "sitting", "standing still", "walking", "running", and "bicycling". The first three activities are stationary and the rest are non-stationary activities. We also make sure that there is no sample in the dataset that indicates two of them happening at the same time (possibly due to misreporting).

Before introducing a hierarchy into the labels, we train a flat classifier and observe the performance of the classifier carefully. If there is a considerable misclassification between two or more distinct sets of classes, a hierarchy might boost the performance of the model.

As shown in Fig. 1, we employ a hierarchical classification with one local classifier per parent node (LCPN) [31]. In this setting, a classifier first classifies activity samples into stationary and non-stationary. In the second (child) level, one algorithm further classifies the stationary samples into sitting, standing, and lying down, and another classifier labels non-stationary activities into walking, running, and biking. This is a two-level hierarchy. The training of the parent nodes are done independently but, in the test phase we use a top-down prediction method to assess the overall performance of the model.

2.3 Deep Neural Networks

Deep neural networks (DNN) are artificial neural networks (ANN) with more than one hidden layer. The schematic architecture of a DNN is shown in Fig. 2. The number of nodes at the input layer is equal to the number of the features and the number of the nodes at the output layer is equal to the number of the classes [19]. DNNs are capable of finding a complex nonlinear relationship between the inputs and outputs. Also, they have shown tremendous power in prediction if designed and tuned well. Tuning of a DNN consist of tuning the number of the layers, number of the nodes per layer, appropriate activation function, regularization factor and other hyperparameters.

2.4 Evaluation

Training and testing of the classifiers in this setup are done independently which calls for creating separate training and test sets for each classifier. Since stationary and non-stationary are not among the labels, we create them using a simple logical OR over the child labels and build a binary classifier to label them. The second and third classifiers are corresponding to stationary and non-stationary nodes at the parent level and both of them are three-class classifiers.

The test data is passed through the first classifier at 0th level and based upon the predicted class it is passed to one of the classifiers at the parent level to make the final prediction out of the six labels. By comparing the actual labels with the predicted ones, we calculate the confusion matrix. Other performance metrics, such as accuracy, precision, recall, and F1 score can be readily obtained from the confusion matrix.

Accuracy score definition can be extended to multi-class classification and can be reported per class or averaged over all classes. However, as data becomes more imbalanced, the accuracy can be misleading. Therefore, the balanced accuracy becomes more relevant.

2.5 Baseline Methods

To better evaluate our model we needed a baseline to compare our results with. To our knowledge, there is no similar study with the same setup (same labels and dataset). Therefore we applied several classification algorithms along with a flat DNN on our data. Decision tree with a max depth of 20, k Nearest Neighbors (kNN) with k = 10, support vector machine (SVM) with "RBF" kernel, random forest with max depth of 10 and 10 estimators, and multi-layer perceptrons (one hidden layer with 64 nodes) were applied on the six class classification problem. The accuracy of these algorithms are compared with our model in the results section.

3   Experiment

3.1   Dataset

The Extrasensory dataset that is used in this work is a publicly available dataset collected by Vaizman et al. [35] at the University of California San Diego. This data was collected in the wild using smartphones and wearable devices. The users were asked to provide the ground truth label for their activities. The dataset contains over 300k samples (one sample per minute) and 50 activity and context labels collected from 60 individuals. Each label has its binary column indicating whether that label was relevant or not. Data was collected from accelerometer, gyroscope, magnetometer, watch compass, location, audio, and phone state sensors. Both featurized and raw data are provided in the dataset. Featurized data has a total of 225 features extracted from six main sensors. We used the featurized data in this work.

3.2   Results and Discussion

As mentioned earlier, we first train a flat classifier and examine the performance of the classifier before deciding to use the hierarchy. Hence, a DNN with 3 hidden layers and 256, 512, 128 nodes respectively were trained. A dropout layer was also added to prevent overfitting [32]. The number of nodes and the number of

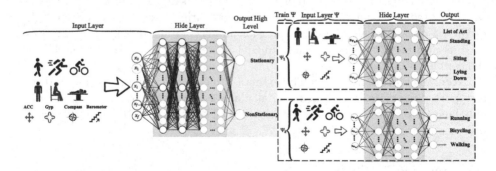

Fig. 2. HHAR-Net: Hierarchical Human Activity Recognition using Neural Networks. This is our Deep Neural Network (DNN) approach for Activity Recognition. The left figure depicts the parent-level of our model, and the right figure depicts child-level models defined by Ψ_i as input activity in the parent level [18].

[Flat DNN]

		Stationary			Non-stationary		
		Standing	Sitting	Lying Down	Running	Walking	Bicycling
Stationary	Standing	1092	277	25	1	45	14
	Sitting	142	3726	19	7	80	13
	Lying Down	33	102	3408	0	7	0
Non-stationary	Running	4	13	0	124	8	1
	Walking	34	130	6	10	432	13
	Bicycling	21	18	0	0	12	415

[Hierarchical DNN]

		Stationary			Non-stationary		
		Standing	Sitting	Lying Down	Running	Walking	Bicycling
Stationary	Standing	1264	149	7	0	16	7
	Sitting	111	3783	23	2	18	3
	Lying Down	22	31	3580	0	2	0
Non-stationary	Running	0	10	1	114	8	2
	Walking	82	142	10	10	373	17
	Bicycling	21	23	1	1	16	383

Fig. 3. Confusion matrix of a) flat classifier and b) hierarchical classifier

hidden layers varied between {2, 3} and {64, 128, 256, 512} respectively. The activation functions for the hidden layers were set to rectified linear unit (ReLU) [26]. In the output layer, softmax was used as the activation function for multi-class classifiers and sigmoid for the $0^t h$ level of the hierarchical model. Class weights were set to the inverse of the frequency of classes to compensate for the class imbalance. The implementation was done in Python using Keras [7] and sklearn [29] packages. From the entire dataset, 80% was selected for the training of the model and 20% was reserved for the evaluation of the performance of the model. Also, 5% of the training data was used as an evaluation set to tune hyper-parameters.

The resulting confusion matrix is shown in Fig. 3a. A quick glance at the confusion matrix reveals that the number of misclassifications between stationary activities and non-stationary activities is not negligible. But, to make a quantitative comparison, we calculated the number of misclassifications, between stationary activities and non-stationary activities and within each of them. To do so, we summarized misclassifications for our flat DNN in Table 1a. The absolute number of misclassifications within the stationary class is 598 which is 57.8% of

Table 1. Miss-classification within and between stationary (S) and non-stationary (Non-S) classes.

(a) Flat DNN. Total misclassifications are 1035.

	S	Non-S
S	598	167
Non-S	226	44

(b) Hierarchical DNN. Total misclassifications are 735

	S	Non-S
S	343	48
Non-S	290	54

all misclassifications. It is not a surprise to our intuition as discerning stationary activities are expected to be a challenge. However, the main takeaway from this table, is that nearly 38% of the total misclassifications are coming from the confusion between stationary and non-stationary classes. This implies that our classifier is struggling with distinguishing between these two types of activities in some cases. Surprisingly, misclassifications between "sitting" and "walking" is contributing to more than 20% of the total misclassifications. This urged us to come up with a hierarchy to reduce the misclassifications between the two sets of classes.

We trained a hierarchical DNN with the same hyper-parameters that were used for the flat DNN. Then we tested our model on the test data as explained in the evaluation section. The resulting confusion matrix is shown in Fig. 3b. Likewise, we calculated the misclassifications and summarized it into Table 1b. By comparing Tables 1a and 1b, we can see that the total misclassifications dropped significantly. Not only the number of misclassifications between stationary activities and non-stationary activities dropped from 393 to 338 but also the number of misclassifications within stationary classes decreased. This could be because in this hierarchical model, we are dealing with three-class classifiers at the parent level whereas the flat classifier has six classes.

Table 2. Performance comparison for several flat classifiers and the hierarchical classifier.

(a) Accuracy.

Classifier	Accuracy
Decision Tree	84.3
k-NN	87.5
SVM	87.7
Random Forest	83.5
MLP	87.8
Flat DNN	89.8
HHAR-Net	**92.8**

(b) Balanced Accuracy

Classifier	Balanced Accuracy
Decision Tree	75.9
k-NN	78.8
SVM	79.2
Random Forest	70.9
MLP	81.4
Flat DNN	84.1
HHAR-Net	**85.2**

Table 3. Detailed performance comparison of flat DNN and HHAR-Net. HHAR-Net exhibits a dominant performance over flat DNN

			Precision	Sensitivity	Specificity	F1-score
Flat DNN	Stationary	Standing	82.35± 1.95	75.10 ± 2.22	97.19± 0.84	78.56 ± 2.10
		Siting	87.34± 1.03	93.45± 0.77	91.02± 0.89	90.29± 0.92
		Lying Down	98.55± 0.39	96.00± 0.64	99.14± 0.30	97.26± 0.54
	Non-Stationary	Running	87.32± 1.10	82.67± 6.06	99.80± 0.19	84.93± 3.25
		Walking	73.97± 1.44	69.12± 7.39	98.30± 1.01	71.46± 4.10
		Bicycling	91.01± 0.94	89.06± 1.03	99.54± 0.22	90.02± 0.99
HHAR-net	Stationary	Standing	84.27± 1.88	87.60± 1.70	97.21± 0.85	85.90± 1.80
		Siting	91.42± 0.87	96.02± 0.61	94.15± 0.73	93.66± 0.76
		Lying Down	98.84± 0.35	98.49± 0.40	99.30± 0.27	98.66± 0.37
	Non-Stationary	Running	89.76± 0.99	84.44± 6.11	99.86± 0.13	87.02± 3.12
		Walking	86.14± 1.12	58.83± 8.30	99.35± 0.63	69.92± 4.26
		Bicycling	92.96± 0.83	86.07± 1.13	99.68± 0.18	89.38± 1.00

Tables 2a and 2b show the accuracy and balanced accuracy comparison of our proposed model, hierarchical DNN, and the baselines respectively. It is evident that our model is performing better than our baselines and also flat DNN. We achieved the accuracy of 95.8% at the $0^{t}h$ level, differentiating between stationary and non-stationary activities. Moreover, the classification accuracy of the stationary and non-stationary activities at the parent level were 92.8% and 93.2% respectively resulting in a 92.8% total accuracy.

To further verify the performance of HHAR-Net, we measured precision, sensitivity, specificity and F1 score for both flat DNN and HHAR-Net in Table 3. All of these metrics can be easily extracted from the confusion matrix nonetheless provided here for a more detailed comparison. One can witness that HHAR-Net shows better precision for all classes and also an improved Sensitivity for all classes except walking and bicycling. The same trend is seen in F1 score as it is a harmonic mean of precision and sensitivity. Another observation is that introducing a hierarchy has led to a significant improvement within the stationary classes which is our abundant class. On the other hand, non-stationary activities suffer from a lack of sufficient training data leading to larger error bars. This makes it hard to draw any conclusion about the slight superiority of the flat DNN F1 score over the HHAR-Net. Nevertheless, we can see a dominant performance of the proposed HHAR-Net over the flat DNN.

4　Conclusion

In this paper, we proposed a hierarchical classification for an HAR system, HHAR-Net. Our model proved to be capable of differentiating six activity labels successfully with a high accuracy surpassing flat classifiers, even a flat DNN with the same architecture. The idea of recognizing a structure in the labels can be applied in many other similar systems to improve the performance of the model

28. Nguyen, N.T., Phung, D.Q., Venkatesh, S., Bui, H.: Learning and detecting activities from movement trajectories using the hierarchical hidden Markov model. In: 2005 IEEE Computer Society Conference on Computer Vision and Pattern Recognition (CVPR 2005), vol. 2, pp. 955–960. IEEE (2005)
29. Pedregosa, F., et al.: Scikit-learn: machine learning in Python. J. Mach. Learn. Res. **12**, 2825–2830 (2011)
30. Ronao, C.A., Cho, S.B.: Human activity recognition with smartphone sensors using deep learning neural networks. Expert Syst. Appl. **59**, 235–244 (2016). https://doi.org/10.1016/j.eswa.2016.04.032
31. Silla, C.N., Freitas, A.A.: A survey of hierarchical classification across different application domains (2011). https://doi.org/10.1007/s10618-010-0175-9
32. Srivastava, N., Hinton, G., Krizhevsky, A., Sutskever, I., Salakhutdinov, R.: Dropout: a simple way to prevent neural networks from overfitting. J. Mach. Learn. Res. **15**(1), 1929–1958 (2014)
33. Su, X., Tong, H., Ji, P.: Activity recognition with smartphone sensors. Tsinghua Sci. Technol. **19**(3), 235–249 (2014). https://doi.org/10.1109/TST.2014.6838194
34. Vail, D.L., Veloso, M.M., Lafferty, J.D.: Conditional random fields for activity recognition. In: Proceedings of the 6th International Joint Conference on Autonomous Agents and Multiagent Systems, p. 235. ACM (2007)
35. Vaizman, Y., Ellis, K., Lanckriet, G.: Recognizing detailed human context in the wild from smartphones and smartwatches. IEEE Pervasive Comput. **16**(4), 62–74 (2017). https://doi.org/10.1109/MPRV.2017.3971131
36. Wang, A., Chen, G., Yang, J., Zhao, S., Chang, C.: A comparative study on human activity recognition using inertial sensors in a smartphone. IEEE Sens. J. **16**(11), 4566–4578 (2016). https://doi.org/10.1109/JSEN.2016.2545708
37. Wang, J., Chen, Y., Hao, S., Peng, X., Hu, L.: Deep learning for sensor-based activity recognition: a survey. Pattern Recogn. Lett. **119**, 3–11 (2019). https://doi.org/10.1016/j.patrec.2018.02.010
38. Zeng, M., et al.: Convolutional neural networks for human activity recognition using mobile sensors. In: 6th International Conference on Mobile Computing, Applications and Services, pp. 197–205, November 2014. https://doi.org/10.4108/icst.mobicase.2014.257786

Analysis of Streaming Information Using Subscriber-Publisher Architecture

Aparajit Talukdar[✉]

Assam Don Bosco University, Guwahati 781017, Assam, India
aparajit.talukdar2019@gmail.com

Abstract. Kafka is a massively scalable publisher-subscriber architecture. Using the features of Kafka this research paper accounts to create an architecture composed of subscriber-publisher environment incorporated with streaming data analysis application like Kafka in order to grab information from IOT based devices and provide summarized data along with prediction to the designated consumers. Through the use of Kafka, streaming huge volumes of data can be accomplished within seconds. In this research paper, Kafka's architecture have been integrated with machine learning algorithms to generate predictions on a variety of use cases which can be advantageous for the society. These algorithms have been examined on the basis of time to calculate speed of data production thereby providing an idea as to which algorithm is advantageous for anticipating the power consumption of a household.

Keywords: Kafka · IOT · Subscriber-publisher architecture

1 Introduction

In our daily lives, we are generating data at an unprecedented rate. This data can be of many types such as numeric, categorical, or free text and produced from all sectors of society. Data analysis and data collection are two very important terms for generating meaningful results [4].

In today's world data analysis plays a crucial role in any individual or organization's day to day activities. Data analysis gives interesting insights, informs conclusions and supports decision making for people and organizations. It is one of the biggest challenges associated with big data [5]. But before that, the specific data must be collected and also it should be able to make the data available to users. Data collection is the technique of collecting and measuring information from innumerable different sources. This is the work Apache Kafka [1]. It can handle trillions of events occurring in a single day. It helps in the transfer of data such that concentration can be given on data and not on how the data can be transferred and shared.

With this information we draw the concept of a subscriber-publisher architecture [4]. In this model, senders called publishers do not intend the messages to be sent to receivers called subscribers, but categorize the messages into classes

© Springer Nature Switzerland AG 2021
M. Singh et al. (Eds.): IHCI 2020, LNCS 12615, pp. 59–69, 2021.
https://doi.org/10.1007/978-3-030-68449-5_7

or topics without the knowledge of which subscribers there are. In a similar fashion, subscribers expose their interest to one or more classes or topics and only receive those messages that are specific to those groups without the knowledge of which publishers there are [9].

A point to note here is that the data collected needs to be error free and have appropriate information for the task required. This is essential because a good data helps to develop high performance predictive models. Predictive models are built by capturing records of past events and analyzing them to find recurring patterns and then using those patterns and machine learning algorithms to look for trends and predict future changes [16].

Apache Kafka has become one of the most widely used frameworks for critical real time applications. It has been integrated with a well organized machine learning framework to support real time data streaming. Kafka is used for building real-time data pipelines and streaming apps. It is horizontally scalable, fault-tolerant, increasingly fast, and runs in production in thousands of companies [12].

The households generally don't have access to a prediction system where they can forecast their weekly consumption of electricity. Hence, they need a system where they can send their previous data and get a forecast about the expected electricity consumption rate. This project aims to address that situation. We have proposed a way to use the features of Kafka in order to analyze streaming information. Since Kafka is a massively scalable publisher-subscriber architecture, it allows us to send entire datasets for analysis. This architecture then pairs up with machine learning algorithms to predict forecast values which the consumer wants. This architecture is fast enough to produce results without long waiting times and is also able to use multiple servers instead of a single one for increased reliability. This project helps the consumer to have an economical understanding of his/her electricity consumption. The consumer will also be able to save electricity when he/she has a previous knowledge about the incoming consumption and thus reduce wastage. This paper also aims to calculate the time required for each machine learning algorithm used to find out which algorithm is faster.

2   Related Works

2.1   Traffic Monitoring System

This is an application that processes the IoT data sent in real time in order to monitor traffic [8,15]. There are three components in this application.

- IoT Data Producer: The data producer sends the messages generated by the connected vehicles for processing. This processing is done by the streaming application.
- IoT Data Processor: This takes in data from the producer and processes them for analysing traffic. This is done by storing the total vehicle count and the total vehicle count for the last 30 seconds for different types of vehicles on

different routes on the Cassandra database. This is also done by storing the details of vehicles withing a certified radius in the Cassandra database.
- IoT Data Dashboard: This receives data from the Cassandra database and sends it to the web page. It displays the data in charts and tables.

2.2 Recipes Alert System

This is an alert system designed to notify if calories in a recipe meet a certain threshold. Allrecipes data is used for this system [7]. There are two topics that are used here:

- raw_recipes: This stores the HTML of each recipe.
- parsed_recipes: This is the JSON format of the parsed data of each recipe.

The producer program first accesses Allrecipes.com and collects the raw HTML data. This raw HTML is stored in raw_recipes topic. The next program acts as the consumer as well as the producer. It consumes data from the raw_recipes topic, transforms the data into JSON format and then sends it to the parsed_recipes topic.

These recipes are stored in JSON and raw HTML format for further use. Then a consumer program connects with the parsed_recipes topic and sends a message if certain criteria of calories match. The calories are counted by decoding the JSON and a notification is issued.

3   Proposed Methodology

3.1 Components Used

In this paper the dataset that is used is the Household Power Consumption dataset which contains the power consumed for four years in a specific household. We have used the first three years to train the model and then the final year for evaluating the model.

This dataset shows the observations of power consumed for everyday between December 2006 and November 2010. The dataset has eight columns and they are as follows:

- Date: Shows the corresponding date.
- Global active power: It shows the total active power consumed by the household in kilowatts.
- Global reactive power: It demonstrates the total reactive power consumed by the household in kilowatts.
- Voltage: It represents the average voltage in volts.
- Global intensity: It gives the average current intensity in amps.
- Sub metering 1: Represents kitchen's active energy.
- Sub metering 2: Represents laundry's active energy.
- Sub metering 3: Represents climate control systems' active energy.

This paper has only dealt with the global active power for analysis and prediction.

- Kafka: Kafka is the backbone of this architecture. This architecture relies on Kafka to send and receive messages. It uses a subscriber publisher model for real time streaming. It is a highly scalable platform which has low-latency and high throughput for handling data feeds. The Household Power Consumption dataset is fed into the Kafka framework so that it can be analyzed and produced to the consumer. Kafka consists of a producer, consumer, topics, broker, cluster [6].
- Kafka Producer: As the name stands the Kafka Producer is used to publish the message by pushing data to brokers. Messages are published into specific topics. These topics are intended for specific consumers who subscribes them [1].
- Kafka Consumer: Consumers are those that consume or receive the messages published by the producer. Consumers have a consumer group name, and the data is ingested by the consumer from the topic whenever it is ready [1].
- Kafka Topic: It is a particular category of a stream of messages where data is stored. Each topic is split into one minimum partition [1].
- Kafka Broker: Kafka brokers use ZooKeeper to maintain clusters since the Kafka Cluster consists of multiple brokers. This helps to support load balance. A cluster can be enlarged without downtime and is stateless. A Kafka broker can handle huge amount of messages without taking a hit in performance [1].
- ZooKeeper: It manages and coordinates the Kafka broker. If any new broker is present in the Kafka architecture ZooKeeper notifies the producer and the consumer. As per the notifications, decisions are taken and tasks are performed [2].

3.2 Algorithms Used

- Linear Regression: It is based on a linear relationship between an input variable (say X) and an output variable (say Y) [11]. Y is calculated from X i.e. Y is dependent on a linear combination of the variable X. When there is a single input variable involved then it is called a simple linear regression and when there are more than one input variables involved then the regression is called multiple linear regression. There are various techniques [10] available for training the dataset with linear regression one of which is the ordinary least square technique.

  Linear regression in simple to perform because of its representation as a linear equation. It gives the predicted output (Y) for a set of input values (X). Therefore, both the input and output values are numeric in nature.

  A coefficient is assigned for each and every input value. This coefficient is represented by the Greek letter Beta. Another coefficient is also added which helps to move up and down the plotted line called the intercept or bias coefficient.

The linear regression equation for a simple problem can be represented as:

$$Y = B0 + B1 * X \tag{1}$$

Here a single X and a single Y value is taken and B0 is the intercept while B1 is the slope of the line taken. In case of more than one input the line becomes a plane or a hyper plane. This takes place in higher dimensions.

A point to note here is that as soon as a coefficient becomes zero, the influence of the input variable becomes useless and consequently the prediction obtained from the model also becomes zero.

- Naive Forecasting: This forecast is suited for time series data and in case of cost effective situations. It uses the last period's value to calculate forecasts that are equal to that observed value. This model provides a benchmark through which more complex models can be compared.

  Based on data recorded for a past period the naive forecasting method [10] evaluates a prediction for a coming period. The naive forecast can be considered as equals to a previous period's actual values, or an average of the actual values of certain specific previous periods. Adjustments for variations and trends of past periods are not considered for this forecasting. It generates a forecast with-out considering these factors. Causal factors are also not a concern for naive forecasting as they are not considered for generating predictions. This helps the naive forecast to check results of much more complicated and sophisticated methods.

- Lasso Regression: The lasso regression model was introduced in 1986. It was again brought to the public view and popularized by Robert Tibshirani in 1996. It is a form of linear regression but uses shrinkage i.e. it involves a process to shrink values towards a mean or a central point. It conducts L1 regularization in which case a penalty is added equivalent to the absolute value of the coefficients' magnitude. The larger the penalty, the more coefficient values nearer to zero. This is idealistic situation for generating simpler models [9].

  It was designed originally for least squares models. This justifies its relationship to ridge regression and best subset selection. It also defines the relationship between lasso coefficient estimates and soft thresholding.

4   Architectural Design

- The first thing done in this research is running the Kafka server which is the main backbone of the entire architecture. The Kafka server acts as the broker between the Producers and the Consumers. The Kafka server also needs the Zookeeper server along with it. Therefore, the Zookeeper server is also started.
- Zookeeper is used to maintain the synchronization with the various components running in the architecture. It keeps information about the status of topics, partitions and cluster nodes used. Within Zookeeper the data is divided among multiple nodes and it helps to achieve reliability. If one node fails, the data is not lost and it can perform failover migration.

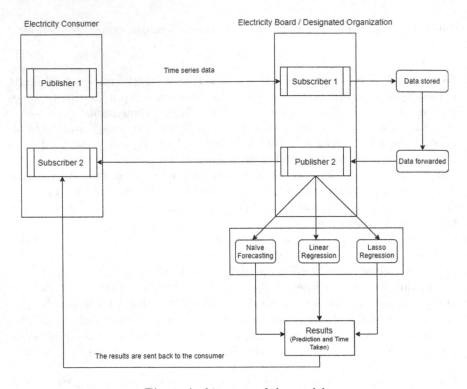

Fig. 1. Architecture of the model.

- In the next step a Kafka topic is created to publish a message or a stream of messages. These topics can be subscribed by multiple subscribers and can have single or multiple partitions but here only single partition is used.
- For initial testing a producer and a consumer are first set up to publish and receive messages. Kafka's command line client is used to read messages and publish them from the producer to the consumer either from a text file or from the command prompt/terminal.
- In an attempt to utilize Kafka's features, instead of a single broker the architecture uses two brokers with two different localhost addresses (but in the same local machine). So another Kafka server is run simultaneously.
- After successfully sending and receiving dummy messages, the main dataset and its results are being operated upon. The main Household Power Consumption dataset is first loaded, cleaned and saved before it is read by a producer and sent to the consumer for analysis. This is done because the dataset contains mixed types of data and also missing values.
- The producer in this case is the local household and the consumer is the electricity board or the entity responsible for analysis. The consumer writes out the dataset to a file to store it after receiving the input from the producer.

- For the analysis part, the electricity board becomes the producer and runs a publisher program to perform analysis on the dataset. The Global active power is predicted for the upcoming week. This is done by splitting the dataset into standard weeks and further splitting them into years. This dataset is then divided into three years for training predictive models and one year for evaluating models. The analysis is carried out by using three machine learning algorithms namely linear regression, lasso regression and naïve forecasting. They all use a technique called walk forward validation, which uses previous weeks' data to predict the subsequent week. The time required for each algorithm to perform their respective forecast is also calculated. This output is delivered to the designated household, which now becomes the consumer.
- The consumer can use this data to optimize the power consumption to mitigate cost and save energy. The electricity provider can also get an idea as to how much power should be supplied to a particular customer based on a daily consumption.

5   Observations

Before setting up the architecture for analysis of electricity consumption, the initial sending and receiving of messages through Kafka is shown below in Fig. 2 and Fig. 3. This is just a dummy message in order to test whether the setup is working properly.

Fig. 2. Sending of a test message.

Fig. 3. Receiving the test message.

The architecture mentioned in this paper uses three different machine learning algorithms [14] namely, naive forecasting, linear regression and lasso regression. These algorithms are used for predicting and sending the electricity consumption forecasts for a week to the specified household. In addition to providing the predicted values, the time taken for prediction are also calculated for the three algorithms. This gives an idea as to which process is faster.

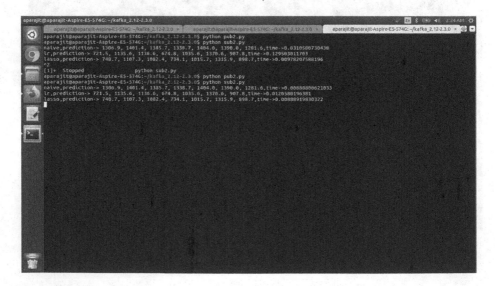

Fig. 4. Output of predicted values and time required.

Figure 4 shows the predicted values for two observations and the time taken for each process for the three different algorithms. These predicted values are received by the consumer. In an effort to see which algorithm works faster ten observations are taken for consideration.

Table 1. Observations for time taken by the algorithms

Sl. No	Naive Forecasting	Linear Regression	Lasso Regression
1	0.008685112	0.011534929	0.00907588
2	0.008874893	0.011316061	0.008983135
3	0.009013891	0.011430025	0.008874178
4	0.013002872	0.011392117	0.008968115
5	0.008614063	0.011363983	0.008756161
6	0.01412487	0.017096996	0.011085033
7	0.009347916	0.013273001	0.009985924
8	0.010995865	0.018764019	0.009129047
9	0.008780956	0.011371136	0.008622885
10	0.010751009	0.016293049	0.009701967

Table 1 shows ten observations that were taken for comparison. The following graph in Fig. 5 shows the results in a more decisive manner.

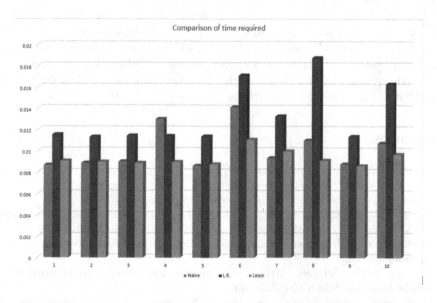

Fig. 5. Graph showing the time comparison between different algorithms.

The above graph in Fig. 5 shows the time required by the three algorithms in giving the predicted values. The x-axis shows the number of observations (in this case ten observations are taken) and the y-axis indicates the time taken.

Some interesting conclusions can be inferred from this graph. It can be seen that out of the three algorithms linear regression takes the highest amount of time to generate the predictions in all except one of the ten observations. Naive forecasting technique takes the shortest time in four observations and the longest time in one of them. Lasso regression takes the shortest time in six observations but does not take the longest time in any of the other ones. It records the shortest average time taken across all observations whereas linear regression takes the longest average time across all observations. Also, in most cases the time difference between naive forecasting method and lasso regression method is minimal indicating that at times both algorithms takes almost the same amount of time to generate predictions.

Thus, considering time it can be concluded that lasso regression is a better choice for our architecture to deliver an optimal performance in swiftness.

6   Conclusion

Through the use of Kafka, streaming large volumes of data can be achieved within seconds. Multiple servers can be established so that data for different consumers do not get mixed up. Kafka's architecture can be integrated with machine learning algorithms to generate predictions on a variety of use cases which will be beneficial for the society. This paper has taken up the issue of electricity consumption by using three algorithms.

These three algorithms have been analyzed on the factor of time to calculate speed of data production. The results have proved that lasso regression takes the shortest time in most occasions and linear regression takes the highest amount of time. These results give an idea as to which algorithm is beneficial for predicting the electricity consumption of a household on the basis of speed of delivery of the output of the respective algorithms.

References

1. Apache Kafka. https://kafka.apache.org/
2. Zookeeper. http://zookeeper.apache.org/
3. Postscapes. https://www.postscapes.com/diy-apache-kafka/
4. Instructable. https://www.instructables.com/id/Subcriber-PublisherEnvironment/
5. TheVerge. https://www.theverge.com/ad/17604188/Big_data_analysis
6. Confluent. https://www.confluent.io/blog/using-apache-kafka-drive-cutting-edge-machine-learning
7. Towards Data Science. https://towardsdatascience.com/getting-started-with-apache-kafka-in-python-604b3250aa05
8. InfoQ. https://www.infoq.com/articles/traffic-data-monitoring-iot-kafka-and-spark-streaming/

9. Karapanagiotidis, P.: Literature review of modern times series forecasting methods. Int. J. Forecasting **27**(4), 11–79 (2012)
10. Towards Data Science. https://towardsdatascience.com/putting-ml-in-producti on-using-apache-kafka-in-python-ce06b3a395c8
11. Rajkumar, L.R., Gagliardi, M.: The real-time publisher/subscriber inter-process communication model for distributed real-time systems: design and implementation. In: Proceedings of the IEEE Real-Time Technology and Applications Symposium (1995)
12. Fusco, G., Colombaroni, C., Comelli, L., Isaenko, N.: Short-term traffic predictions on large urban traffic networks: applications of network-based machine learning models and dynamic traffic assignment models. In: Proceedings of the 2015 IEEE International Conference on Models and Technologies for Intelligent Transportation Systems (MT-ITS), Budapest, Hungary, 3–5 June 2015, pp. 93–101 (2015)
13. Machine Learning Mastery. https://machinelearningmastery.com/naive-methods-for-forecasting-household-electricity-consumption/
14. Albanese, D., Merler, G.S., Jurman, Visintainer, R.: MLPy: high-performance python package for predictive modeling. In: NIPS, MLOSS Workshop (2008)
15. Research Gate. https://www.researchgate.net/publication330858371_Adaptive_Traffic_Management_System_Using_IoT_and_Machine_Learning
16. Schaul, T.: PyBrain. J. Mach. Learn. Res. 743–746 (2010)

Fraudulent Practices in Dispensing Units and Remedies

Undru Vimal Babu[1,4], M. Nagamani[2(✉)], Shalam Raju[3(✉)],
and M. Rama Krishna[4(✉)]

[1] Legal Metrology, Hyderabad, India
uvbabu43@gmail.com
[2] SCIS, UOH, Gachibowli, Hyderabad, India
nagamanics@uohyd.ac.in
[3] Legal Metrology, Guntur, Andra Pradesh, India
shalemraju.tambala@gmail.com
[4] Vignan Foundation for Science, Technological University,
Guntur, Andra Pradesh, India
drmramakrishnaa@gmail.com

Abstract. Growth of Artificial Intelligence and its rational thought "doing things right" makes handling fraudulent detection practices by implementing Machine Learning and Deep Learning methods. The detection needs to analyze regular patterns with its anomalies that are easy and speedy manner by machine than human cognitive process. The humans perceive and predict suspicious situation but machines analyze and detect. Analyzing regular patterns to anomalies enable machine to detect rationally than human inspections in various data. Hence human cognition capabilities incorporating in machine is trend of technology to prevent fraudulent practice. This work explores re-engineering on discovering fraudulent practice in dispensing unit. Initially short delivering is established, understand dispensing unit in the retail outlet of Petroleum products in responding to short volume, which is used for fraudulent practice and triggering methods. Then explored concepts are structured as theory for the short delivery which is to propose the secure system to prevent fraud.

Keywords: Artificial Intelligence · Machine learning · Deep learning · Patterns · Petroleum product · Dispensing unit · Fraudulent

1 Introduction

The petroleum products are dispensed through the dispensing units in the retail outlets of Oil companies shown in Fig. 1. The dealers are resorting to unethical practices while dispensing the Fuel through the dispensing units such that the deliveries through the unit are given short than the expected. This short delivery

Supported by Legal Metrology, Telangana.

can be achieved through the insertion of external devices into the hardware parts of the Dispensing Pump. Some of the electronic devices can be controlled through the remote devices in India [19].

1.1 Introduction to Artificial Intelligence Fraud Detection

Artificial Intelligence era explores as various domain applications with exercising and applying reverse engineering and re-engineering methods to build intelligent system that prevent the fraudulent system in society. The present paper work focus on the exploring the fraudulent practices in Dispensing pumps in terms of their short delivery of fuel from the retail outlet of petroleum products with less volume than intended. The present intelligent users malpractices and their detection process using various physical and AI methods are used to identify the context of the fraudulent situation and their prevention method proposals through data engineering applications are main context presented in this work. Various electronic systems and their visual inspection as cognitive method and the system detected methods explored [1].

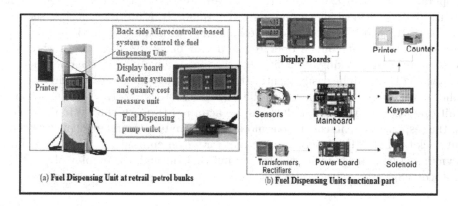

Fig. 1. Dispensing unit at retail outlets of petrol products fraudulent practices

1.2 Petroleum Product Dispensing Units at Retail Outlet

The petroleum products are dispensed through the dispensing units in the retail outlets of Oil companies. The dealers are resorting to unethical practices while dispensing the Fuel through the dispensing units such that the deliveries through the unit are given short than the expected. This short delivery can be achieved through the insertion of external devices into the hardware parts of the Dispensing Pump [12]. Some of the electronic devices can be controlled through the remote devices in India [13].

2   Review Works on the Existing Systems

This is a practical approach and is presented in this article, without going in detail into the architectures of the hardware Dispensing Pump. The android mobiles can be utilized to control the process simulators and controllers, and these can be shared with other [1]. Normally the Lead acid batteries are used to give power input to the devices. There shall be renewable power sources to the devices such that there shall not be any gap in the detection of source through the sensors [2]. Handheld devices can also be utilized to control the home appliances and for computers [3]. Similarly, the hardware devices of the Dispensing Units can be easily controlled through remote. The devices which can interact with the people with the recognized voice system otherwise called pervasive computing. These can be used to lock the Dispensing System or unlock the same even when the password is forgotten [4]. These kinds of devices help the owner of the Dispensing units to control the delivery through the unit. The person operating the unit can be easily identified. This will also help in the recognition of the person who is involved in the process of manipulation if any at and the time of manipulation. The error logs can be traced to identify the time and person who is operating the dispensing unit. Here the authors have given description about the irradiated parts which can be controlled remotely, Materials and Life science Facility (MLF) [5] The person can be identified with the bio-metric system [6]. The dispensing unit can be operated and can be accessed to the authorized persons through this king of facility. So, the person involving in the operation of Dispensing Unit can be easily identified. The real analogue electronic circuits are validated through the internet with the help of remote lab [7,12]. The Dispensing Unit operation and real time data can be acquired through this system to perk up the response time in concurrent mode and to analyse the reliance of system to put back the Lab-VIEW algorithms so also for better employment. The electronic devices and be managed with the help of internet through the employ of TCP-IP protocols [8].

2.1   Fraudulent Practices in Dispensing Units Causes

To gain more money unlawfully, some of the dealers of retail outlets generally resort to fraudulent practices in Dispensing Units. Two methods the vender following to cheat the user

- By delivering short volume [18]
- By adjusting totalizer reading [15,19]

First Method. The first methods isolated circuit allow the intended to manipulate the dispens-ing pump by external control incorporated to manipulate the speed and flow of the fuel delivery can be altered without effecting other module functionality hence the user cannot able to detect the fraud. The chances of alteration possible in the system are Metering Unit, Pulsar/Sensor, Mother Board, Keypad, Display Unit.

Second Method. The second method is through programming alteration of the pulsar [12] functionality with remote operation. Whenever the verification they set back the original program. With this it is difficult to notice the fraud as it was fully control of the in tender. In the second case the protection can me made such that the software should be strong enough that it shall not allow by passing the original parts. Any by pass of the original system should be notified and detected as fraudulent condition identification. The following Fig. 2 will describe the system functionality [21] who describe in his patent about the detection process with flow diagram with the parts in-tended to prone are Mother board and totalizer [15, 20].

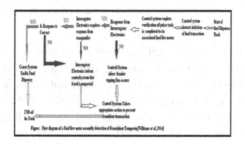

Fig. 2. Fuel dispenser protection method for malpractices [15, 21]

3 Observation and Analysis

3.1 Fraudulent Practices

For achieving the above two goals offenders resorts to fraudulent practices.

The physical inspection and exploring the complaints and data of the dispensing pumps by legal enforcement team identify the fraudulent methods from the me-mechanical, semi-automated and fully automated system are the major malpractice with external system of dispensing unit. [13] To gain more money unlawfully, some of the dealers of retail outlets generally resort to fraudulent practices in Dispensing Units. Two methods the vender following to cheat the user (1) By delivering short volume [16], (2) By adjusting totalizer reading [15]. In the first methods isolated circuit allow the intended to manipulate the dispensing pump by external control incorporated to manipulate the speed and flow of the fuel delivery can be altered without effecting other module functionality hence the user cannot able to detect the fraud. The chances of alteration possible in the system are Metering Unit, Pulsar/Sensor, Mother Board, Keypad, Display Unit.

The second method is through programming alteration of the pulsar [9] functionality with remote operation. Whenever the verification they set back the

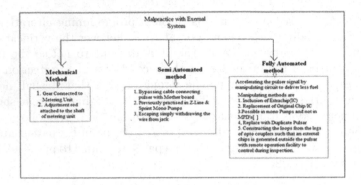

Fig. 3. Fraudulent practice types with various methods [15].

original program. With this it is difficult to notice the fraud as it was fully control of the in-tender. In the second case the protection can me made such that the software should be strong enough that it shall not allow by passing the original parts. Any by pass of the original system should be notified and detected as fraudulent condition identification. The following Fig. 3 will describe the system functionality [20] who de-scribe in his patent about the detection process with flow diagram with the parts in-tended to prone are Mother board and totalizer.

3.2   Method of Adjusting K-Factor

Initially the present K-Factor is recorded. With the help of the formula given below, new K-Factor is arrived to get the desired output (short delivery of the product).

$$NewK - Factor = PresentK - Factor * \left(\frac{DispensedVolume}{DisplayedVolume} \right) [15] \qquad (1)$$

Here is interface is in ON position and the emergency switch is in OFF position. To run the Dispensing Unit with modified K-Factor for getting the short delivery (Output) of the product, the plug-in device will be attached with modified K-Factor and the emergency switch will be in ON position. If any suspicion arose in the mind of the customer or in any inspections of en-forcement agencies like Legal Metrology Department, Police, Civil Supplies Depart-ments or by Oil Marketing Companies, they put OFF the emergency switch or de-taches the battery cable so that the unit runs as usual.

3.3   Reminiscences Possibilities

There will not be any Reminiscences like soldering or planting any extra fittings or extra wires found for the reason that after the Dispensing Unit was started with modified K-Factor the operator removes the external plug-in device and connects the same original cable. Hence it is possible to notice the reminiscences (Fig. 4).

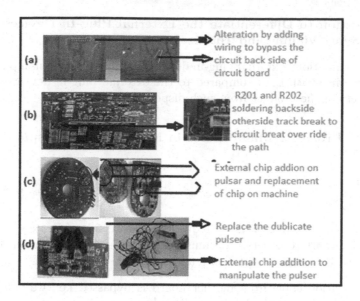

Fig. 4. Visual inspection to identify the fraud in dispensing units [15]

4   Experimental Practices Visual Inspection of System by Human Cognition

4.1   Identify Malpractices in GVR Make Dispensing Units

Totalizer Related Malpractices. Here the 20 pin connectors on the left side middle portion of mother board pertain to GVR make Dispensing Unit. This is original unused for manipulation one. Defaulters prepares an external plug-in device which acts as second totalizer and inserts this external plug in device into the unused connector. To make running this second total-izer just above the 20-pin connector, two resistors namely R201, R202 must be sol-dered and other side of the mother board original communication is disturbed, by cutting the track path and alternate communication has arranged. Just opposite side of the 20-pin connector on mother board some soldering takes place.

4.2   Short Delivery Methods

- *A.* If possible, the targeted Dispensing Unit shall be allowed for running by detaching cable of battery or putting OFF the emergency switch.
- *B.* Then if possible, try to take the print of K-Factor of the unit. This will enable to get modified K-Factor with date and time. By analysing the calibration counts and K-Factor we can guess the quantum of short deliveries they have attained.

4.3   Detection to Differentiate the External Plug-In Devices for Short Volume and Totalizer

Physical examination: External plug-in device used for totalizer manipulation is will be more in width than compared to the external plug-in devices used for short delivery of the product from such dispensing units.

4.4   Parts Prone for Malpractices in MIDCO Make Dispensing Units

– Keypad
– Display Unit

Keypad: Additional fitting will come on the opposite side of the keypad, which is not vis ible part to observe, we must unscrew the screws to open the user interface board (Key-pad) [4] (Figs. 5 and 6).

– Communication cable from mother board is bypassed i.e., we can observe second cable. So, communication is disturbed
– We can observe external fitting
– Extra wire is used for voltage impedance

Fig. 5. By passing calibration cable & chip soldered to manipulate the circuit path

Fig. 6. Bypassing the wires to change the circuit path

4.5   Process to Identify Malpractices

Family integrity concept should be broadened by including every peripheral device into family. When there is tamper to integrity of family, alert messages should be there (Figs. 7 and 8).

- Dispensing Units shall avoid using weak software.
- While issuing Model Approvals, International Organization of Legal Metrology (OIML - D31) directives shall be followed
- Original Equipment Manufacturers research and development shall not be leaked.
- Unused connectors shall not there be on mother board.
- In place of metallic cover plates, better to use transparent covers.
- Fool proof automation is also one of the solutions to malpractices.
- Oil Marketing Companies shall prefer quality tamper proof Dispensing Units, rather than cost effective Dispensing Units.
- There shall be high standard Recruitment process for Original Equipment **Manufacturer technicians to contain/maintain standards as specified to the Unit**.
- There shall be strict mandate and following of statutory rules for the **authorities-conducting the inspecting against the Dispensing Unit such as Legal Metrology officials.**

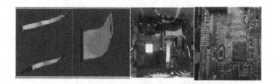

Fig. 7. External plugins and mother board connections [15]

Fig. 8. Remote controllers using key chain option to control []

5   Discussions and Suggestion

With Human cognition the observation and inspections on the fraudulent practices by the petroleum product dispensing outlets and their system of malpractices the following conclusion drawn with research support and the law enforcement authority.

5.1 Suggestions

The statutory authorities put seals to the different parts of the Dispensing unit so that there shall not be any scope for manipulation. The businessmen have adopted new device such that the seals put by the statutory are intact and the hardware parts are changed to the benefit of businessmen. This kind of tactics are undetected both by the Statutory and the Oil Companies. With this kind of techniques of the traders the end consumers are losing the product what they are paid for. In order to prevent such insertions, there shall be a greater number of error logs to be recorded and made available to the inspecting authority. With this error logs the kind of cheating can be easily detected. One shall look at the repeated errors. If there are about 10 to 15 errors of E09 in the time span of 30 to 60 min there shall be serious view. The error log E09 will occur when there is fluctuation in the power supply. To check this is a normal error or not one shall cross check the same error in other dispensing units situated in the same retail outlet. If there is a variation from one dispensing unit to other unit, one should doubt the manipulation. The error logs E09 coupled with E28 is even more serious (Fig. 9).

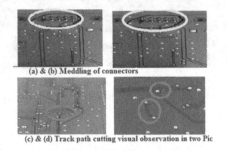

(a) & (b) Meddling of connectors

(c) & (d) Track path cutting visual observation in two Pic

Fig. 9. Meddling and tract cutting with visual observations [21]

6   Conclusion and Future Scope

The software of the Dispensing Unit shall detect the insertion of external devices into the main hardware parts like Mother Board, Calibration Card, and Display Board. The software is either failed to detect the external devices inserted to the various hardware parts or there may be change of software in the Dispensing Unit. There must be fool proof system to automatically identify the malpractices; else there shall be automatic system not to function automatically. Incorporation of Artificial intelligence methods [12–14, 16] with visual inspection using image processing methods and signal bypass detection methods using the deep learning and machine learning techniques, and prevention of alteration by incorporation secure algorithms in programming of the microprocessor that monitor the flow and any signal by pass of the original mother board immediate detection of communication breaking symptom detection should cause the functional down of the system helps to come out of these fraudulent. Exploration of the various secure algorithms [17] that signals the remote alteration of the pulsar.

Acknowledgements. We would like to acknowledge Legal Metrology Departments from Telangana and Andra Pradesh states for their support and opportunity to work on Fuel Dispensing pump outlet system fraud analysis by understanding thee entire system functionality. We would also like to acknowledge the Mr. Ajay kumar BPCL, Mumbai willing to collaborate and opportunity for the University students to work in the area of Machine learning and Deep learning methods to explored practical observation data. Finally acknowledgement to School of Computer and Information Sciences, University of Hyderabad for environment to work in this domain with resource support.

References

1. Papadopoulos, A.V.: Politecnico di Milano, Dipartimento di Electronica, Informazione e Bioingegneria, Labouratories over the network: from remove mobile, 1 ISSN0th IFAC symposium Advances in Control Education, The International Federation of Automatic Control August 2 2249-684X, Sheffield, UK, pp. 8–30 (2012)
2. Dewan, A., Ay, S.U., Karim, M.N., Beyenal, H.: Alternative power sources for remote sensors: a review. J. Power Sources **245**, 129–143 (2014)
3. Myers, B.A.: Using handhelds for wireless remote control of PCs and appliances. Interact. Comput. **17**(3), 251–264 (2005)
4. Zhu, M., He, T., Lee, C.: Technologies toward next generation human machine interfaces: from machine learning enhanced tactile sensing to neuromorphic sensory systems. Appl. Phys. Rev. **7**(3), 031305 (2020)
5. Jeong, H.D.J., Lee, W., Lim, J., Hyun, W.: Utilizing a Bluetooth remote lock system for a smartphone. Pervasive Mob. Comput. **24**, 150–165 (2015)
6. Kinoshita, H., et al.: Remote handling devices in MLF. Nuclear Instrum. Methods Phys. Res. A **600**, 78–80 (2009)
7. Limbasiya, T., Doshi, N.: An analytical study of biometric based remote user authentication schemes using smart cards. Comput. Electr. Eng. **59**, 305–321 (2017)
8. Hernandez-Jayo, U., Garcia-Zubia, J.: Remote measurement and instrumentation laboratory for training in real analog electronic experiments. Measurement **82**, 123–134 (2016)
9. Soto-Cordova, M.: A remote control and supervision device through internet, research and development division national. In: IFAC Telematics Applications in Automation and Robotics, Institute for Research and Training in Telecommunications (INICTEL), Weingarten, Germany (2001)
10. Vimal Babu, U., Ramakrishna, M., Nagamani, M.: Review on utilization of nanomaterials and polymers in fruit ripening, food packing and its hazards to human health (2015)
11. Limbasiya, T., Doshi, N.: An analytical study of biometric based remote user authentication schemes using smart cards. Comput. Electr. Eng. **305–321**, 59 (2017)
12. Vimal Babu, U., Krishna, R., Mani, N.: Review on the detection of adulteration in fuels through computational techniques. Mater. Today: Proc. **4**(2), 1723–1729 (2017)

13. Vimal Babu, U., Mani, M.N., Krishna, M.R., Tejaswini, M.: Data preprocessing for modelling the audulteration detection in gasoline with BIS. Mater. Today: Proc. **5**(2), 4637–4645 (2018). https://doi.org/10.1016/j.matpr.2017.12.035. http://www.sciencedirect.com/science/article/pii/S2214785317330109
14. Vimal Babu, U., Ramakrishna, M., Nagamani, M., et al.: Detection of fuel adulteration through multivariate analysis using python programming. IOSR J. Comput. Eng. (IOSR-JCE) **20**(5), 23–26 (2018)
15. Nagamani, M., Babu, U., Ramakrishna, M., Kumar, S.: Manipulation of electronic devices and data in dispensing pumps. Int. J. Innovative Technol. Exploring Eng. **8**, 1–13 (2019). https://doi.org/10.35940/ijitee.K1287.0981119
16. Vimal Babu, U., Rama Krishna, M., Naga Mani, M., Tejaswini, M: Environmental and health hazards of fuel adulteration and its detection, safeguards. Int. J. Electrical Eng. Educ. The Author(s) 2020 Article reuse guidelines:https://sagepub.com/journals-permissions. https://doi.org/10.1177/0020720920940609. https://journals.sagepub.com/home/ije
17. Times of India Hyderabad Petrol Pump Owners cheating by inserting chips in pumps. https://timesofindia.indiatimes.com/city/hyderabad/petrol-pump-owners-install-cheat-chips-trick-motorists-in-telangana-and-andhra-pradesh/articleshow/77957198.cms
18. 6th September 2020, Seized 11 petrol pumps for cheating. https://www.newindianexpress.com/states/telangana/2020/sep/06/eleven-fuel-pumps-in-telangana-seized-more-than-13-arrested-for-duping-people-2193046.html/
19. Reference manual of C4000 processor based dispensing pumps. http://www.compac.biz/vdb/document/551
20. MobileMark Homepage. https://www.thehindu.com/news/national/andhra-pradesh/racket-pilfering-fuel-at-petrol-pumps-busted/article32526270.ece/
21. US patent: Williams 2014. https://patentimages.storage.googleapis.com/9c/d1/0d/cd215c91b5bebc/US8844587.pdf

Biomedical Signal Processing and Complex Analysis

Digital Processing of Blood Image by Applying Two-Dimensional Haar Wavelets

H. N. Zaynidinov[1](✉) ⓘ, I. Yusupov[1](✉) ⓘ, J. U. Juraev[2](✉) ⓘ,
and Dhananjay Singh[3] ⓘ

[1] Tashkent University of Information Technologies named after Muhammad al Khwarizmi,
Tashkent, Uzbekistan
tet2001@rambler.ru, ibrohimbek.211_10@mail.ru
[2] Samarkand State University, Samarkand, Uzbekistan
jurayevju@mail.ru
[3] ReSENSE Lab, Department of Electronics Engineering, Hankuk University of Foreign
Studies, Seoul, South Korea

Abstract. In this work, compression and zero coefficients have been determined by using digital processing of the blood image that is 1 µl (approximately one drop of blood) in the Haar wavelet. Also using this amount, an algorithm is proposed to define the number of leukocytes in human blood. We know that there are currently two types of methods for determining the number of leukocytes: grease and counting by using a microscope. This is a loss of time, causes a number of errors and inconveniences. By using this algorithm, a program has been developed to calculate the number of leukocytes with high accuracy in a blood image of 1 µl (about one drop of blood).

Keywords: Two-dimensional signals · Haar wavelet · Degree of decomposition · Scaling function · Wavelet function · Wavelet transform · Compression coefficients · Zero coefficients

1 Introduction

Nowadays, two-dimensional wavelets are used in digital processing of various signals, for example, in the analysis of blood cells, in the digital processing of radiographic images, to determine blood groups, fingerprints and other similar issues [2]. Decreasing the number of leukocytes in the blood leads to a variety of diseases in humans. In this study, an algorithm for calculating the number of leukocytes with high accuracy using a special program has been developed. This algorithm was implemented by applying two-dimensional Haar wavelets [3]. It is known that one-dimensional Haar wavelet is constructed with the help of the scaling function and the wavelet function. The process of compressing and filtering signals using one-dimensional Haar wavelet has been performed depending on the scaling function [2]. The smaller we scale the filtration process is carried out with such high quality. Conversely, as you zoom in the quality of the filtration process decreases so much. Images are made up of pixels (dots). As we enlarge the image, the pixels in it begin to appear in the form of a right rectangle. The two-dimensional Haar wavelet processes signals into a rectangular shape [1]. This reflects the fact that image processing in two-dimensional Haar wavelets gives better results than other types [7]. 1 μl = one micro liter. Let's make a clear what 1 μl drop of blood. In the laboratory, 1 μl (1 drop) of blood is taken from the patient. It is an image and this image is converted into numbers by performing 1, 2, 3, 4, 5, 6, 7-degree fractions in a two-dimensional Haar wavelet. The numbers that we got are the coefficients of two-dimensional Haar wavelets. With these coefficients, zero coefficients are discarded as a result of each compression. This process is continued until the number of compression coefficients is equal. Non-zero coefficients are counted and it gives the number of leukocytes [9].

Digital processing of images in two-dimensional Haar wavelets also uses the scaling function, but it is called 1, 2, 3, 4, 5, 6, 7-degree fragmentation (compression) as we mentioned above. Wavelet coefficients that are close to zero are discarded when each compression is performed. But, coefficients close to zero will not always be dropped out. Because it is interpreted differently depending on the type of disease [2, 9]. We know that the shaped elements of the blood include leukocytes, erythrocytes, and platelets. Using Wavelet coefficients to determine the number of erythrocytes and platelets that has different approach.

2 Building a Two-Dimensional Haar Wavelet

For example, suppose a 2×2-dimensional image is given in array view [12],

$$[x_{i,j}], \quad i = 1, \ldots, 2^n; \quad j = 1, \ldots, 2^n \tag{1}$$

In that case $[0, 1] \times [0, 1]$ defined in the unit area $f(s, t)$ the unit field can be expressed as a function of two variables, invariant. For given two-variable function,

$$f(s, t) = \sum_{i=1}^{2^n} \sum_{j=1}^{2^n} x_{i,j} H_{I_i \times I_j}(s, t) \tag{2}$$

Let the equality be fulfilled, here

$$I_i \times I_j = \left[\frac{i-1}{2^n}, \frac{i}{2^n}\right] \times \left[\frac{j-1}{2^n}, \frac{j}{2^n}\right] = \left\{(s, t) : s \in \left[\frac{i-1}{2^n}, \frac{i}{2^n}\right), \ t \in \left[\frac{j-1}{2^n}, \frac{j}{2^n}\right)\right\}$$

and

$$H_{I_i \times I_j}(s, t) = \begin{cases} 1, & ((s, t) \in I_i \times I_j) \\ 0, & ((s, t) \notin I_i \times I_j) \end{cases} = H_{I_i}(s) H_{I_j}(t) = \frac{\phi_{n,i-1}(s)}{\sqrt{2^n}} \frac{\phi_{n,j-1}(t)}{\sqrt{2^n}} \tag{3}$$

The parameter s entered here represents a line along the vertical, $x_{i,j}$ are the indexed rows of the matrix i. Substituting (3) into (2),

$$f(s,\,t) = \frac{1}{2^n} \sum_{i=1}^{2^n} \sum_{j=1}^{2^n} x_{i,j}\phi_{n,i-1}(s)\phi_{n,j-1}(t) = \frac{1}{2^n} \sum_{i=1}^{2^n} \left\{ \sum_{j=1}^{2^n} x_{i,j}\phi_{n,j-1}(t) \right\} \phi_{n,i-1}(s)$$

$$= \frac{1}{2^n} \sum_{i=1}^{2^n} z_i(t)\phi_{n,i-1}(s) \tag{4}$$

here

$$z_i(t) = \sum_{j=1}^{2^n} x_{i,j}\phi_{n,j-1}(t) \tag{5}$$

For each i step Eq. (5) is similar to (3) and the first phase of the one-dimensional Wavelet Transformation takes place [7]. Then we have an equation of a different form for $z_i(t)$, $\;i = 1, 2, \ldots, 2^n$ (see formula 2)

$$z_i(t) = \sum_{j=0}^{2^{n-1}-1} a^i_{n-1,j}\phi_{n-1,j}(t) + \sum_{j=0}^{2^{n-1}-1} d^i_{n-1,j}\psi_{n-1,j}(t) \tag{6}$$

Now we add (6) to (4) and get the following

$$f(s,\,t) = \frac{1}{2^n} \sum_{i=1}^{2^n} z_i(t)\,\phi_{n,i-1}(s) = \frac{1}{2^n} \sum_{i=1}^{2^n} \left(\sum_{j=0}^{2^n} a^i_{n-1,j}\phi_{n-1,j}(t) + \sum_{j=0}^{2^n} d^i_{n-1,j}\psi_{n-1,j}(t) \right) \phi_{n,i-1}(s)$$

$$= \frac{1}{2^n} \left(\sum_{j=0}^{2^{n-1}-1} \left\{ \sum_{i=1}^{2^n} a^i_{n-1,j}\phi_{n-1,j}(s) \right\} \phi_{n-1,j}(t) + \sum_{j=0}^{2^{n-1}-1} \left\{ \sum_{i=1}^{2^n} d^i_{n-1,j}\phi_{n,i-1}(s) \right\} \psi_{n-1,j}(t) \right)$$

$$= \sum_{j=0}^{2^{n-1}-1} \alpha_j(s)\phi_{n-1,j}(t) + \sum_{j=0}^{2^{n-1}-1} \beta_j(s)\psi_{n-1,j}(t) \tag{7}$$

here

$$\alpha_j(t) = \sum_{i=0}^{2^n} a^i_{n-1,j}\phi_{n,i-1}(s) \text{ and } \beta_j(s) = \sum_{i=1}^{2^n} d^i_{n-1,j}\phi_{n,i-1}(s).$$

Expressions α_j and β_j for j are constant and are similar to (3). You can apply a one-dimensional WT to each column of the image. By doing this, we get [13, 14],

$$\alpha_j(t) = \sum_{i=0}^{2^n} a_{n-1,j}^i \phi_{n,i-1}(s) = \sum_{i=0}^{2^{n-1}-1} \tilde{a}_{n-1,i}^j \phi_{n-1,i}(s) + \sum_{i=0}^{2^{n-1}-1} \tilde{d}_{n-1,i}^j \psi_{n-1,i}(s)$$

$$\beta_j(t) = \sum_{i=0}^{2^n} d_{n-1,j}^i \phi_{n,i-1}(s) = \sum_{i=0}^{2^{n-1}-1} \tilde{\tilde{a}}_{n-1,i}^j \phi_{n-1,i}(s) + \sum_{i=0}^{2^{n-1}-1} \tilde{\tilde{d}}_{n-1,i}^j \psi_{n-1,i}(s)$$

here

$$f(s,\ t) = \tfrac{1}{2^n}\left(\sum_{i=1}^{2^{n-1}-1} \alpha_j(s)\,\phi_{n-1,j}(t) + \sum_{i=1}^{2^{n-1}-1} \beta_j(s)\,\psi_{n-1,j}(t)\right)$$
$$= \tfrac{1}{2^n}\left(\sum_{j=0}^{2^{n-1}-1}\left\{\sum_{i=0}^{2^{n-1}-1} \tilde{a}_{n-1,i}^j \phi_{n-1,i}(s) + \sum_{i=0}^{2^{n-1}-1} \tilde{d}_{n-1,j}^j \psi_{n-1,i}(s)\right\}\phi_{n,j}(t)\right.$$
$$\left.+ \sum_{j=0}^{2^{n-1}-1}\left\{\sum_{i=0}^{2^{n-1}-1} \tilde{\tilde{a}}_{n-1,i}^j \phi_{n-1,i}(s) + \sum_{i=0}^{2^{n-1}-1} \tilde{\tilde{d}}_{n-1,j}^j \psi_{n-1,i}(s)\right\}\psi_{n-1,j}(t)\right)$$

Considering the equations given, we obtain the following,

$$f(s,\ t) = \sum_{i=0}^{2^{n-1}-1}\sum_{i=1}^{2^{n-1}-1} a_{i,j}^{n-1}\,\phi_{n-1,j}(t)\phi_{n-1,i}(s) + \sum_{i=0}^{2^{n-1}-1}\sum_{i=1}^{2^{n-1}-1} h_{i,j}^{n-1}\,\phi_{n-1,j}(t)\psi_{n-1,i}(s)$$

$$+ \sum_{i=0}^{2^{n-1}-1}\sum_{j=0}^{2^{n-1}-1} v_{i,j}^{n-1}\,\psi_{n-1,j}(t)\phi_{n-1,i}(s) + \sum_{i=0}^{2^{n-1}-1}\sum_{i=1}^{2^{n-1}-1} d_{i,j}^{n-1}\,\psi_{n-1,j}(t)\psi_{n-1,i}(s)$$

$$(8)$$

here

$$a_{i,j}^{n-1} = \frac{1}{2^n}\tilde{a}_{n-1,i}^j, \quad h_{i,j}^{n-1} = \frac{1}{2^n}\tilde{d}_{n-1,i}^j, \quad v_{i,j}^{n-1} = \frac{1}{2^n}\tilde{\tilde{a}}_{n-1,i}^j, \quad d_{i,j}^{n-1} = \frac{1}{2^n}\tilde{\tilde{d}}_{n-1,i}^j$$

As a result $f(x,\ y)$ functions are divided into the following functions [4],

$$\phi_{n-1,j}(t)\phi_{n-1,i}(s), \quad \phi_{n-1,j}(t)\psi_{n-1,i}(s), \quad \text{and}$$
$$\psi_{n-1,j}(t)\phi_{n-1,i}(s), \quad \psi_{n-1,j}(t)\psi_{n-1,i}(s)$$

3 Digital Signal Processing

As a result of discrete wavelet transform, the processed signal is divided into two pieces of equal size [5]. One is the average value view a_n or approximation of the signal, and the other is the different value view d_n or detail of the signal [7,11].

They are represented in the following form,

$$a_n = \frac{f_{2n-1} + f_{2n}}{\sqrt{2}}, \quad n = 1, 2, 3, \ldots, N/2 \tag{9}$$

Here is $a = \{a_n\}$, $n \in Z$ - the formula for determining the average values.

If the signal has a different value,

$$d_n = \frac{f_{2n-1} - f_{2n}}{\sqrt{2}}, \quad n = 1, 2, 3, \ldots, N/2 \tag{10}$$

Here is $d_i = (d_1, d_2, \ldots, d_{N/2})$ - the formula for determining the different values.

These values create two new signals: one to restore the original signal $a = \{a_n\}$, $n \in Z$ and the other to restore the original signal, $d = \{d_n\}$, $n \in Z$ indeed [6]

$$f_{2n-1} = a_n + d_n$$

$$f_{2n} = a_n - d_n \tag{11}$$

To understand how Haar fast transform works, let's look at a simple example below [1, 3, 4]. Suppose

$$I = \begin{pmatrix} 1 & 2 & 3 & 4 \\ 4 & 5 & 6 & 7 \\ 8 & 9 & 1 & 2 \\ 3 & 4 & 5 & 6 \end{pmatrix}$$

Based on the approximation coefficients (9) when applying one-dimensional HT along the first line

$\frac{1}{\sqrt{2}}(1 + 2)$ and $\frac{1}{\sqrt{2}}(3 + 4)$

and the difference coefficients are based on (10)

$\frac{1}{\sqrt{2}}(1 - 2)$ and $\frac{1}{\sqrt{2}}(3 - 4)$

An array I can be applied to other rows by the same modification. By placing the approximation coefficients of each row in the first two columns and the difference coefficients corresponding to the next two columns, we obtain the following results [10, 11].

$$I = \begin{pmatrix} 1 & 2 & 3 & 4 \\ 4 & 5 & 6 & 7 \\ 8 & 9 & 1 & 2 \\ 3 & 4 & 5 & 6 \end{pmatrix} \xrightarrow{\text{on one-dimensional HT line}} \frac{1}{\sqrt{2}} \begin{pmatrix} 3 & 7 & : & -1 & -1 \\ 9 & 13 & : & -1 & -1 \\ 17 & 3 & : & -1 & -1 \\ 7 & 11 & : & -1 & -1 \end{pmatrix}$$

The approximation coefficients and the difference coefficients in a given ratio are separated by a dot on each line. If we apply a one-dimensional HT to the column of the array formed in the next step, we have the resulting array of the first level [9, 16].

$$\frac{1}{\sqrt{2}} \begin{pmatrix} 3 & 7 & : & -1 & -1 \\ 9 & 13 & : & -1 & -1 \\ 17 & 3 & : & -1 & -1 \\ 7 & 11 & : & -1 & -1 \end{pmatrix} \xrightarrow{\text{on one-dimensional HT column}} \frac{1}{2} \begin{pmatrix} 12 & 20 & : & -2 & -2 \\ 24 & 14 & : & -2 & -2 \\ \dots & \dots & : & \dots & \dots \\ -6 & -6 & : & 0 & 0 \\ 10 & -8 & : & 0 & 0 \end{pmatrix}$$

So we have the following arrays,

$$A = \begin{pmatrix} 12 & 20 \\ 24 & 14 \end{pmatrix}, \quad H = \begin{pmatrix} -2 & -2 \\ -2 & -2 \end{pmatrix}$$

$$V = \begin{pmatrix} -6 & -6 \\ 10 & -8 \end{pmatrix}, \quad D = \begin{pmatrix} 0 & 0 \\ 0 & 0 \end{pmatrix}$$

Each array shown in the example above has a size of 2x2, and they are called A, H, V, and respectively D [11].

A (approximation area) is the area that contains information about the global properties of the image. Loss of coefficients close to zero from this area leads to the greatest distortion of the original image. H (horizontal area) contains information about the vertical lines hidden in the image. Subtracting coefficients close to zero from this area removes the horizontal details in the original image. V (vertical area) contains information about the horizontal lines hidden in the image. Subtracting coefficients close to zero from this area eliminates the vertical details in the original image. D (diagonal area) contains information about the diagonal details hidden in the image. Subtracting coefficients close to zero from this area results in minimal distortions in the original image [10]. Thus, the Haar Fast Transform (HFT) is applied to arrays where the number of rows and columns in the image array is equal to two [19].

In order to split an image, you must first apply a one-dimensional HFT to each line denoting the pixel value. A one-dimensional HFT is then applied to each column.

Thus, the first stage of two-dimensional HT is continued by applying the first stage of one-dimensional HT to each column of the image and then applying the first stage of one-dimensional HT to each row of the array [20].

The above (8) is the two-dimensional Haar wavelet transform, which requires finding a large number of coefficients. The use of a long chain of coefficients and signal values allows to improve the quality of signal recovery. Signal filtering is performed using two types of filters, high frequency (HF) and low frequency (LF). As a result, the image is divided into four parts: LFLF, LFHF, HFHF and HFLF. As we know, because of the image is two-dimensional, filtering the pixel values is done first by columns, then by rows [7]. While the filtering process, the pixel color values are multiplied by Haar wavelet coefficients and are the sum of the result. So this process of change continues until the last pixel of the image is etched [11] (Fig. 1).

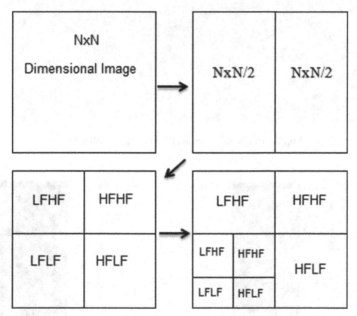

Fig. 1. 1st and 2nd degree fragmentation scheme of images in two-dimensional Haar wavelet

C ++ Builder and Matlab programs were used based on the formula given in (8) to find the number of leukocytes in a 1 μ l blood image in Fig. 2. The following results were obtained after 1, 2, 3, 4, 5, 6, 7 degree fractions [8, 15, 17] (Figs. 3, 4, 5, 6 and 7).

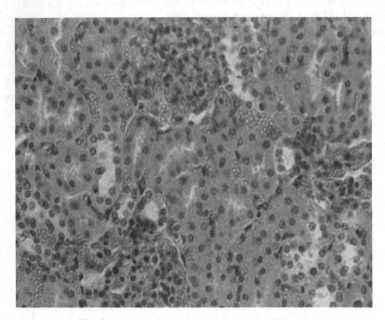

Fig. 2. Image of blood in 1 μl (1 drop of blood)

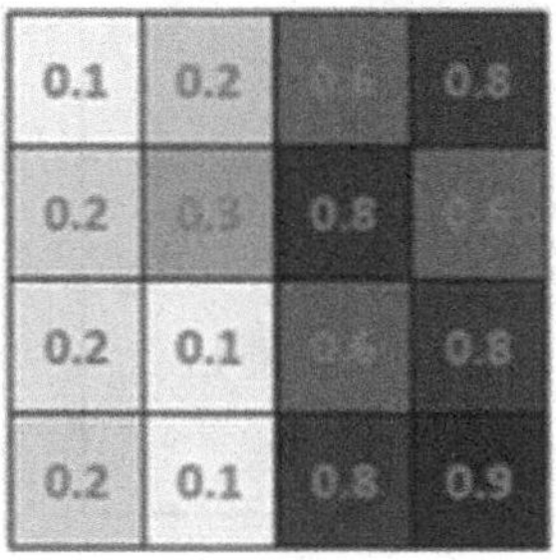

Fig. 3. Representation of the image in pixels 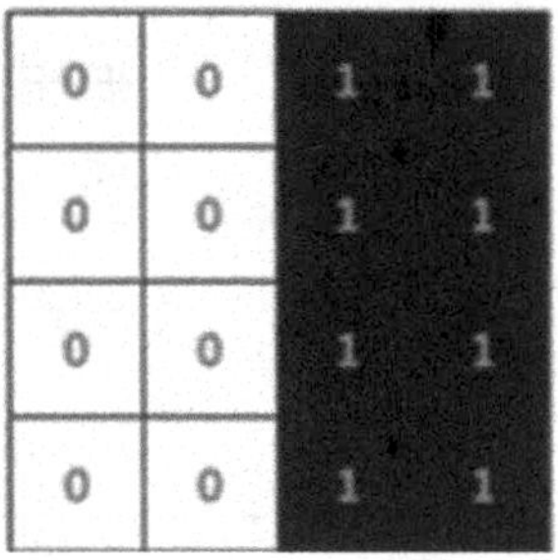   **Fig. 4.** Compressed image

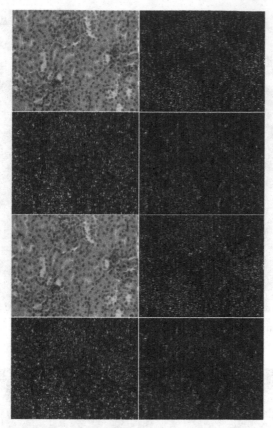

Fig. 5. Level 1 fragmentation of digital processing of two dimensional signals using Haar wavelets

Table presentation of the amount of coefficients after degree fractions (Table 1).

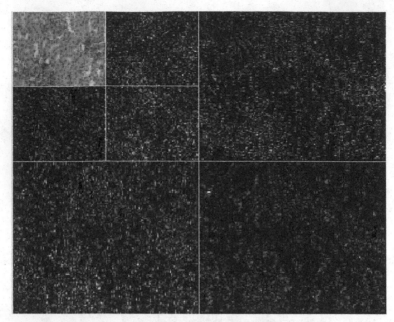

Fig. 6. Level 2 fragmentation of digital processing of two-dimensional signals using Haar wavelets

Fig. 7. Level 5 fragmentation of digital processing of two-dimensional signals using Haar wavelets

Table 1.

Number of fractions	The number of total coefficients	The number of non-zero coefficients	Number of compression coefficients	The number of zero coefficients	The percentage of zero coefficients relative to the compression coefficients (%)
Level 1 fragmentation	213120	53280	4	159840	75,000
Level 2 ragmentation	213120	12787	17	200333	94,000
Level 3 ragmentation	213120	4262	50	208858	98,000
Level 4 fragmentation	213210	4254	50	208956	98,005
Level 5 fragmentation	213315	4216	51	209099	98,024
Level 6 ragmentation	213315	4216	51	209099	98,024
Level 7 fragmentation	213327	4208	51	209119	98,027

N - Number of total coefficients (equal to the number of transmitted signals).
$N/(N - N1)$ - Number of compression coefficients.
N1 - Number of zero coefficients.
$(N - N1)$ - The number of non-zero coefficients.
Note that the number of compression coefficients increases after each level of fragmentation, while the number of non-zero coefficients decreases. What interests us is the declining number of non-zero coefficients. Because the number of spots (leukocytes) in the resulting image depends on this parameter [19].

4   Conclusion

As a result of digital processing of two-dimensional Haar wavelet transform, 1, 2, 3, 4, 5, 6, 7-degree fractions (compression) of 1 μl blood image were performed. An algorithm for determining the number of leukocytes in 1 μl of blood was developed based on the amount of coefficients compressed image. The number of leukocytes according to Table 1 is N = 4208. And, the accuracy of the number of leukocytes was 98.027%. Generally, in the system of Si international units, it is assumed that the average number of leukocytes in a healthy person is 4000–9000 leukocytes in 1 μl (about 1 drop) of blood, and the number of leukocytes are estimated based on this 1 μl blood volume.

References

1. Zaynidinov, H., Juraev, J., Juraev, U.: Digital image processing with two-dimensional haar wavelets. Int. J. Adv. Trends Comput. Sci. Eng. **9**(3), 2729–2734 (2020)
2. Zaynidinov, H.N., Mallayev, O.U., Yusupov, I.: Cubic basic splines and parallel algorithms. Int. J. Adv. Trends Comput. Sci. Eng. **9**(3), 3957–3960 (2020)
3. Zaynidinov, H.N., Yusupov, I., Juraev, J.U., Jabbarov, J.S.: Applying two-dimensional piecewise-polynomial basis for medical image processing. Int. J. Adv. Trends Comput. Sci. Eng. **9**(4), 5259–5265 (2020)
4. Zayniddinov, H.N., Mallayev, O.U.: Paralleling of calculations and vectorization of processes in digital treatment of seismic signals by cubic spline. In: IOP Conference Series: Materials Science and Engineering, vol. 537, no. 3 (2019)
5. Fang, Y.: Sparse matrix recovery from random samples via 2D orthogonal matching pursuit. arxiv Preprint arXiv11015755, no. 61001100, pp. 1–11 (2011)
6. Singh, D., Singh, M., Hakimjon, Z.: Requirements of MATLAB/simulink for Signals. In: Singh, D., Singh, M., Hakimjon, Z. (eds.) Signal Processing Applications Using Multidimensional Polynomial Splines. SAST, pp. 47–54. Springer, Singapore (2019). https://doi.org/10.1007/978-981-13-2239-6_6
7. Sharma, A., Zaynidinov, H., Lee, H.J.: Development and modelling of high-efficiency computing structure for digital signal processing. In: 2009 International Multimedia, Signal Processing and Communication Technologies, IMPACT 2009, pp. 189–192 (2009)
8. Xiao, L.L., Liu, K., Han, D.P.: CMOS low data rate imaging method based on compressed sensing. Opt. Laser Technol. **44**(5), 1338–1345 (2012)
9. Yan, C.L., Deng, Q.C., Zhang, X.: Image compression based on wavelet transform. In: Applied Mechanics and Materials, vol. 568–570, pp. 749–752 (2014)
10. Khamdamov, U., Zaynidinov, H.: Parallel algorithms for bitmap image processing based on daubechies wavelets. In: 2018 10th International Conference on Communication Software Networks, pp. 537–541, July 2018

Development of the Method, Algorithm and Software of a Modern Non-invasive Biopotential Meter System

J. X. Djumanov[1]([✉]) [iD], F. F. Rajabov[1]([✉]) [iD], K. T. Abdurashidova[1]([✉]) [iD],
D. A. Tadjibaeva[2]([✉]) [iD], and N. S. Atadjanova[1]([✉]) [iD]

[1] Tashkent University of Information Technologies Named After Muhammad al-Khwarizmi, Tashkent, Uzbekistan
jamoljon@mail.ru, faekhad63@mail.ru, kamolabdurashidova@mail.ru, nozmaatadjanova@gmail.com
[2] Tashkent Chemical - Technological Institute, Tashkent, Uzbekistan
difuzatadjibayeva79@gmail.com

Abstract. In this article discusses the possibilities of creating hardware and software for a non-invasive computer bio-meter. The main goal of this work is to create a computer bio-meter. To achieve this goal proposed a new approach to the development of adequate mathematical models and algorithms for an automated system for processing health diagnostics data. It has also shown that the effective functioning of a non-invasive hardware becomes needs are primarily being utilized to control relatively simple computer interfaces, with the goal of evolving these applications to more complex and adaptable devices, including small sensors.

Keywords: Computer bio meter · Non-invasive method · Software · Interference suppression

1 Introduction

According to the World Health Organization (WHO), 17.5 million people die from cardiovascular disease (CVD) every year. More than 75% of CVD deaths occur in low- and middle-income countries. But 80% of premature heart attacks and strokes can be prevented. One of the most effective ways to prevent death from CVD is timely diagnosis based on electrocardiography (ECG) and electroencephalography (EEG). The issues of timely diagnosis of CVDs are especially relevant now in connection with the global pandemic COVID19, since coronavirus primarily dies from CVD [1].

As we know, it is no secret that positive changes in all spheres of society are associated with the level of direct application and implementation of information and communication technologies (ICT) in practical activities. Scientific research aimed at solving problems and tasks arising in the integration process in ICT with the area of research under consideration are associated with the use of system analysis tools, data processing and management. In many cases, the integration process is manifested in the implementation of methods and means of automated system management. This is especially in

M. Singh et al. (Eds.): IHCI 2020, LNCS 12615, pp. 95–103, 2021.
https://doi.org/10.1007/978-3-030-68449-5_10

demand in the field of medicine, where operational processing of the results of experiments and the adoption of a scientific - informed decision is required. And this is achieved using the below tools. In this connection, the task of developing a software shell for the effective functioning of a non-invasive glucometer apparatus becomes relevant. Here it is necessary to emphasize that the apparatus of non-invasive bio-meters will also be improved [3, 15].

2 Related Work

An analysis and review of the methods of acquisition and digital processing of biomedical signals based on computer technology shows that the discoveries by Richard Caton [13] determine the electrical signals on the surface of the brain of animals. Later, in 1924, Hans Berger using a galvanometer on paper recorded a curve describing the biopotentials taken from the surface of the head using needle electrodes. To suppress interference associated with electromyographic signals, an atremor low-pass filter (LPF) is desirable, limiting the input signal range to 60–70 Hz, and to combat network interference a notch filter at 50 Hz (60 Hz) [4]. On the other hand, in some studies [9, 14], for example, stress tests, a shorter time constant is consciously chosen for better retention of the contour [5]. Thus, we can conclude that according to the survey of biomedical signals, in particular the ECG signal, the useful signal must be taken into accordance with standard leads, such a reference potential formation scheme is called the right leg driver circuit or Right Leg Drive Circuit (RLD) [6–10]. Bio potential meters using non-invasive equipment's have been a major focus of research and development. The use of conductive methods to produce results capable of monitoring heart rate and generating electrocardiogram waveforms has been reported by numerous researchers [11]. In order to obtain adequate bio potential signals, intimate small skin-sensor contact is required. For this purpose, the current state of the development of non-invasive bio-measuring devices has been analyzed and the problems of creating computer bio-measuring devices have been revealed [12–17].

- Setting goals and basis objectives are the questions of creating a computer bio-meter. To achieve this goal, the following theoretical and practical problems were solved:
- The study a complex of methods analog and digital filtering of cardio signals;
- analysis of the spectral-temporal characteristics of electrocardiogram signals by wavelet function methods;
- development of methods for increasing the efficiency of adaptive noise filtering algorithms in cardiac signals;
- creation of hardware and software for implementing the filtering method in cardiac signals;
- study the influence of the parameters of the proposed digital filter circuit on its characteristics in order to increase the filtering efficiency in cardiac signals;
- study the possibility of using the developed means of recording and processing cardiac signals to control the operating modes of medical complexes.

When processing an ECG, it is very important to accurately determine the heartbeat, because it is the basis for further analysis, and can also be used to obtain information about the heart rate. The energy of the heartbeat is mainly found in the QRS complex, so an accurate QRS detector is the most important part of ECG analysis.

So, the basis of functional diagnostics devices based on computer technology should be a computer bioelectric meter built on a modern elemental base - multichannel low-noise operational amplifiers, multi-bit and multichannel integrated analog-to-digital converters (ADC), programmable logic matrices and/ or microcontrollers [1–3]. The basis of this digital bio-meter is an ADC - the main characteristics of the entire system largely depend on its parameters. One of the advanced methods is the use of a multi-bit ADC (22–24 bit) as an ADC, which allows you to measure a bio signal directly from electrodes located on a bio object.

This implementation of the bio-meter has the following advantages:

A. No need for multi-channel low-noise amplifiers
B. Downsizing of the system
C. The possibility of applying the technology of saving the signal "as is"

The "as is" signal preservation technology allows you to save a bio signal as removed from the electrodes located on the bio-object without post-processing. This technology allows you to change the parameters of the system after measuring, for example, the sensitivity or frequency range of the measured signal.

3   Significance of the System

The microcontroller used to register bio signals must have the ability to process digital signals (DSP), high speed, large memory and rich peripherals. This criterion is well met by 32-bit STM32 processors with the Cortex M4 (3) core of ST Microelectronics (Fig. 1). In this implementation, in addition to a rich communication interface (USB, SPI, etc.), there is a real-time clock timer, an interface for the LCD module, an interface for the touch screen (touch panel), and an automatic direct memory access device (DMA) and the priority system of singing (Interrupt). The number of channels of the bio-meter can vary from 4–8 (ECG, EMG and Holter ECG) to 32 (EEG), depending on the type of bio-meter. Based on the above considerations means, the authors are propose the following structure of a modern multi-channel bio-meter consisting of several (3 pieces) of a fully-equipped (Front-End) ADC and ARM Cortex microprocessor. Of course with the possibility of autonomous (battery power), LCD touch screen, micro CD memory and also having a wireless connection. The ADS1299 ADC (8) is connected to the microprocessor via the SPI serial port, in cascade mode, in the Daisy-Chain mode [1, 2]. At the same time, 216 bits = 24 status bits + 24 bits * 8 channels = 3 bytes + 3 bytes * 8 channels = 27 bytes for each ADC are read in the DMA device by direct memory access. The code of this algorithm in C++ is given below. This EXTI9_5_IRQ Handler subroutine interrupt service is called by the readiness signal (Data Ready (DRDY)) of the ADC and allows using the DMA2 root channel to write 27 + 1 bytes of data through SPI1 to the memory received from the ADC. After receiving all 27 bytes, the DMA device generates a preference signal to proceed to the maintenance of the DMA2_Stream0_IRQ Handler routine. This subroutine creates a stream of 25 samples of 32 bit bio signal values over the entire 24 channels (Fig. 2).

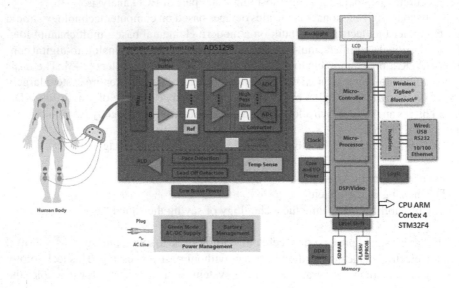

Fig. 1. The structure of a modern bio-meter based on a multi-channel 24-bit ADC Analog Front End (AFE) IC from Texas Instruments (TI) and a single-chip STM32 CortexM4 microcontrollers.

To adequately display the ECG signal and an ECG device operation signal pre-filtering is needed. The first step is to remove baseline drift (zero) so that the signal is always displayed within the screen. This is necessary because of the features of the ADS1298 ADC, which digitizes the signal with a DC. Mathematical view and algorithm this can be achieved [1] by a simple IIR digital filter HPF (1):

$$H(z) = \frac{z - 1}{z - 0.995} \qquad (1)$$

The difference equation for such a filter is expressed by the formula (2):

$$Y(n) = [X(n) - X(n - 1)] + 0.995 \times Y[n - 1] \qquad (2)$$

The amplitude and phase characteristics of the filter specified by Eq. (1 or 2) to remove the drift of the signal of the display in Fig. 3. After digitization, the ECG signal has HF noise and interference. They can be loosened by a simple 3-order low pass FIR Hanning, defined by the expression (3):

$$Y(n) = \frac{1}{4}[X(n) + 2 \times X(n - 1) + X(n - 2)] \qquad (3)$$

Hanning filter transfer function defined by the expression (4):

$$H(z) = \frac{1}{4} \times \left[1 + 2 \times Z^{-1} + Z^{-2}\right] \qquad (4)$$

The result of high-pass filtering based on formula (1) is shown in Fig. 4.

4   Results

For transferring the ECG data packet from the controller to the computer, the optimal asynchronous serial transmission method is USART (universal serial asynchrony receiver transmitter). In computers running under MS Windows, such ports are called COM (communication port - COM1, COM2 ...) and real such ports work in the industry standard. RS232, but modern USB exchange devices also create virtual COM ports. Therefore, the use of such a port for transmission, which has wireless and USB versions of the converters (controller-computer bridge), makes the ECG controller universally modular. Figure 4 shows the result of the filter LPF Hanning. Determination of the exchange interface and calculation of the data transfer traffic of the ECG bio-meter (Fig. 4). By default, one byte in this method is transmitted in 10 consecutive bits - 1 start bit +8 data bit + 1 stop + 0 bit parity/odd check. Then, for example, to transmit 16 bit ECG samples, 16 bits = 2 bytes => 2 * 10 = 20 bits are required.

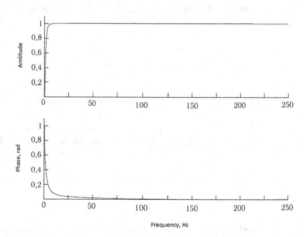

Fig. 2. Amplitude and phase characteristics of the filter specified by Eq. (1) to remove the signal drift.

The amplitude and phase characteristics of the filter specified by Eq. (3 or 4) and results to Hanning LPF of the display in Fig. 4. We are calculate the minimum required ECG data rate via USART: 500 times per second x 20 x 8 channels = 80,000 bot. The nearest traffic, advising the transmission of this information from a number of standard USART (COM-RS232) port exchange rates, is 115200 bot. Data sent to the PC via UART contains eight ECG leads [7]. These signals are sent at a rate of 500 packets per second Fig. 5. The program installed in the computer calculates the remaining four ECG leads using data from the 1 lead and 2 lead [7].

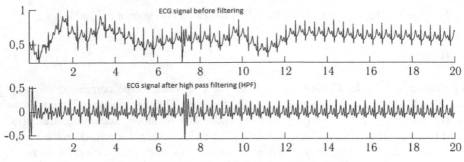

Fig. 3. ECG signal before and after filtering HPF.

For transmission to a computer, two types of ECG data packet are generated: the first packet type is repeated every second once and is called a full packet, where there is a header consisting of 2 bytes of synchronization, 1 byte with heart rate (HR), 1 byte of lead status (is there an electrode connection or not skin) and 16 bytes of data from eight ECG leads (Table 1), the second type of packet, called a shortened packet, has 1 byte packet number and 16 bytes of data from eight ECG leads (Table 2), and the packet number increases with each new packet.

Table 1. Format of a complete ECG data packet (where, SB is the most significant byte, LB is the least significant byte).

0x55	0xAA	HR	Lead On/off	I		II		C1		C2		C3		C4		C5		C6	
Sync (2 bytes)		1 bytes	1 bytes	SB	LB	SB	LB	SB	LB	SB	LB	SB	LB	SB	LB	SB	LB	SB	LB
Package header				ECG data 8 leads															

Thus, 1 full and 499 reduced packet is transmitted within one second, and at the same time traffic can be calculated as follows: $1 \times (4$ bytes header $+16$ bytes of data$) \times 10 + 499 \times (1$ bytes header $+16$ bytes of data$) \times 10 = 100 + 84\,830 = 84930$ bot (or bits/sec). Therefore, the UART configuration is set to 115200 bps, 8 data bits, 1 stop bit and no parity is enough to transmit ECG data. On the basis of the above considerations, the following functional diagram of a modern multichannel biometer (consisting of several fully equipped (Front-End) ADCs and an ARM Cortex microprocessor, with the possibility of autonomous (battery-powered), touch-sensitive LCD display and micro-CD memory, with wireless communication) is proposed. It also generates a bias signal BIAS - the potential of the patient "Block MK" (microcontroller) controls the operating modes of the device. The signal in digital form comes here from the signal block, is processed, buffered and sent to the PC through the "Communication block" and the signal can be saved to a Micro SD memory card using the "Memory block".

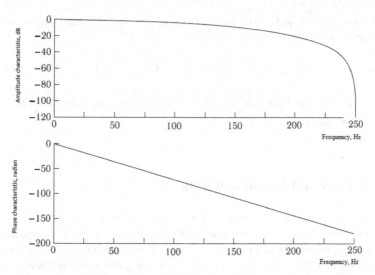

Fig. 4. Amplitude and phase characteristics of the Hanning LPF filter

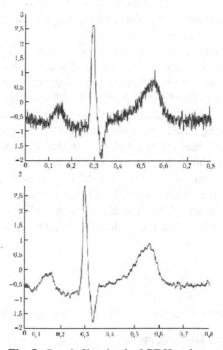

Fig. 5. Result filtration by LPF Hanning.

Table 2. The format of the reduced ECG data packet (where, SB is the most significant byte, LB is the least significant byte).

Number of Package	I		II		C1		C2		C3		C4		C5		C6	
1 bytes	SB	LB	SB	LB	SB	LB	SB	LB	SB	LB	SB	LB	SB	LB	SB	LB
Package header	ECG data 8 leads															

5   Conclusions and Recommendations

In embedded systems, such as portable ECG systems, it is desirable to use algorithms based on differentiation and time analysis based on a priori data of ECG signals. The Pan and Tompkins algorithm passed the test of time; it is popular and high-speed with good quality analysis of QRS complexes. In stationary computer ECG systems, algorithms for analyzing ECG based on neural networks and wavelet transforms are promising. To find the R peaks of the ECG signal, it is promising to use the wavelet transform with the correct choice of the threshold for determining the peaks and the window within which these peaks are sought. The results of research on the development of adequate mathematical models and software and software for improving the functions of processing removable experimental data and diagnostics, automated ECG meter, will be announced in subsequent publications.

Heart beat the article describes the main aspects of the development of functional diagnostics devices based on computer technology. The device for functional diagnostics of a bio-measuring, built on a modern element base (multi-channel low-noise operational amplifiers, multi-bit and multi-channel integrated analog-to-digital converters (ADC), programmable logic arrays and/or microcontrollers), is considered in detail. A structure of a modern bio-measuring based on a multi-channel 24-bit ADC.

A controller for a multichannel digital ECG signal has been developed that meets the modern requirements of a digital biometer: with the possibility of autonomous power supply; graphical input and output interface; ADC with high accuracy (4 μV) at a frequency of 500 times per second; using VLSI with a complete analog part; with the ability to save ECG data on micro-CDs; having wired USB and wireless Bluetooth communication with the host computer; drivers for ADC, graphic screen and touch screen were developed.

References

1. Cardiovascular Diseases. https://www.who.int/health-topics/cardiovascular-diseases/#tab=tab_1
2. ADS1298ECG-FE/ADS1198ECG-FE ECG Front-End Performance Demonstration Kit – User's Guide/ Texas Instruments [Digital resource]. http://www.ti.com/lit/ug/sbau171c/sbau171c.pdf. Accessed 23 Feb 2014
3. GOST 19687-89: General technical requirements and test methods. Rossiya, Moskov (1989). (in Russian)

4. Led, S., Fernandez, J.: Design of a wearable device for ECG continuous monitoring using wireless technology. USA, 3318p. (2004)
5. Rodriguez, C. Borromeo, S.: Wireless ECG based on Bluetooth protocol: design and implementation. Greece, 48p. (2006)
6. Rajabov, F.F., Ubaydullayev, S.A., Mukxamedjanov, N.Z.: Development and production of a multifunctional medical diagnostic complex based on a modern elemental base. Rossiya, Moskva: International scientific-practical conference "Private partnership in the field of medical technology, pp. 54–55 (2009). (in Russian)
7. Rajabov, F.F., Abdurashidova, K.T.: Typical solutions for the construction of modern electrocardiographs (ECG). Tashkent TUIT -BULLETIT 2(46), 42–55 (2018)
8. Rajabov, F.F., Abdurashidova, K.T., Salimova, H.R.: The issues of creating a computer bio-measuring and noise suppression methods. Muhammad al-Khwarazmiy Descendants. Sci. –Pract. Inf. Anal. J. Tashkent 1(3), 23–27 (2019)
9. Bokeria, L.A.: Modern non-invasive diagnostic tool in cardiac surgery: a rational approach a brief way to diagnosis. Federal guide. Reducing mortality caused by socially significant diseases. Ann. Arrhythmol. (4) (2010)
10. Ruttkay-Detski problems electrocardiology evaluation of the impact of the autonomic nervous system to the heart. Bull. Arrhythmia. 56–60 (2001)
11. Zaynidinov, H.N., Yusupov, I., Juraev, J.U., Jabbarov, J.S.: Applying two-dimensional piecewise-polynomial basis for medical image processing. Int. J. Adv. Trends Comput. Sci. Eng. 9(4),156, 5259–5265 (2018)
12. Singh, D., Singh, M., Hakimjon, Z.: Signal Processing Applications Using Multidimensional Polynomial Splines. Springer, Singapore (2019)
13. Heart-vascular prevention. National recommendations. 2011. Heart-vascular therapy and prevention. 2011. N 10(6). Annex 2. www. scardio.ru. Accessed 12 Apr 2014
14. Haas, L.F.: Hans Berger (1873–1941), Richard Caton (1842-1926) and electroencephalography. J. Neurol. Neurosurg. Psychiatr. 74 (2003). https://doi.org/10.1136/jnnp.74.1.9
15. Man, D.: Paszkiel Sczepan, Sadecki Jan (eds): Współczesne problemy w zakresie inżynierii biomedycznej i neuronauk, Studia i Monografie - Politechnika Opolska, vol. 434 (2016). Oficyna Wydawnicza Politechniki Opolskiej, ISBN 978-83-65235-33-6, 108p
16. Singh, D., Zaynidinov, H., Lee, H.: Piecewise-quadratic Harmut basis functions and their application to problems in digital signal processing. Int. J. Commun Syst 23, 751–762 (2010)
17. Zaman, S., Hussain, I., Singh, D.: Fast computation of integrals with fourier-type oscillator involving stationary point. Mathematics 7, 1160 (2019)

Deception Detection Using a Multichannel Custom-Design EEG System and Multiple Variants of Neural Network

Ngoc-Dau Mai[1], Trung-Hau Nguyen[2], and Wan-Young Chung[1(✉)]

[1] Department of AI Convergence, Pukyong National University, Busan, South Korea
wychung@pknu.ac.kr
[2] Department of Electronic Engineering, Pukyong National University, Busan, South Korea

Abstract. This study aims to develop a BCI system using Electroencephalo-gram (EEG) for lie detection. EEG signal is obtained in a non-invasive manner with eight electrodes mounted at designated positions on the scalp employing a hand-made design EEG headset. The selected input for classification in this research comprises 11 features extracted across all three domains: time, frequency, and time-frequency. In which nine time-domain features include: Mean, Variance, Standard Deviation, Skewness, Kurtosis, Permutation Entropy, SVD Entropy, Approximate Entropy, and Sample Entropy. Over the frequency domain and time-frequency domain, we have the Spectral Entropy and an 8-dimensional image computed by Continuous Wavelet Transform. Modern researches indicate that Neural Networks have an extraordinary capacity to classify patterns with diverse inputs such as data sequences and pictures. Four variations are applied in this study, including the Multilayer Perceptron (MLP), Long Short-Term Memory (LSTM), Gated Recurrent Unit (GRU), and Convolutional Neural Network (CNN). In which MLP, LSTM, GRU are worked on the first ten time and frequency features, while the 8-dimensional image composed after the Continuous Wavelet Transform is adopted as the input of the CNN network. The evaluation results demonstrated the feasibility of our EEG-Based BCI system in Deception recognition.

Keywords: Lie detection · BCI · EEG · Neural networks · CNN · LSTM · GRU · MLP · CWT

1 Introduction

"Brain-Computer Interface (BCI)" is an approach that empowers humans to interact with the outside world through the utilization of potential signals generated from the inner brain instead of other peripheral ones such as voiced communication or muscle movement. EEG is one of the most frequently practiced techniques that captures brain activity through electrodes settled on the scalp in a non-invasive way. Using EEG, BCI systems have been implemented in multiple applications and diverse disciplines, such as robot

This work was supported by an NRF grant of Korea Government (2019R1A2C1089139).

M. Singh et al. (Eds.): IHCI 2020, LNCS 12615, pp. 104–109, 2021.
https://doi.org/10.1007/978-3-030-68449-5_11

control, clinical trials, emotion recognition, entertainment, and deception detection. The purpose of lie detection is to identify the truth perceived and covered by others. The standard Polygraphy examination is performed to define the truth or lie: The Concealed Information Test (CIT) [1]. CIT is also known as the Guilty Knowledge Test, designed to discover a person's criminal awareness of a crime. The mechanism for detecting deception lies inside the individual body itself [2]. The own body responds to external influences, including lie tests, through biological signals. Studies in lying discovery have been conducted based on various physiological signals such as skin conductivity, body temperature, eye behavior, pupil size, voice, and brain activity. However, many of the Polygraphy examinations previously done on criminals by analyzing their sensitive activity replies like sweating, facial expressions, or heart rate have failed because they are incredibly proficient in converting their responses [3]. Hence, the performance of these body feedbacks is not reliable enough. Instead, using EEG-based BCI gives a higher compatible and natural solution to the issue by recording brain activity by its unbiased evaluation and great accuracy. Up-to-date studies have shown that machine learning techniques can help better interpret neural signals more accurately. Over traditional approaches, neural networks are one of State-of-the-Art machine learning techniques in various fields, including EEG task classification [4]. Four variants of neural networks were applied in this study for deceit detection.

2   Experimental Framework

2.1   Custom-Design Headset

In this study, we utilized a wearable 8-channel custom-design headset to collect EEG signals from the scalp. Here, dry sensors are adopted to obtain the signals instead of wet ones because of their convenience and benefits in measuring through hair. The flexible dry EEG electrodes sensor with $100–2000 \, K\Omega$ on unprepared skin used in this research is a product of Cognionics Inc. Additionally, to minimize ambient noise, cable movement artifacts, and interference caused by electrical effects, active sensors are employed. Those sets of active sensors require a pre-amplification circuit for dealing with the high electrode and skin interfacial impedances. Figure 1a shows our EEG custom-design headset.

2.2   Experiment Protocol

The people who participated in lie detection experiments consist of three males and two females (aged 24 to 40). Individual participants must read an informational document about the experiment and then sign a consent form under ethical principles for medical study on human subjects. All stayed in steady physical and mental wellness, principally do not have optical errors and eye-related disorders. Before the examination, volunteers were obliged not to use alcohol and drugs to impact the measurement outcomes negatively. Participants would carry the device on their heads and are seated pleasantly in front of a computer screen. Each person would watch 30 short videos, and each lasting 44 s will match to a trial. The stimulus consists of three types of events: target, irrelevant,

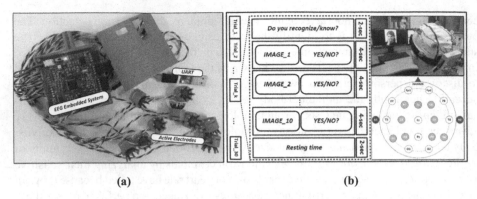

(a)   (b)

Fig. 1. (a) The 8-channel custom-design EEG circuit, (b) Experiment procedure.

and probe in any order. The subject can identify Target photos, Irrelevant images that they did not know before, and Probe ones related to the crime [5]. According to multiple diverse topics, the experiment's selected pictures comprise celebrities, relatives, advertisements, movies, or famous events. Figure 1b gives more details about the steps to conduct the experiment and the measuring locations according to the International 10–20 System. This study's measuring areas contain F3, F4, C3, C4, P3, P4, T5, T6, and Fz is the reference electrode.

3 Proposed Method

3.1 Pre-processing and Feature Extraction

Data preprocessing is achieved with the task of reducing the adverse effects of signal artifacts due to power-line noise and cross-talk. They can be handled applying filters: a notch filter to eliminate 50 Hz power-line noise and a band-pass filter to allow only wanted signals in the frequency range of 0.5–60 Hz passing through. Because the data collection is continuous, to guarantee the temporal correlation among data points, a sliding window of 128 lengths with an overlap rate of 50% is adopted to segment the data.

The preprocessed signals will be extracted rich features across both the time and frequency domains before using them as inputs for future classification models. These features are obtained according to each segmentation mentioned in the previous section. In which, time-domain features include: (1) Mean, (2) Variance, (3) Standard Deviation, (4) Skewness, (5) Kurtosis, (6) Permutation Entropy, (7) SVD Entropy, (8) Approximate Entropy, and (9) Sample Entropy. The frequency-domain feature is (10) Spectral Entropy. The final one is (11) taken after applying Continuous Wavelet Transform on the time-frequency domain. The success of the CWT over traditional techniques stems from the unique attributes of the wavelet template. The wavelet is proper for EEG signal analysis, in which the spectral features of the signal differ over time.

3.2 Classification

Neural networks are considered one of the most modern Machine learning algorithms in countless different fields today [6]. Many various studies and applications have proven their superiority over other traditional algorithms in efficiency and speed. This study utilizes the most optimal four variants, namely The Multilayer Perceptron (MLP), Long Short-term Memory (LSTM), The Gated Recurrent Unit (GRU), and Convolutional Neural Network (CNN), for lie detection based on EEG signals. Figure 2 shows the proposed architectures we used in this research.

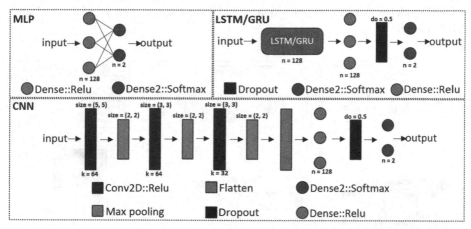

Fig. 2. The architectural proposed models.

Multilayer perceptron (MLP) is a class of feed-forward networks widely used in many applications, precisely pattern classification. MLP uses a backpropagation, which is a generally accepted supervised learning algorithm in model training. Long Short-Term Memory (LSTM) is a more advanced variant of the Recurrent Neural Network (RNN) capable of working with sequence data that eliminates the vanishing gradient that the RNN suffers. LSTM has a great deal of accomplishment in feature extraction of a data sequence than MLP can do, obtained thanks to its specialized memory cells. Gated Recurrent Unit (GRU) is comparable to the LSTM but possesses some variation within the structure. It blends the forget gate and input gate into one and performs some modifications by combining the cell state and hidden state. Therefore, the design of this type is more straightforward, with fewer parameters than LSTM. Convolutional Neural Network (CNN) is known for its ability to work exceptionally well when learning features from images and bio-signals. Continuous Wavelet Transform was applied to the EEG data to extract time-frequency domain information as images by computing the result as a scalogram and then taking advantage of CNN's outstanding image classification effort.

4   Results and Discussion

Figure 3b shows an overview of the entire process from pre-processing, characteristic extraction methods, and ultimately depending on the features that the most suitable models would apply. The ten features extracted per time and frequency domain are used for three classification models: MLP, two variations of RNN: LSTM, and GRU. If we go into each potential model's details, the MLP model has the highest accuracy for S2 is 95.56%, the lowest with S4 is 90.85%, and the average accuracy is 94.01%. Meanwhile, LSTM also achieved the highest accuracy on S2 with 96.54%; the lowest is 92.24% on S5 and 94.23% for average accuracy. Finally, the average accuracy when applying the GRU model is 94.64%, with the highest and lowest accuracy of 96.35% on S1 and 92.48% on S5.

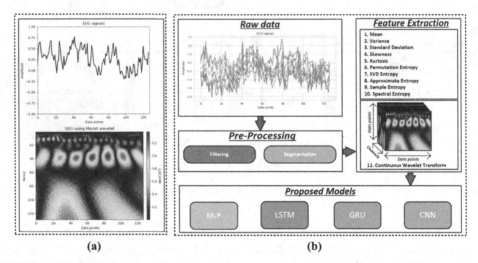

(a)   (b)

Fig. 3. (a) An EEG signal and its composed image using Morlet-CWT, (b) Block diagram of the proposed algorithm.

The remaining feature is extracted on the time-frequency domain using Morlet-Continuous Wavelet Transform to assemble an 8-dimensional image corresponding to eight EEG channels, shown in Fig. 3a. With the remarkable ability to work with images, CNN would apply here. Compared to the average highest accuracy achieved when used with the previous three models across ten types of features, the CNN Network with this feature as multidimensional images outperformed with an average accuracy of 96.51%, the highest on S2 is 98.02%, and the lowest is 94.52% on S4.

We have applied the neural networks for deceit detection on each individual, and that classification is called subject-dependent. However, to become more suitable in real-world applications, the subject-independent using the data set mixed with all from the subjects will be performed. Table 1 compares the detection accuracy of all models with the subject-dependent and subject-independent. In general, the average accuracy for each model with subject-independent is higher than the subject-dependent implementation.

Table 1. Classification results using the proposed models for both subject-dependent and subject-independent strategies across five subjects

Subject	Accuracy (%)			
	Multi-features (10 features)			8-D images
	MLP	LSTM	GRU	CNN
S1	94.74	95.14	96.35	97.58
S2	95.56	96.54	95.93	98.02
S3	94.91	94.97	95.02	96.78
S4	*90.85*	92.66	93.43	94.52
S5	93.97	92.24	92.48	95.67
AVG	94.01	94.23	94.64	**96.51**
Subject-independent	91.34	92.43	92.92	**94.62**

The CNN model with features generated after the CWT process also provides us higher accuracy than the rest of the models with multiple features, 96.51% and 94.62% on subject-dependent and subject-independent, respectively.

5 Conclusion

In this study, we have developed a deception detection system using EEG signals. The technique combines multiple methods in feature extraction as well as multi-neural networks in the lie classification. Our proposed approach has been proven feasible and practical in real-life applications by comparing achieved accuracy results.

References

1. Ben-Shakhar, G.: Current research and potential applications of the concealed information test: an overview. Front. Psychol. **3**, 342 (2012)
2. Sebanz, N., Shiffrar, M.: Detecting deception in a bluffing body: the role of expertise. Psychon. Bull. Rev. **16**(1), 170–175 (2009). https://doi.org/10.3758/PBR.16.1.170
3. Haider, S.K., et al.: Evaluation of p300 based lie detection algorithm. Electr. Electron. Eng. **7**(3), 69–76 (2017)
4. Jiao, Z., et al.: Deep convolutional neural networks for mental load classification based on EEG data. Pattern Recogn. **76**, 582–595 (2018)
5. Cutmore, T.R.H., et al.: An object cue is more effective than a word in ERP-based detection of deception. Int. J. Psychophysiol. **71**(3), 185–192 (2009)
6. Schmidt, J., et al.: Recent advances and applications of machine learning in solid-state materials science. NPJ Comput. Mater. **5**(1), 1–36 (2019)

A Study on the Possibility of Measuring the Non-contact Galvanic Skin Response Based on Near-Infrared Imaging

Geumbi Jo[1] , Seunggeon Lee[2] , and Eui Chul Lee[3]([✉])

[1] Department of Computer Science, Graduate School, Sangmyung University,
Seoul, South Korea
[2] Department of AI & Informatics, Graduate School, Sangmyung University, Seoul, South Korea
[3] Department of Human-Centered AI, Sangmyung University, Seoul, South Korea
eclee@smu.ac.kr

Abstract. Galvanic skin response (GSR) is the change in the electrical conduction of the skin caused by an emotional response to a stimulus. This bio-signal appears to be the action of the autonomic nervous system as a change in sweat gland activity and reflects the intensity of emotion. There have been various technologies for detecting autonomic nerve reactions to bio-signals in image processing; research on non-contact detection of these technologies has become a major issue. However, research on GSR for detecting signals in a non-contact method is lacking compared to other bio-signals. In this paper, we propose a non-contact GSR detection method from the human face region using a near-infrared camera. The proposed method used a 20-min visual stimulus video that caused tension or excitement among the participants and set up an isolated experimental environment to minimize emotion induction caused by external factors. Afterward, a GSR sensor was attached to the inside of the participant's finger, and a visual stimulus video was shown. The participant's face region was shot using a near-infrared camera, which also stored GSR sensor data and infrared face images that were input during the experiment. The face region was detected from the acquired face images, and the region of interest inside the face was compared with the data obtained through the GSR sensor. Finally, the correlation between the image intensities inside the designated region of interest and the data from the GSR sensor were analyzed.

Keywords: Bio-signal · Galvanic Skin Response (GSR) · Non-contact · Image processing · Near-infrared camera

1 Introduction

There are several kinds of bio-signals, such as electroencephalogram (EEG), galvanic skin response (GSR), skin temperature, and respiratory rate. Recently, research on non-contact-based bio-signal detection has been steadily progressing. Research includes detecting heart rate using a webcam [1–4]; measuring pulse rate using a webcam [5]; measuring respiration volume through a depth camera [6]; and detecting such as blood

© Springer Nature Switzerland AG 2021
M. Singh et al. (Eds.): IHCI 2020, LNCS 12615, pp. 110–119, 2021.
https://doi.org/10.1007/978-3-030-68449-5_12

volume, heart rate, and respiratory rate through a webcam [7]. Conversely, GSR measures sweat on the surface of the skin. Given that GSR is measured using a sensor attached to the body, research on non-contact signal detection is insufficient compared to other bio-signals. Therefore, in this study, we propose a non-contact method for detecting the GSR signal among various bio-signals.

Sweat gland activity increases with emotional changes, which may be emotions such as joy or fear [8]. Sweat glands are distributed between the subcutaneous fat under the dermis, as shown in Fig. 1, and are surrounded by adipose tissue; these secrete moisture to the skin surface through pores [8]. The amount of positive and negative ions in the secreted moisture changes, and current flows make it possible to measure the conductivity of the skin [8]. As more sweat glands activity occurs, skin resistance and conductivity decrease and increase, respectively [9]. GSR measures these changes through skin conductivity. Therefore, GSR is a measure of the change in moisture secreted from the sweat glands due to changes in a person's mental or psychological state [10].

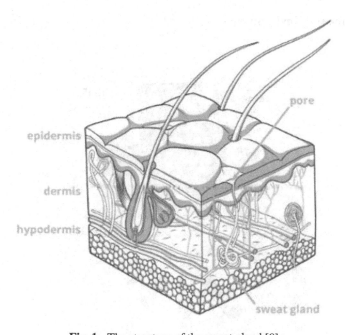

Fig. 1. The structure of the sweat gland [8]

In this paper, we propose a non-contact method for detecting GSR signals using a near-infrared camera. To experimentally validate the method, a 20-min visual stimulus video was prepared to elicit nervousness or excitement in the participant, and an isolated experimental environment was configured to minimize emotion induction caused by other environmental factors. Then, after attaching the GSR sensor to the inside of the index finger and ring fingers, the prepared video was played, and the face region of the participant viewing the video was captured using a near-infrared camera. During the experiment, the GSR data and face images were captured. Next, the face region was detected from the facial images, and the region of interest (ROI) inside the face was specified. The average intensity of the ROI was compared with the data obtained through the contact GSR sensor. Finally, the correlation between the intensities of the ROI and the GSR sensor data were analyzed.

2 Proposed Method

2.1 Experimental Environment

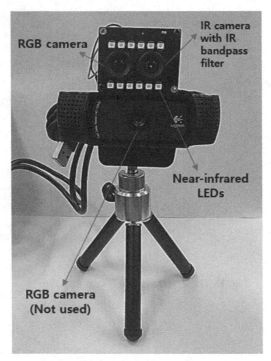

Fig. 2. A camera device used in an experiment

To capture face images, a webcam with an infrared passing filter—with an 850 nm wavelength—was fixed using a tripod, as shown in Fig. 2. The webcam also contained 12 near-infrared wavelength LEDs around the lens. Near-infrared wavelengths penetrate the skin better than visible wavelengths. Therefore, it is suitable for observing changes caused by sweat near the endothelium.

The Neulog GSR sensor, as shown in Fig. 3, was equipped to measure the change in sweat gland activity for emotions induced by stimuli. This sensor was attached to the inside of the index finger and ring finger, which allows for easy measurement of GSR because sweat is well secreted [11].

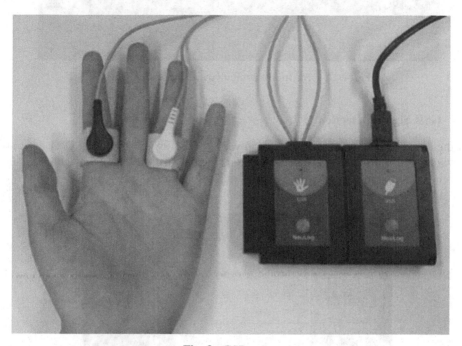

Fig. 3. GSR sensor

An isolated experimental environment was prepared so that participants could focus on the experiment and that emotions were not induced by stimuli other than the experiment. A 20-min visual stimulation video that caused excitement and tension was produced. Afterward, 20 participants were asked to watch the visual stimulus video through the computer monitor screen. The near-infrared images and GSR data were stored for 20-min. The following scene was obtained from a 20-min visual stimulation video (Fig. 4).

Fig. 4. A scene from a 20-min visual stimulation video

2.2 GSR Signal Detection

In the saved face images, the face region was tracked using OpenCV's Deep Neural Network (DNN) Face Detector, and the ROI inside the face was compared with the data obtained through the GSR sensor. Figure 5 shows the tracked face region.

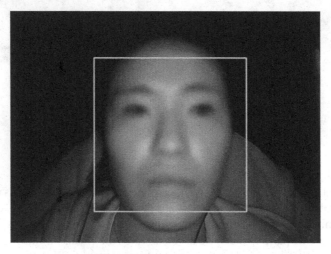

Fig. 5. Tracked face region

The specified ROI was the forehead, as shown in Fig. 6, because the forehead secretes more sweat than other parts of the face when one is emotionally tense or excited [12, 13].

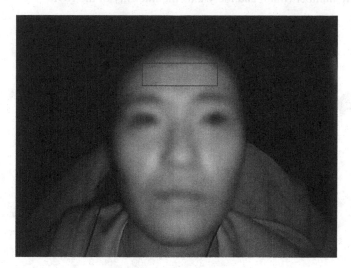

Fig. 6. ROI in face images

To analyze the correlation with the stored GSR data, the average intensity of the ROI was successively extracted. Figure 7 is an example of time series GSR data, while Fig. 8 shows the average intensity of the ROI. In Fig. 7, the GSR time series data is graphed,

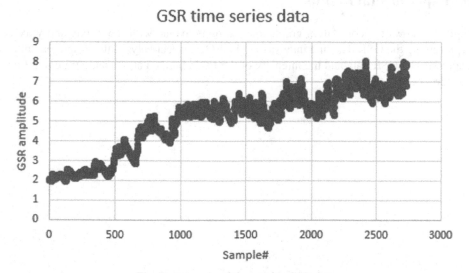

Fig. 7. Example of time series GSR data

with the x-axis and the y-axis representing the successive number (time) and amplitude, respectively.

In Fig. 8, the average intensity of the ROI graph is illustrated, with the x-axis being the successive number (time) and the y-axis the intensity of the ROI.

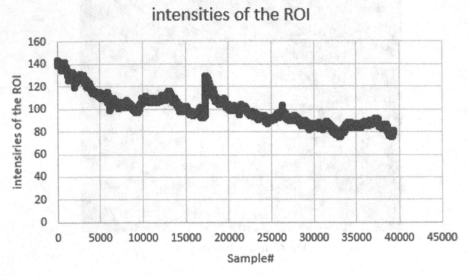

Fig. 8. Example of the ROI intensity

3   Experimental Results

Figure 9 shows the correlation graphs of five participants, with the x-axis and y-axis representing the GSR data and intensity of the ROI, respectively. In the graph, one plot expresses the GSR data and the intensity of the ROI obtained at the same time point in the form (x, y).

The results of GSR data and ROI intensity obtained simultaneously showed a decreasing trend from left to right. The R2 suggested a high correlation, with an average of 0.5931. This may have been caused by the increase in the amount of infrared light components scattered or absorbed in the skin area due to sweat generated inside the skin, and the amount of light reflected and acquired as images decrease. As a result of the experiment, the possibility of replacing the GSR sensor was confirmed through infrared facial images.

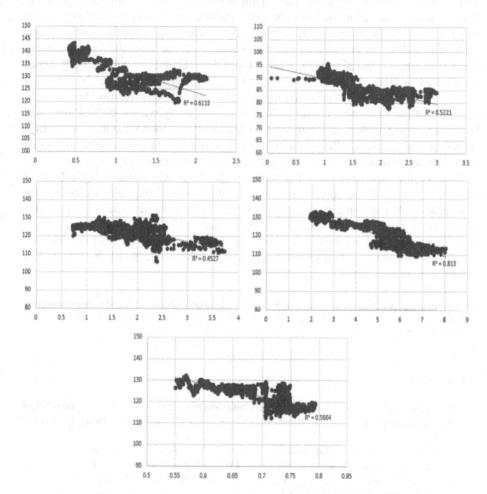

Fig. 9. Correlation Analysis between GSR data and ROI intensities for five participants

4 Conclusion

There are various bio-signals, such as PPG, ECG, GSR, skin temperature, and respiratory rate. Research on detecting these bio-signals using image processing has been steadily progressing, and in recent years, research on detecting bio-signals using non-contact methods has drawn attention. However, unlike other bio-signals, GSR measures sweat on the surface of the skin; since GSR has always detected signals through a sensor attached to the body, research cases for detecting GSR signals in a non-contact method are insufficient.

In this paper, we proposed a non-contact method for detecting GSR from a human face region using a near-infrared camera. A 20-min visual stimulus video that could cause tension or excitement in the participant was shown, and an isolated experimental environment was set to minimize emotion induction due to factors other than the experiment. After attaching a GSR sensor to the inside of the participant's index and ring fingers and having them watch a 20-min visual stimulus video, the face region was filmed with a near-infrared camera. It also stored data entered through the GSR sensor and infrared face images during the experiment. The ROI was specified by detecting the face region in the acquired facial images. Afterward, the correlation between the intensities of the ROI and the GSR sensor data were analyzed; the correlation graph showed a tendency to decrease from left to right, and it was found that there was a high correlation with an average R2 value of 0.5931. This may have been due to the reduction of the light in the images by light reflected by sweat from the participant's skin.

Through the experiment, it was confirmed that GSR was induced by feelings of excitement or tension caused by a visual stimulation video. It was also found that the GSR value can be estimated through a non-contact method, namely the analysis of skin images using a near-infrared camera. In other words, the possibility of replacing the GSR sensor could be confirmed through the analysis of the near-infrared skin images. In future works, we plan to research performance stabilization that is no affected by external light or movement.

Acknowledgements. This work was supported by the NRF (National Research Foundation) of Korea and funded by the Korean government (Ministry of Science and ICT) (NRF-2019R1A2C4070681).

References

1. Monkaresi, H., Calvo, R.A., Yan, H.: A machine learning approach to improve contactless heart rate monitoring using a webcam. IEEE J. Biomed. Health Inform. **18**(4), 1153–1160 (2013)
2. Wei, L., Tian, Y., Wang, Y., Ebrahimi, T., Huang, T.: Automatic webcam-based human heart rate measurements using Laplacian Eigenmap. In: Lee, K.M., Matsushita, Y., Rehg, J.M., Hu, Z. (eds.) ACCV 2012. LNCS, vol. 7725, pp. 281–292. Springer, Heidelberg (2013). https://doi.org/10.1007/978-3-642-37444-9_22
3. Suh, K.H., Lee, E.C.: Contactless physiological signals extraction based on skin color magnification. J. Electron. Imaging **6**(3), 063003-1–063003-9 (2017)
4. Madan, C.R., Harrison, T., Mathewson, K.E.: Noncontact measurement of emotional and physiological changes in heart rate from a webcam. Psychophysiology **55**(4), 1–2 (2018)
5. Lewandowska, M., Rumiński, J., Kocejko, T., Nowak, J.: Measuring pulse rate with a webcam—A non-contact method for evaluating cardiac activity. In: 2011 Federated Conference on Computer Science and Information Systems (FedCSIS), pp. 405–410. IEEE, Szczecin (2011)

6. Yu, M.C., Liou, J.L., Kuo, S.W., Lee, M.S., Hung, Y.P.: Noncontact respiratory measurement of volume change using depth camera. In: 2012 Annual International Conference of the IEEE Engineering in Medicine and Biology Society, pp. 2371–2374. IEEE, San Diego (2012)
7. Poh, M.Z., McDuff, D.J., Picard, R.W.: Advancements in noncontact, multiparameter physiological measurements using a webcam. IEEE Trans. Biomed. Eng. **58**(1), 7–11 (2010)
8. IMOTION Homepage. https://imotions.com/blog/galvanic-skin-response. Accessed 28 Oct 2020
9. Nourbakhsh, N., Chen, F., Wang, Y., Calvo, R.A.: Detecting user's cognitive load by galvanic skin response with affective interference. ACM Trans. Interactive Intell. Syst. (TiiS) **7**(3), 1–20 (2017)
10. Vijaya, P.A., Shivakumar, G.: Galvanic skin response: a physiological sensor system for affective computing. Int. J. Mach. Learn. Comput. **3**(1), 31–34 (2013)
11. van Dooren, M., Janssen, J.H.: Emotional sweating across the body: comparing 16 different skin conductance measurement locations. Physiol. Behav. **106**(2), 298–304 (2012)
12. Drummond, P.D., Lance, J.W.: Facial flushing and sweating mediated by the sympathetic nervous system. Brain **110**(3), 793–803 (1987)
13. Kamei, T., Tsuda, T., Kitagawa, S., Naitoh, K., Nakashima, K., Ohhashi, T.: Physical stimuli and emotional stress-induced sweat secretions in the human palm and forehead. Anal. Chim. Acta **365**(1–3), 319–326 (1998)

EEG Motor Classification Using Multi-band Signal and Common Spatial Filter

Tan Yu Xuan[1], Norashikin Yahya[2(✉)] (iD), Zia Khan[2] (iD), Nasreen Badruddin[2] (iD), and Mohd Zuki Yusoff[2] (iD)

[1] Intel Malaysia Sdn Bhd, Bayan Lepas, Malaysia
[2] Centre for Intelligent Signal and Imaging Research (CISIR),
Department of Electrical and Electronic Engineering,
Universiti Teknologi PETRONAS (UTP), Seri Iskandar, Malaysia
norashikin_yahya@utp.edu.my

Abstract. Electroencephalography (EEG) signal is one of the popular approaches for analyzing the relationship between motor movement and the brain activity. This is mainly driven by the rapid development of Brain-Computer-Interface BCI devices for applications like prosthetic devices, using EEG as its input signal. The EEG is known to be highly affected by artefact and with more motor events, this may result in low classification accuracy. In this paper, classification of 3-class hand motor EEG signals, performing grasping, lifting and holding using Common Spatial Pattern (CSP) and pre-trained CNN is investigated. Thirteen electrodes capturing signals related to motor movement, C3, Cz, C4, T3, T4, F7, F3, Fz, F4, F8, P3, Pz and P4 are utilized and signal from α, β, Δ and θ bands are selected in the pre-processing stage. CSP filters utilizing the scheme of pair-wise are used to increase the discriminative power between two classes whereby the signals extracted by the CSP filter are converted into scalograms by utilizing Continuous Wavelet Transform (CWT). The accuracy of the proposed multi-band and CSP based classification algorithm tested using DenseNet giving average accuracy values of 97.3%, 93.8% and 100%, for GS, LT and HD movements, respectively. These results indicate that the classification framework using CSP filters and pre-trained CNN can provide a good solution in decoding hand motor movement from EEG signals.

Keywords: CNN · Motor EEG · Filterbank · CSP · CWT · EVD

1 Introduction

In 1924, Hans Berger, a psychiatrist from Germany invented a method for recording brain waves, known as electroencephalography (EEG). Olejniczak in [18] stated that EEG is a realistic portrayal of the distinction in voltage between two diverse cerebral areas captured over time. As the cells of the human brain

© Springer Nature Switzerland AG 2021
M. Singh et al. (Eds.): IHCI 2020, LNCS 12615, pp. 120–131, 2021.
https://doi.org/10.1007/978-3-030-68449-5_13

communicate with one another through electrical driving forces, resulting in electrical activity of the brain cell. The EEG signal measurement is described by what is known as volume conduction, which is the effect of current movement through the tissues between the electrical generator and the recording electrodes [18]. An EEG can be utilized to help distinguish potential issues related to this movement in different brain areas.

An EEG tracks cerebrum wave data and records data using little flat metal plates called cathodes. The cathodes capture the brain's electrical driving forces and send a sign to a PC which logs the results. In an EEG recording, the electrical driving forces will have the pattern of wave lines with valleys and peaks. These lines empower doctors to assess unusual patterns quickly with inconsistencies can be an indication of spasms or different unsettling influences in the cerebrum. It should be noted that the nature of EEG which is non-invasive and can be utilized easily and this has positioned EEG as the most appropriate method for BCI application [23]. Smith in [22] claimed that EEG can show significance abnormality in patients with altered or unclear mental states.

During the recording of EEG imagery motor movement signals, the subjects will be concentrating on the tools positioned in front of them, for example a pair of chopsticks [16]. Next, when it is the phase of performing the task, subjects will be asked to imagine the motor movement that is required to utilize the tool and the subjects will be entering the resting phases by relaxing and not thinking about the movement [16]. On the other hand, when the executed motor movement is performed, the subject is required to operate tools which are the chopsticks in this case and physically perform the movement that was imagined previously.

Based on the results in [21], the theory of the imagery motor movement and executed motor movement are neurological alike phenomena. However, Nakano et al. in [16] claimed that there are no existing method to do direct measurement on ability to imagine for each of the subjects that performed imagery motor movement and this might cause different in results among the subjects due to the different ability to imagine [16]. The second issue is the differences in time taken for activities to be operated in terms of imagined and executed [12]. Lotze et al. in [12] claimed that the duration of imagery motor movement might be overestimated if it is a difficult task like lifting a heavy load while walking, but the duration of the simple task have high similarity in terms of duration. This theory is supported by Fitt's law which claims that more time will be taken to perform a difficult motor movement physically as compared to an easy motor movement and this theory can be applied to the imagery motor movement too [9, 12].

The paper outline is as follows. In Sect. 2, papers related to CSP and filter-bank techniques are reviewed. Then, Sect. 3 detail out how multi-band signal, CSP and CNN is put together as a framework for classification of 3-class EEG motor signals. Section 4 presents result and discussion. Lastly, Sect. 5 concludes the paper and suggests further work in this area.

2 Related Work

The low signal-to-noise ratio (SNR) nature of EEG will have a detrimental effect on the performance of EEG classification algorithms, hence Common Spatial Pattern (CSP) is often utilized for improving EEG classification performance. CSP is a feature extraction technique that can transform the data into a new time series resulting in maximizing variance of one event while minimizing variance of the other event. CSP is widely utilized in the area of BCI for 2-class event classification from the EEG signal [24].

In [8], Khan et al. utilized CSP filter to perform spatial features extraction for motor function EEG signal efficiently as CSP maximizes ratio of variances between data of different events [8]. The maximized variance in one event while minimizing the variance in another event can improve the accuracy of classification. The features of the CSP are in the form matrix while the electrodes will weigh on each of the rows of the data matrix. There are several researches completed and proven that CSP is an efficient technique and widely used to perform features extraction from EEG signal [7].

The important point of the CSP method is that it is based on two-class classification, so multi-class problems will be broken into several classification two-class problems. In [4,14], two well known methods which are One-versus-the-Rest (OVR) and pair of two-class classification problems, pair-wise were presented and each method has its strength and weakness. For the OVR method, an assumption is made that the covariance matrix of all the events are approximately similar, but this assumption is not applicable in the realistic world data.

In [7], Ang et al. utilized Filter Bank Common Spatial Pattern (FBCSP) for classification of EEG signal of imagery motor movement for BCI. FBCSP is claimed to have higher generalization compared to the approach of Sub-band Common Spatial Pattern (SBCSP). Their results indicate that the FBCSP approach has generated higher accuracy of classification as compared to CSP and SBCSP when the frequency bands are manually selected [7].

Selection of the frequency bands played an important role when CSP is utilized for feature extraction of EEG signal since it carries the information related to specific event [10]. In addition, correct band selection will ensure that the band-power changes generated by the motor function are well captured before being input to the classifier [7].

There are four progressive steps performed by the FBCSP, the first phase is filtering by filter bank to filter the EEG in the desired frequency range only, the second phase is utilizing the CSP algorithm for spatial filtering, the third phase is select the features and the last phase is perform the classification for the motor function based on the selected features only [3].

Several variants of CSP approaches will be discussed briefly in this paragraph. In [10], Common Spatio-Spectral Pattern (CSSP) is proposed to improve the performance of the CSP and this is implemented by having a temporary delay inserted to enable the frequency filters have the ability to be tuned individually [10]. Next, a further improvement is done by searching common spectral pattern for all of the channels instead of searching the different spectral pattern for each of the channels in CSSP, the improved algorithm was known as Common Sparse Spectral Spatial pattern (CSSSP) [10].

As another alternative, an algorithm Sub-band Common Spatial Pattern (SBCSP) is introduced to filter the EEG signals at several sub-bands and the feature extraction is performed at each sub-bands. Even though the SBCSP has better accuracy compared to the CSSSP, the method ignores the possible association of the CSP features from various sub-bands. Therefore to account for association of the CSP features from various sub-bands, FBCSP is introduced [10]. The key feature of FBCSP is the algorithm estimate the mutual data of the CSP features from various sub-bands to achieve the objective of getting the most discriminative features while the features that are selected will be fed into the classifier as input [10]. It also can be said that FBCSP will only deploy the effective spatial filter pairing with selected CSP features while the SBCSP will deploy all the spatial filters [8]. Other works that also utilize FBCSP include EEG classification of motor imagery [3] and executed motor movement for a single limb [20].

3   Methodology

The overall machine learning framework of this work is illustrated by the block diagram in Fig. 1. The stages include signal pre-processing for artefact removal, followed by multi-band filtering, data normalization, feature extraction using CSP filter-wavelet transform, training and testing of the classifier model for all 12 subjects. The details for each stage will be explained in the subsequent sections.

3.1   Three-Class EEG Dataset

The data collection was done by WAY consortium (wearable interfaces for hand function recovery) recorded using an EEG recording device, ActiCap that consists of 32 electrodes 500 Hz sampling frequency [13]. The dataset consists of EEG recordings of 6-motor tasks collected from 12 healthy subjects where each subject performed 10 series of recordings, having 30 trials within each recording. In this work, only a 3 motor tasks are used to investigate the performance of classification using multi-band EEG and CSP method.

3.2   Electrode Selection and Pre-processing

The original EEG dataset was recorded using 32-channel electrodes [13] but in this study, only 13 electrodes, known to be active for executed motor function are selected. In [1] Alomari et al. established that the electrodes C3, CZ and C4 are the prominent channels for the executed hand movement of both right and left side. In this paper, adapting electrode selection as in [1,2] where 13 electrodes T3, T4, F7, F3, Fz, F4, F8, C3, Cz, C4, P3, Pz and P4 are chosen. The electrode T7,T8 of the WAY dataset is equal to the electrode T3 and T4

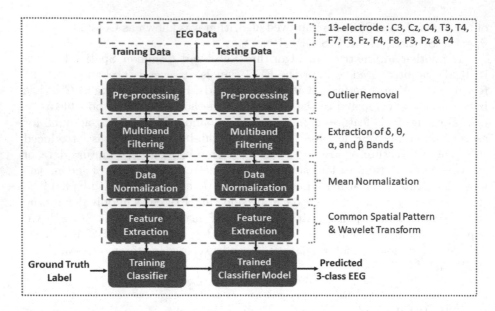

Fig. 1. General methodology for training and testing machine learning algorithm in classifying motor EEG signals using CSP filter and wavelet transform.

[2,13]. The pre-processing of the EEG signal was by utilizing a 50-Hz notch filter to remove the unwanted signals coming from the power line [11,13,20]. For removal of eye-blink artefact method based on empirical mode decomposition (EMD), and canonical correlation analysis (CCA), known as FastEMD–CCA is used to preserve the neural information in EEG signal [5].

3.3 Multi-band Filtering

The selection of multi-band EEG is considered based on the range frequency related to motor movement. In [10], it was stated that the frequency range 4 Hz to 40 Hz covered most of the features that are related to the imagery motor movement. In [11], Li et al. utilized 5 bands of frequencies which are Δ 1 Hz to 3 Hz, θ 4 Hz to 7 Hz, α 8 Hz to 12 Hz, β 13 Hz to 30 Hz and γ 31 Hz to 50 Hz. Sleight et al. [21] mentioned that the frequency waves of α, β and θ are representing the adult's activity of the brain. Therefore, based on these work, the frequency bands selected are Δ, θ, α and β and the frequency range will be 0 to 3.5 Hz, 4 to 7 Hz, 8 12 Hz and 13 to 30 Hz, respectively. The γ band was not chosen to avoid the noise from the power line which will affect the accuracy of the classification result.

The multi-band signal extraction is achieved using filter bank technique. The set of filters used includes a Butterworth lowpass filter for Δ band, while a 12th-order Butterworth Bandpass Filter will be used to filter the EEG data at the frequency band of θ, α and β. An illustrative diagram detailing on multi-band extraction from the 13-electrode is presented in Fig. 2.

3.4 Data Normalization

In general, normalizing the data will speed up learning and leads to faster convergence of the trained network. Here,

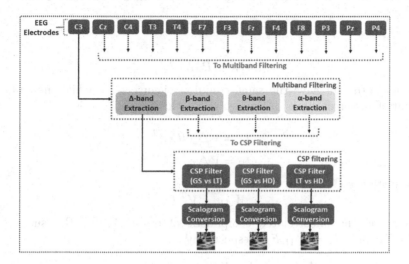

Fig. 2. Multi-band extraction, CSP filtering and scalogram conversion of EEG signal from 13 electrodes.

3.5 Common Spatial Filter

Consider 2 motor events, A and B, with their EEG signals given as $E_A \in \mathbb{R}^{N \times T}$ and $E_B \in \mathbb{R}^{N \times T}$, where T is the number of samples per electrode and N is the number of electrodes. The steps in performing the CSP filtering are as follows:

1. Determine the normalized spatial covariance, for event A and B.

$$\overline{R}_A = \frac{E_A E_A^T}{trace(E_A E_A^T)} \tag{1}$$

$$\overline{R}_B = \frac{E_B E_B^T}{trace(E_B E_B^T)} \tag{2}$$

where E^T is the transpose of E while the $trace(E)$ is the sum E's diagonal elements.

2. If $\overline{R}_A$ and $\overline{R}_B$ are the average normalized covariance of event A and B respectively, next step is to perform the eigendecomposition of the composite spatial covariance matrix

$$R_{AB} = \overline{R}_A + \overline{R}_B = \mu \triangle \mu^T \tag{3}$$

where μ is the eigenvectors and $\triangle$ is the diagonal matrix of the eigenvalues.

3. Determine the whitening transformation matrix

$$P_w = \triangle^{-\frac{1}{2}} \mu^T \tag{4}$$

4. Determine the transformed average covariance matrix of event A and B

$$Q_A = P_w R_A P_w^T \tag{5}$$

$$Q_B = P_w R_B P_w^T \tag{6}$$

Q_A and Q_B will have the same common eigenvectors, while the summing result of the eigenvalues of the two matrices, $\triangle_A$ and $\triangle_B$ is 1.

$$Q_A = V\triangle_A R_A V^T \tag{7}$$

$$Q_B = V\triangle_B R_B V^T \tag{8}$$

5. Calculate the CSP filter

$$W_{AB} = V^T P_w \tag{9}$$

6. Determine the uncorrelated component of event A and B, using $Z_A = W_{AB}E_A$ and $Z_B = W_{AB}E_B$, respectively.

The 2-class CSP filter is used to improve the discriminative power between the 2-class EEG. Details on the pair-wise scheme utilized for every 2-class, GS vs LT, GS vs HD, LT vs HD is illustrated in Fig. 2.

3.6   Scalogram Conversion Using Continuous Wavelet Transform

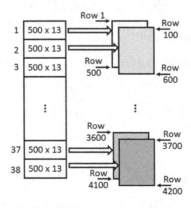

Fig. 3. Windowing operation of 13-electrode EEG for conversion to scalogram using CWT.

Conversion to the scalogram of signal from each electrode is performed using window size of 500 samples with overlaps of 100 samples as illustrated in Fig. 3.

Continuous wavelet transform (CWT) will take in the signals of 500-sample for conversion to grayscale scalograms. Since every event has EEG data size 4200-sample, the overlapped sliding window technique will increase the EEG data size to 19000 for each of the events. Therefore, for one event, each electrode will produce 38 scalograms (19000/500 = 38). It is given that the GS, LT and HD movement EEG recording for one subject was repeated for 8 series. Hence, the grayscale scalograms from 13 electrodes are generated for 8 series. Then, the same process is applied on 8 series data of one subject is repeated for four frequency bands of α, β, θ and Δ, as illustrated in Fig. 2.

With 13 electrodes, the RGB format scalogram will be based on 5 combination of electrodes as follows; (C3, Cz and C4), (T3, T4 and F7), (F3, Fz and F4), (F8, P3 and Pz) and (P3, Pz and P4) as illustrated in Fig. 4. Also, Fig. 4 shows the formation of RGB scalogram from 5 sets of 3-electrodes from α band signals. The same steps are applied for signals from β, θ and Δ band. In total, the number of RGB scalograms for each GS vs LT, GS vs HD and LT vs HD is 24320 images.

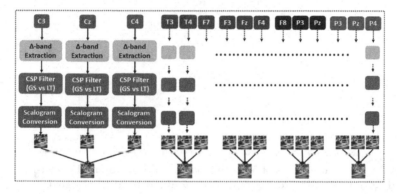

Fig. 4. Illustration of RGB scalogram formation from set of three electrodes in α band EEG signals. Same steps are repeated for β, θ and Δ bands.

3.7 Implementation Details

The RGB scalograms will be split as testing, validation and training data with the ratio of 0.1:0.1:0.8, respectively. The training data is used for training the pre-trained CNN whereas the testing data is used as unseen input to the trained-CNN to evaluate the performance of the network. The validation data used during training of CNN is for assessing how well the network is currently performing hence can be used to detect over-fitting problem. The CNN models training and testing were carried out on MATLAB platform, using a 32 GB memory PC, Intel Core i7-8700k CPU, Geforce RTX 2080Ti with 32 GB RAM graphic card, and 10.1 Cuda version. The network is trained at 30 epoch with SGDM optimizer, learning rate of 0.0001 and minibatch size of 16.

4   Result

4.1   Common Spatial Pattern Filtering

Samples of scatter plots of EEG signals before and after CSP filtering from the 4 bands are shown in Fig. 5. CSP filtering essentially transformed the data into a new subspace having better inter class separation for the 3-class signal. This trend is clearly visible for signals, especially from Δ and β bands. Notably, change of signal variance after CSP filtering has resulted in signals from the same class to be closely clustered together, giving good separation for the 3-class data. This indicates that the systematic role of Δ band signals on motor responses which is in line with the findings in [6,19].

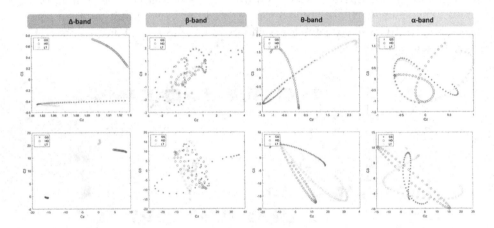

Fig. 5. C3 vs Cz scatter plots of 50-sample EEG for GS, HS and LT (1st-row) before and (2nd-row) after CSP filtering, from Δ, β, θ and α band EEG signals.

4.2   3-Class EEG Motor Classification

Table 1 shows the performance of the EEG motor classification using DenseNets expressed in terms of accuracy for 12 subjects. The overall accuracy of classifying the 3-class motor function, GS vs LT vs HD varies between 94.9% to 99.5%. Specifically, GS classification accuracy for 12 subjects varies from 94.5% to 99.6% with the best one recorded by Subject 4. Accuracy for LT, is the lowest for the 3 classes with a minimum value of 88% to maximum value of 99.5%. From the confusion matrix shown in Fig. 6, this indicates the lifting movement, LT are often being wrongly classified as grasping, GS and vice versa. This may be due to the fact that the 2 movements occurred in consecutive manner, GS, followed by LT. Besides, the change of movements as the subjects change from GS to LT movement is relatively small and this may have resulted in low classification accuracy (<95%) for LT class. Perfect classification is achieved with hold,

HD movement for all subjects and this may be attributed to the unique finger movement of HD, where the subject has to press both thumb and index fingers on the object. As expected, the variation in performance of different subjects is subject-dependent [15,17] because EEG varies from one subject to the other.

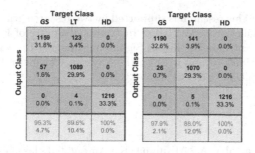

Fig. 6. Confusion matrix of Subject 6 and 8.

Table 1. Accuracy (in %) of 3-Class EEG Motor Classification using DenseNet for 12 subjects.

	S1	S2	S3	S4	S5	S6	S7	S8	S9	S10	S11	S12
GS	97.5	97.9	98.4	97	99.6	97.9	99.1	95.3	98.1	94.5	97.3	94.6
LT	90.7	88	99.1	92.9	98.8	88	99.0	89.6	99.5	91.0	99.0	90.0
HD	100	100	100	100	100	100	100	100	100	100	100	100
Average	96.1	95.3	99.2	96.6	99.5	95.3	99.4	95.0	99.2	95.2	98.8	94.9

5   Conclusion

Analysis of EEG signal motor function plays an important role in the development of practical BCI systems commonly developed for the neurological diseases patients and amputees. Classification of motor EEG signals based on Δ, β, θ and α and CSP filtering has shown promising results with average subject accuracy between 94.9% and 99.5%, tested on DenseNet. The main framework of the algorithm uses generalized eigenvalue decomposition, namely CSP filtering that has resulted in maximizing the variance in one class while minimizing the variance in the other class. Besides, the utilization of multi-band EEG provides the classifier, DenseNet with more enriching data needed for a well trained CNN. The combination of multi-band and CSP filtering offer the advantage of giving good classification accuracy for the 3-class EEG data. This result has demonstrated the effectiveness of the multi-band CSP based method for classification of EEG motor functions. The implementation of CSP in the EEG classification algorithm is made under the assumption that the EEG signal is wide-sense stationary having minimal intra-session and inter-session variation. This means the CSP based algorithm is sensitive to outliers, and can be addressed using regularization method for more realistic BCI system.

Acknowledgment. This research is supported by two research grants: (1) by the Ministry of Education Malaysia under Higher Institutional Centre of Excellence (HICoE) Scheme awarded to Centre on Intelligent Signal and Imaging Research (CISIR) under grant number 015MA0-050(6), and (2) by the Yayasan Universiti Teknologi Petronas under grant number YUTP-FRG 015LC0-031.

References

1. Alomari, M.H., Samaha, A., AlKamha, K.: Automated classification of L/R hand movement EEG signals using advanced feature extraction and machine learning. arXiv preprint arXiv:1312.2877 (2013)
2. AydemiR, Ö.: Common spatial pattern-based feature extraction from the best time segment of BCI data. Turk. J. Electr. Eng. Comput. Sci. **24**(5), 3976–3986 (2016)
3. Bentlemsan, M., Zemouri, E., Bouchaffra, D., Yahya-Zoubir, B., Ferroudji, K.: Random forest and filter bank common spatial patterns for EEG-based motor imagery classification. In: 2014 5th International Conference on Intelligent Systems, Modelling and Simulation, pp. 235–238 (2014). https://doi.org/10.1109/ISMS.2014.46
4. Dornhege, G., Blankertz, B., Curio, G., Muller, K.R.: Boosting bit rates in non-invasive EEG single-trial classifications by feature combination and multiclass paradigms. IEEE Trans. Biomed. Eng. **51**(6), 993–1002 (2004)
5. Egambaram, A., Badruddin, N., Asirvadam, V.S., Begum, T., Fauvet, E., Stolz, C.: FastEMD-CCA algorithm for unsupervised and fast removal of eyeblink artifacts from electroencephalogram. Biomed. Sig. Process. Control **57**, 101692 (2020). https://doi.org/10.1016/j.bspc.2019.101692
6. Hamel-Thibault, A., Thénault, F., Whittingstall, K., Bernier, P.M.: Delta-band oscillations in motor regions predict hand selection for reaching. Cereb. Cortex **28**(2), 574–584 (2016). https://doi.org/10.1093/cercor/bhw392
7. Ang, K.K., Chin, Z.Y., Zhang, H., Guan, C.: Filter bank common spatial pattern (FBCSP) in brain-computer interface. In: 2008 IEEE International Joint Conference on Neural Networks (IEEE World Congress on Computational Intelligence), pp. 2390–2397 (2008). https://doi.org/10.1109/IJCNN.2008.4634130
8. Khan, J., Bhatti, M.H., Khan, U.G., Iqbal, R.: Multiclass EEG motor-imagery classification with sub-band common spatial patterns. EURASIP J. Wirel. Commun. Netw. **2019**(1), 1–9 (2019). https://doi.org/10.1186/s13638-019-1497-y
9. Kourtis, D., Sebanz, N., Knoblich, G.: EEG correlates of Fitts's law during preparation for action. Psychol. Res. **76**(4), 514–524 (2012). https://doi.org/10.1007/s00426-012-0418-z
10. Kumar, S., Sharma, A., Tsunoda, T.: An improved discriminative filter bank selection approach for motor imagery EEG signal classification using mutual information. BMC Bioinform. **18**(16), 545 (2017). https://doi.org/10.1186/s12859-017-1964-6

11. Li, T., Xue, T., Wang, B., Zhang, J.: Decoding voluntary movement of single hand based on analysis of brain connectivity by using EEG signals. Front. Hum. Neurosci. **12**, 381 (2018)
12. Lotze, M., Halsband, U.: Motor imagery. J. Physiol.-Paris **99**(4–6), 386–395 (2006)
13. Luciw, M.D., Jarocka, E., Edin, B.B.: Multi-channel EEG recordings during 3,936 grasp and lift trials with varying weight and friction. Sci. Data **1**(1), 1–11 (2014)
14. Meisheri, H., Ramrao, N., Mitra, S.: Multiclass common spatial pattern for EEG based brain computer interface with adaptive learning classifier. arXiv preprint arXiv:1802.09046 (2018)
15. Melnik, A., et al.: Systems, subjects, sessions: to what extent do these factors influence EEG data? Front. Hum. Neurosci. **11** (2017). https://doi.org/10.3389/fnhum.2017.00150
16. Nakano, H., Ueta, K., Osumi, M., Morioka, S.: Brain activity during the observation, imagery, and execution of tool use: an fNIRS/EEG study. J. Nov. Physiother. **S1**, 1–7 (2012)
17. Nishimoto, T., Higashi, H., Morioka, H., Ishii, S.: EEG-based personal identification method using unsupervised feature extraction and its robustness against intra-subject variability. J. Neural Eng. **17**(2), 026007 (2020). https://doi.org/10.1088/1741-2552/ab6d89
18. Olejniczak, P.: Neurophysiologic basis of EEG. J. Clin. Neurophysiol. **23**(3), 186–189 (2006)
19. Saleh, M., Reimer, J., Penn, R., Ojakangas, C., Hatsopoulos, N.: Fast and slow oscillations in human primary motor cortex predict oncoming behaviorally relevant cues. Neuron **65**, 461–71 (2010). https://doi.org/10.1016/j.neuron.2010.02.001
20. Shiman, F., et al.: Classification of different reaching movements from the same limb using EEG. J. Neural Eng. **14**(4), 046018 (2017)
21. Sleight, J., Pillai, P., Mohan, S.: Classification of executed and imagined motor movement EEG signals, vol. 110. University of Michigan, Ann Arbor (2009)
22. Smith, S.: EEG in neurological conditions other than epilepsy: when does it help, what does it add? J. Neurol. Neurosurg. Psychiatry **76**(Suppl 2), ii8–ii12 (2005)
23. Thiyam, D., Rajkumar, E.: Common spatial pattern algorithm based signal processing techniques for classification of motor imagery movements: a mini review. IJCTA **9**(36), 53–65 (2016)
24. Yahya, N., Musa, H., Ong, Z.Y., Elamvazuthi, I.: Classification of motor functions from electroencephalogram (EEG) signals based on an integrated method comprised of common spatial pattern and wavelet transform framework. Sensors **19**(22), 4878 (2019)

Segmentation of Prostate in MRI Images Using Depth Separable Convolution Operations

Zia Khan[1] , Norashikin Yahya[1($\boxtimes$)] , Khaled Alsaih[1] ,
and Fabrice Meriaudeau[2]

[1] Centre for Intelligent and Imaging Research (CISIR),
Department of Electrical and Electronic Engineering,
Universiti Teknologi PETRONAS (UTP), Seri Iskandar, Malaysia
norashikin_yahya@utp.edu.my
[2] ImViA/IFTIM, University of Bourgogne Franche-Comté,
Besançon, France

Abstract. The segmentation of the prostate gland into two sub-regions, namely, the central gland (CG) and the peripheral zone (PZ) is crucial for the prostate cancer (PCa) diagnosis. The nature and occurrence of cancer occurred in the prostate is substantially different in both zones. Magnetic resonance imaging modality (MRI) is a clinically primary tool for computer-based assessment and remediation of various cancer types such as PCa. In this paper, we evaluated DeeplabV3+ model on T2W MRI scans using the I2CVB dataset, which is designed in an encoder-decoder style for the zonal segmentation of prostate regions. An important feature of DeeplabV3+ is the depth-wise separable convolutions, which allow more information to be extracted from images as it uses filters with different dilation rates. Prior to being fed to the deep neural network, image pre-processing techniques are applied, including image resizing, cropping, and denoising. The DeeplabV3+ model performance is evaluated using the Dice similarity coefficient (DSC) metric and compared with the vanilla U-Net architecture. Results show that the encoder-decoder network having depth-wise separable convolutions performed better prostate segmentation than the network with standard convolution operations with the DSC value of 70.1% in PZ and 81.5% in CG zone.

Keywords: Zonal segmentation · MRI · Encoder-decoder · U-Net · DeeplabV3+

1 Introduction

Prostate cancer is considered the second-largest diagnosed cancer after skin cancer and one of the leading causes of death among the men over the world [31]. According to the American cancer society, the prostate cancer ratio is the largest

© Springer Nature Switzerland AG 2021
M. Singh et al. (Eds.): IHCI 2020, LNCS 12615, pp. 132–141, 2021.
https://doi.org/10.1007/978-3-030-68449-5_14

in the U.S. For 2019, and there are estimates of 174,650 new cancer cases and 31,620 cancer deaths in the United States [28]. Detecting prostate cancer at early stages can help in saving human life and minimizing treatment costs. Different imaging modalities have been used for prostate cancer scannings, such as computed tomography (CT), transrectal ultrasound (TRUS), and magnetic resonance imaging (MRI). However, MRI plays a vital role in medical abnormalities diagnosis because of its advantage of giving good soft-tissue contrast, and blood circulation visualization [17]. For this reason, MRI carries critical information on prostate anatomy and the prostate tumour level for diagnosis of prostate cancer (PCa) [30]. In particular, T2W MRI imaging remains the primary scanning technique, and it can show a boundary between the central gland (CG) and peripheral zone (PZ) [26]. Manual delineation of the prostate zones is time-consuming due to the longer time taken for the grader to delineate each image manually [22], and it is prone to inter-observer and inter-observer variability. There has been a continuous research effort in improving the computer-aided detection of PCa by implementing robust machine learning algorithms.

The prostate gland consists of three different zones, namely, transition zone, peripheral zone, and central zone. The central zone and transition zone look similar in medical imaging and collectively named as a central gland, as shown in Fig. 1. The occurrence rate of prostate cancer in the PZ is 60–80%, whereas, in the transitional zone (TZ) and CG, the rate is at 30% and 10%, respectively [9]. The severity of cancer in CG is far less than in TZ and PZ, but there is a high chance of extra-capsular expansion and seminal vesicle incursion. Given that the risk of PCa is different in each zone, zonal prostate segmentation is crucial for early detection and improved prostate cancer diagnosis. Therefore, a robust and automated zonal segmentation approach is needed for accurate localization of prostate zones [4].

Over the last decade, before deep learning regained prominence, various prostate segmentation algorithms were proposed for better detection of PCa in computer-aided diagnosis systems. In [29], Toth et al. (2008) developed a level-set based on the gradient, intensity, statistical and geometric information for segmentation of the prostate gland. Atlas-based approach for prostate segmentation was proposed by Klein et al. (2008) [15] by which several images with the related mask are registered to the target image through geometric transformation. The deformable labels are fused using different voting techniques to generate required labels for prostate segmentation. Martin et al. [21] have introduced the automatic segmentation of the prostate gland 3D MRI. Probabilistic segmentation is achieved by registering a new image to a probabilistic atlas and applying a statistical shape model and feature model to refine prostate gland MRI segmentation. Paper by Langerak et al. (2010) [16] presented an atlas-based approach using geometric transformation to register prostate images with related masks to the target images. Then, using the selective and iterative method for performance level estimation (SIMPLE) for deformable labels is a fusion method to generate required labels for prostate segmentation. Qui et al. [24] developed

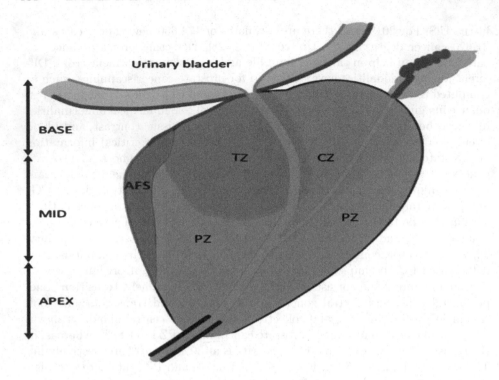

Fig. 1. The prostate gland partition into zonal and regional zones. The PZ symbolizes the peripheral zone. Also, the central zone (CZ) and transitional zone (TZ) together form the CG. The apex region contributes to 30% of the image sequence, mid gland region has 40%, and base region share 30% of the total size of the sequence [11].

a global optimization technique to perform multi-region prostate segmentation numerically stably and efficiently.

Another work by Chi et al. (2014) [3] combined the Gaussian mixture model with atlas probability maps and performed the CG and TZ segmentation with DSC values of 0.83 and 0.52, respectively. In [20], Litjens et al. (2015) presented a segmentation method using a pattern recognition approach based on anatomical, intensity, and texture features. The method is evaluated with the majority voting rule and simultaneous truth and performance level estimation (STAPLE) algorithm. Zhang et al. (2016) [32] performed the whole gland (WG) and TZ segmentation of prostate diffusion-weighted (DW) MRI using atlas semi-automatic, recording DSC values of 0.85 and 0.77, respectively. Zonal-based prostate segmentation using multi-parametric MRI was proposed in 2017 by Lemaitre et al. [19], by which features from the multi-parametric MRI are registered for better segmentation result.

Moreover, Clark et al. [6] developed a deep neural network for prostate whole gland (WG) and transitional zone (TZ) with DSC of 0.93 and 0.88. Khan et al. [13] implemented a DeepLabV3+ network with Resnet18 [10] as a backbone using NCI-ISBI2013 dataset and obtained a DSC value of 73.2 and 89.2 for PZ and CG. Aldoj et al. [1] developed a novel neural network for segmentation of whole gland (WG), Peripheral zone (PZ), and central gland (CG) and obtained DSC values of 89.2, 87.4 and 74.0. Jensen et al. [11] performed zonal prostate segmentation using a U-Net based CNN on z-score normalized data obtained from two different vendors, namely, 1.5T GE scanner and 3T Siemens scanner. The performance of the technique was assessed using DSC, and values of 0.794 and 0.692 were achieved for CG and PZ, correspondingly. In this study, we have applied DeepLabV3+ with the Xception model as a backbone [5]. The encoder-decoder network consists of depth separable convolutions that generate feature maps with different dilation rates, which reduces the computational time and enhances the overall network performance.

The article is outlined as follows: in Sect. 2 the materials and methods are described. Next, Sect. 2.3 shows the evaluation of networks. Section 3 presents results and discussions, and finally, Sect. 4 sums up the paper.

2 Materials and Methods

The encoder-decoder networks are evaluated on a publicly available dataset of the initiative for collaborative computer vision benchmarking (I2CVB), published by Lemaitre [18]. I2CVB contains a multi-parametric MRI of 40 patients acquired using two different scanners with 21 and 19 patients data were collected, respectively, by 1.5T General Electric (GE) scanner and 3T Siemens MRI scanner. The expert radiologist provided the ground truth of the WG, PZ, and CG. There are 6 different image size for this dataset, 256×256, 384×308, 448×336, 448×360, 448×368 and 512×512, with a thickness of 1.5 mm for Siemens and 3 mm for GE. The range of voxel spacing among the patients is 0.27 mm to 0.78 mm. Of the 40 patients, 34 patients are PCa positive, have different lesion sizes (0 to $36c^3$), and the rest are negative cases. The general pipeline of the method utilized for zonal prostate segmentation is shown in Fig. 2.

2.1 Image Pre-Processing

Image Resizing. The I2CVB dataset images are in 6 different dimensions, therefore resizing using the nearest-neighbour interpolation method is performed to obtain one standard image size 320×320.

Image Denoising. The T2W MRI images are filtered using the BM3D filter [7] with a sigma value of 25 to remove noisy pixels.

Image Cropping. In prostate MRI scan, the size of the prostate gland is relatively small compared to the background. Training the deep neural network without considering this can be detrimental to the segmentation result. This is due to the fact that the background pixels will be competing over the prostate

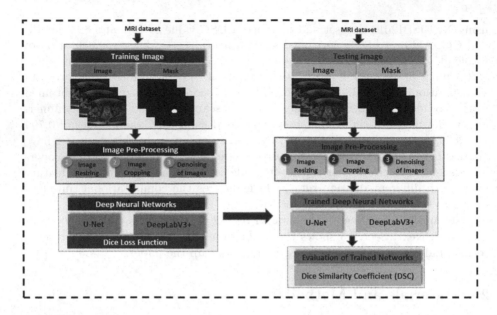

Fig. 2. General pipeline for development of prostate segmentation method.

pixels resulting in poor segmentation of the prostate. Therefore, removing the redundant pixels will reduce this effect and reduce the training time [12, 27]. However, since the image resizing output size is (320 × 320)-pixels, the images are cropped to the (256 × 256)-pixel dimensions.

Image Augmentation. In medical imaging datasets, large numbers of training labels are not available for several reasons. The image delineation in the data set requires an expert (radiologist), which is time-consuming and expensive. When deep neural networks are trained from limited training data, the over-fitting problem transpires [23]. In the training phase of deep neural networks, data augmentation is performed to increase the amount of dataset. The data augmentation substantially increases the spatial variation of the images, improving the trained model's ability to generalize what they have learned to new images. In this paper, two data augmentation approaches are utilized, which are random mirroring and random translation by ±10 pixels in the x and y directions.

2.2 CNN Architecture for Prostate Segmentation on MRI Images

In this part, two encoder-decoder deep neural networks were considered, namely, DeepLabV3+, and U-Net.

The DeepLabV3+ [2] is used with the Xception model as the backbone architecture [8]. Another key feature of this network is the use of depth separable convolutions in place of standard convolution operations that segregates the operation into depth-wise and point-convolution. Feature maps extracted by depth separable convolutions gather more information from filters at different dilation rates. The DeepLabV3+ model utilized depth-wise separable convolution operations instead of max-pooling layers.

In the decoder path, 1 × 1 convolution is used, which effectively performing channel-wise pooling. This technique will decrease the number of feature maps while retaining their salient features and reducing computational time. The output of the 1 × 1 convolution is then concatenated with encoder features and upsampled by 4. Finally, feature maps are refined by 3 × 3 convolution and then again upsampled by 4 to get the same size output.

U-Net [25] is a fully convolutional neural network, which employs the encoder-decoder style. The network is called U-Net because of its U-shaped structure, which interconnects the downsampling part with the upsampling operation. The encoder part consists of successive convolution followed by activation function(ReLU) which store some part of the feature map after each convolution. The feature maps from the encoder part are linked up with the decoder part for precise localization. Skip connections are utilized between the encoder and decoder sides to recover the information lost from the pooling process. On the upsampled feature maps, a 1 × 1 convolution operations are performed to reduce the number of features maps to match the number of wanted classes. Except for sigmoid activation, which is used by the last convolution layer, U-Net uses couple of rectified linear units (ReLU) as the activation function for each convolution layers.

2.3 Evaluation Metric

The prostate segmentation on MRI images is assessed by utilizing the Dice similarity coefficient (DSC). If TP is the true positive, FP is the false positive, and FN is a false negative, then the DSC is

$$DSC = \frac{2|TP|}{2|TP| + |FP| + |FN|}. \tag{1}$$

2.4 Implementation Details

The CNN models in training phase as well as testing phase were carried out on MATLAB platform, on a RAM with the size of 32 GB memory PC, Intel Core i7-8700k CPU, Geforce RTX 2080Ti with 32 GB RAM graphic card, and 10.1 Cuda version. The network is trained with Adam optimizer [14] with a learning rate of 0.0001, minibatch size of 16, and Dice loss function.

3 Result and Discussion

3.1 Quantitative and Qualitative Comparison

The performance of various CNN models is quantitatively evaluated in terms of DSC values as depicted in Table 1. The DSC values in PZ and CG for U-Net and

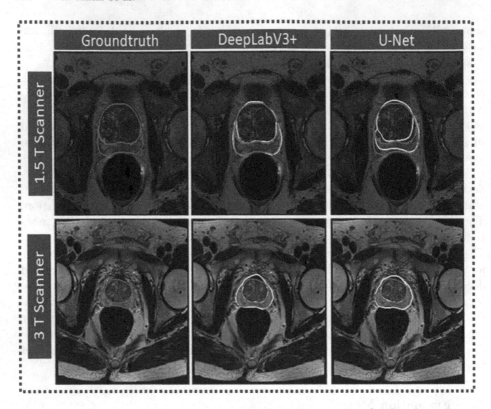

Fig. 3. Segmentation of PZ and CG using DeepLabV3+ and U-Net using an image from dataset of 1.5T and 3T MRI scanner. The white boundary is the predicted boundary, overlay on the red ground truth boundary. (Color figure online)

DeepLabV3+ is 68.9%, 78.6% and 70.5%, 81.3%. This indicates that relative to U-Net, the DSC for DeepLabV3+ is respectively 1.2% and 2.9% higher in the PZ and CG zone. In addition, DeepLabV3+ performed better segmentation in the PZ and CG region as compared to the method in [11]. The DSC value of our work is higher than [11] by 0.9% and 2.1% in the PZ and CG region, respectively.

As for qualitative analysis, we present a sample of segmented images, as shown in Fig. 3 for PZ and CG regions. The visual comparison indicates that DeepLabV3+ can achieve more accurate segmentation. It is clear from Fig. 3 that the mask generated by DeepLabV3+ is able to segment the region around the prostate boundary better than U-Net. This indicates that the use of depth separable convolution extracts feature maps with sufficient information, which helps in accurate localization of prostate boundary compared to standard convolution. Moreover, the use of the Dice loss function during training maximizes the DSC value by increasing the overlapping ratio between the reference and segmented images. The comparison of our work with other prostate segmentation techniques is clearly presented in Table 1.

Table 1. Network performance evaluation using DSC metric values.

	Evaluation	
	PZ	CG
U-Net	68.9%	78.6%
DeepLabV3+	70.1%	81.5%
Jenson [11]	69.2%	79.4%

4   Conclusion

To sum up, encoder-decoder networks are evaluated to perform an automatic zonal segmentation for T2W prostate MR images of two different MRI scanners. The neural networks are tested with the I2CVB MRI dataset and have achieved excellent performance, measured in terms of DSC metric. The better performance achieved by DeepLabV3+ with Xception backbone mainly due to depth separable convolutions. The accuracy of segmentation can further be increased via data augmentation by injecting a generative network.

Acknowledgment. This project is supported by the Yayasan Universiti Teknologi PETRONAS (YUTP) research fund under grant number 015LC0-292.

References

1. Aldoj, N., Biavati, F., Rutz, M., Michallek, F., Stober, S., Dewey, M.: Automatic prostate and prostate zones segmentation of magnetic resonance images using convolutional neural networks (2019)
2. Chen, L.-C., Zhu, Y., Papandreou, G., Schroff, F., Adam, H.: Encoder-decoder with atrous separable convolution for semantic image segmentation. In: Ferrari, V., Hebert, M., Sminchisescu, C., Weiss, Y. (eds.) ECCV 2018. LNCS, vol. 11211, pp. 833–851. Springer, Cham (2018). https://doi.org/10.1007/978-3-030-01234-2_49
3. Chi, Y., et al.: A compact method for prostate zonal segmentation on multiparametric MRIs. In: Medical Imaging 2014: Image-Guided Procedures, Robotic Interventions, and Modeling, vol. 9036, p. 90360N. International Society for Optics and Photonics (2014)
4. Choi, Y.J., Kim, J.K., Kim, N., Kim, K.W., Choi, E.K., Cho, K.S.: Functional MR imaging of prostate cancer. Radiographics **27**(1), 63–75 (2007)
5. Chollet, F.: Xception: deep learning with depthwise separable convolutions. In: Proceedings of the IEEE Conference on Computer Vision and Pattern Recognition, pp. 1251–1258 (2017)
6. Clark, T., Zhang, J., Baig, S., Wong, A., Haider, M.A., Khalvati, F.: Fully automated segmentation of prostate whole gland and transition zone in diffusion-weighted MRI using convolutional neural networks. J. Med. Imaging **4**(4), 041307 (2017)
7. Dabov, K., Foi, A., Katkovnik, V., Egiazarian, K.: Image denoising by sparse 3-D transform-domain collaborative filtering. IEEE Trans. Image Process. **16**(8), 2080–2095 (2007)

8. Dai, J., et al.: Deformable convolutional networks. In: Proceedings of the IEEE International Conference on Computer Vision, pp. 764–773 (2017)
9. Haffner, J., et al.: Peripheral zone prostate cancers: location and intraprostatic patterns of spread at histopathology. The Prostate **69**(3), 276–282 (2009)
10. He, K., Zhang, X., Ren, S., Sun, J.: Deep residual learning for image recognition. In: Proceedings of the IEEE Conference on Computer Vision and Pattern Recognition, pp. 770–778 (2016)
11. Jensen, C., et al.: Prostate zonal segmentation in 1.5 T and 3T T2W MRI using a convolutional neural network. J. Med. Imaging **6**(1), 014501 (2019)
12. Khan, Z., Yahya, N., Alsaih, K., Ali, S.S.A., Meriaudeau, F.: Evaluation of deep neural networks for semantic segmentation of prostate in T2W MRI. Sensors **20**(11), 3183 (2020)
13. Khan, Z., Yahya, N., Alsaih, K., Meriaudeau, F.: Zonal segmentation of prostate T2W-MRI using atrous convolutional neural network. In: 2019 IEEE Student Conference on Research and Development (SCOReD), pp. 95–99. IEEE (2019)
14. Kingma, D., Adam, B.J.: A method for stochastic optimization. arxiv preprint arxiv: 14126980 (2014). Cited on p. 50
15. Klein, S., Van Der Heide, U.A., Lips, I.M., Van Vulpen, M., Staring, M., Pluim, J.P.: Automatic segmentation of the prostate in 3D MR images by atlas matching using localized mutual information. Med. Phys. **35**(4), 1407–1417 (2008)
16. Langerak, T.R., van der Heide, U.A., Kotte, A.N., Viergever, M.A., Van Vulpen, M., Pluim, J.P.: Label fusion in atlas-based segmentation using a selective and iterative method for performance level estimation (SIMPLE). IEEE Trans. Med. Imaging **29**(12), 2000–2008 (2010)
17. Leake, J.L., et al.: Prostate MRI: access to and current practice of prostate MRI in the united states. J. Am. Coll. Radiol. **11**(2), 156–160 (2014)
18. Lemaître, G., Martí, R., Freixenet, J., Vilanova, J.C., Walker, P.M., Meriaudeau, F.: Computer-aided detection and diagnosis for prostate cancer based on mono and multi-parametric MRI: a review. Comput. Biol. Med. **60**, 8–31 (2015)
19. Lemaitre, G., Martí, R., Rastgoo, M., Mériaudeau, F.: Computer-aided detection for prostate cancer detection based on multi-parametric magnetic resonance imaging. In: 2017 39th Annual International Conference of the IEEE Engineering in Medicine and Biology Society (EMBC), pp. 3138–3141. IEEE (2017)
20. Litjens, G.J.S.: Computerized detection of cancer in multi-parametric prostate MRI. Ph.D. thesis, Radboud University, Nijmegen, Netherlands (2015)
21. Martin, S., Troccaz, J., Daanen, V.: Automated segmentation of the prostate in 3D MR images using a probabilistic atlas and a spatially constrained deformable model. Med. Phys. **37**(4), 1579–1590 (2010)
22. Muller, B.G., et al.: Prostate cancer: interobserver agreement and accuracy with the revised prostate imaging reporting and data system at multiparametric MR imaging. Radiology **277**(3), 741–750 (2015)
23. Perez, L., Wang, J.: The effectiveness of data augmentation in image classification using deep learning. arXiv preprint arXiv:1712.04621 (2017)
24. Qiu, W., Yuan, J., Ukwatta, E., Sun, Y., Rajchl, M., Fenster, A.: Dual optimization based prostate zonal segmentation in 3D MR images. Med. Image Anal. **18**(4), 660–673 (2014)
25. Ronneberger, O., Fischer, P., Brox, T.: U-Net: convolutional networks for biomedical image segmentation. In: Navab, N., Hornegger, J., Wells, W.M., Frangi, A.F. (eds.) MICCAI 2015. LNCS, vol. 9351, pp. 234–241. Springer, Cham (2015). https://doi.org/10.1007/978-3-319-24574-4_28

26. Scheenen, T.W., Rosenkrantz, A.B., Haider, M.A., Fütterer, J.J.: Multiparametric magnetic resonance imaging in prostate cancer management: current status and future perspectives. Invest. Radiol. **50**(9), 594–600 (2015)
27. Sekou, T.B., Hidane, M., Olivier, J., Cardot, H.: From patch to image segmentation using fully convolutional networks-application to retinal images. arXiv preprint arXiv:1904.03892 (2019)
28. Siegel, R.L., Miller, K.D., Jemal, A.: Cancer statistics, 2019. CA: Cancer J. Clin. **69**(1), 7–34 (2019)
29. Toth, R., Madabhushi, A.: Multifeature landmark-free active appearance models: application to prostate MRI segmentation. IEEE Trans. Med. Imaging **31**(8), 1638–1650 (2012)
30. Villeirs, G.M., De Meerleer, G.O.: Magnetic resonance imaging (MRI) anatomy of the prostate and application of MRI in radiotherapy planning. Eur. J. Radiol. **63**(3), 361–368 (2007)
31. Wang, Z., Liu, C., Cheng, D., Wang, L., Yang, X., Cheng, K.T.: Automated detection of clinically significant prostate cancer in MP-MRI images based on an end-to-end deep neural network. IEEE Trans. Med. Imaging **37**(5), 1127–1139 (2018)
32. Zhang, J., Baig, S., Wong, A., Haider, M.A., Khalvati, F.: A local ROI-specific atlas-based segmentation of prostate gland and transitional zone in diffusion MRI. J. Comput. Vis. Imaging Syst. **2**(1), 1–3 (2016)

Bone Age Assessment for Lower Age Groups Using Triplet Network in Small Dataset of Hand X-Rays

Shipra Madan[1]([✉]), Tapan Gandhi[1], and Santanu Chaudhury[1,2]

[1] Indian Institute of Technology Delhi, New Delhi, India
shipra.madan@ee.iitd.ac.in
[2] Indian Institute of Technology Jodhpur, Jodhpur, India

Abstract. Skeletal Bone age assessment is a routine clinical procedure carried out by paediatricians and endocrinologists for investigating a variety of endocrinological, metabolic, genetic and growth disorders in children. Skeletal maturity advances with change in structure and size of the skeletal bones with respect to age. This is commonly done by radiological investigation of the left hand due to its non dominant use. Dissent in the skeletal age and bone age values indicates abnormality. In this study, a bone-age assessment model using triplet loss for children in 0–3 years of age is proposed. Furthermore, this is the first automated bone age assessment study on lower age groups with comparable results, using one tenth of the training data samples as opposed to conventional deep neural networks. We have used small number of radiographs per class from Digital Hand Atlas Database System (DHA), a publicly available comprehensive x-ray dataset. Model trained achieves an AUC of 0.92 for binary and 0.82 for multi-class classification with visible separation in embedding clusters; thereby resulting in correct predictions on test data set.

Keywords: Bone-age assessment · Metric learning · Triplet loss · Hand radiographs · Skeletal dysplasias

1 Introduction

Bone age assessment is a standard clinical procedure carried out to determine the skeletal maturity of a child by comparing the bone age with the chronological age [1]. Age estimation based on individual's skeletal maturity is termed as bone age, whereas chronological age is simply assessed from date of birth records. Cases found with discrepancies in bone age and chronological age are referred for further diagnostics for possible skeletal, orthodontics, endocrine or metabolic disorders. This technique finds application in various cases where birth records are not available, people seeking refuge, criminal investigations and accidental deaths.

© Springer Nature Switzerland AG 2021
M. Singh et al. (Eds.): IHCI 2020, LNCS 12615, pp. 142–153, 2021.
https://doi.org/10.1007/978-3-030-68449-5_15

Another important concept that relates to skeletal maturity are the disorders characterized by abnormalities of cartilage and bone called Skeletal Dysplasias. Bone age assessment for children in initial years can present a very exhaustive assessment of such disorders [2,3]. These abnormalities are typically present in the childhood and emerge at later stage due to asymmetric short stature. The onset of skeletal manifestations in many of these conditions is well defined, and therefore, early detection is of paramount importance to establish a right diagnosis. Also radiographs available for children are very less due to reasons like radiation exposure and ethical challenges, so there's a compelling need to devise an automated model which works well on small-scale dataset for lower ages.

Most regularly used methods for skeletal-age determination like Greulich-Pyle (GP) [4] and Tanner-Whitehouse (TW) [5] are primarily based on visual evaluation where left hand scan of the child is compared manually with atlases containing reference images. These methods are used quite extensively but have many limitations. First, they are tedious and time consuming, secondly, require high radiologist experience and more importantly are susceptible to inter-rater and intra-rater variability which further affects the diagnosis process.

BoneXpert [6], a commercial software authorized for clinical use in Europe, applies active appearance model to match the images based on appearance and shape. It considers hand and wrist bones for computing the age and does not use carpal bones, which affects the age estimation for lower age groups where diaphysis and epiphysis are not fused yet. Also, this method requires high quality images and can't handle noisy or poor contrast images. Figure 1 shows the anatomy of an immature hand and skeletal maturity at different ages.

With the advent of technology, several automated approaches have been proposed in the past to assess the skeletal maturity by capturing key morphological features. However, the development of computer aided models for automatically estimating bone age has been hampered by the complexity and large alterations in mineralization of bones, structure and size of the ossification centers of the hand. Many deep learning techniques have been explored lately, mainly based on segmenting the regions of interest and then applying convolutional networks for classification.

Most of these existing methodologies leverage deep learning networks for boneage assessment using large size dataset and rest of the studies, based on other machine learning techniques predict bone-age for higher age groups and exclude lower age groups from their experiments due to unavailability of sufficient data. Table 1 summarizes the related studies in same domain. As evident from the table, most of the studies have excluded smaller age groups and the ones which have included lower age groups used large training data. Our framework proposes an automated model using metric learning for bone-age assessment in hand radiographs of children in 0–3 years of age using small number of samples per class.

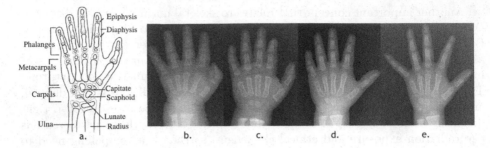

Fig. 1. Example images for hand anatomy and skeletal maturity at various age groups (a.) Hand anatomy of immature hand and ossification centers (b.) 0 years of age (c.) 1 years of age (d.) 2 years of age (e.) 3 years of age

Table 1. A comparative study of similar techniques w.r.t dataset size and age group used.

Proposed method	Dataset size	Age range (years)	Additional comments and metrics used
O'Connor et al. (2014) [7]	221	9–19	Mean Absolute Error
Cunha et al. (2014) [8]	887	7–19	Mean Absolute Error
Urschler et al. (2015) [9]	102	13–20	Mean Absolute Error
Franklin et al. (2015) [10]	388	10–35	Ossification stages w.r.t bone-age
Pinchi et al. (2016) [11]	274	6–17	Percentage classification accuracy
Hyunkwang et al. (2017) [12]	8325	5–18	Mean Absolute Error
Shie et al. (2017) [13]	124	6–15	Mean Absolute Error
Tang et al. (2018) [14]	79	12–17	Mean disparity as metric
Ren et al. (2018) [15]	12480	0–18	Mean Absolute Error, RSNA bone age data
Iglovikov et al. (2018) [16]	11600	0–19	Mean Absolute Error, RSNA bone age data
Zhao et al. (2018) [17]	12611	0–19	Mean Absolute Error, RSNA bone age data
Ours	75	0–3	DHA Digital Hand Atlas

To summarize, our main contributions are:

- Proposing a simple triplet loss based model for estimating bone age in children of 0–3 years of age.
- The visualization of the distributions of the features learned by the triplet loss model shows good class separations indicating the discriminative nature of data.
- Results demonstrate comparable state-of-art performance in bone age prediction using small number of training samples, as opposed to data hungry convolutional neural networks.

2 Prior Work

Deep Neural Networks have been extensively used in varied domains in the recent past owing to their near human level performance in capturing features from data. Deep learning is extensively used in bio-medicine like retinopathy, cancer detection, lung disorders and brain mapping.

Vladimir et al. [16] used VGG style CNN on whole hand as well as particular parts of the hand as regression and classification model, using data consisting of 12.6k hand radiographs from Radiological Society of North America(RSNA) 2017 bone-age challenge. They pre-processed the input images by first detecting the key points of the hand and then registering the images to align them to a common coordinate space and finally used segmentation to extract the hand.

Hyunkwang et al. [10] in their proposed bone-age prediction model, used pre-processed hand radiographs from 5–18 years of age and excluded data for 1–4 years. Pre-trained models like GoogleNet, AlexNet and VGG16 were used to train the model and assigned bone age within one year for female 90.39% of the time and 94.18% of the time for male radiographs. Spampinato et al. [18] also used the whole hand for Bone age assessment using pre-trained models and developed a customised network for training with an error rate of approximately 0.8 years. A regression CNN network proposed by Pengyi et al. [19] extracted carpal bones as the region of interest using boundary extraction method for predicting skeletal age. Chen et al. [20] used VGG as the base network and employed augmentation of dataset using random flips, scaling, rotation and cropping. Bone age assessment was further performed using GoogleNet.

All these methods lead to automation of bone age assessment but they were not complete for assessing skeletal age for all ages (1–18 years), because all of them presented their results on selected age groups due to unavailability of sufficient data particularly for lower age groups. Bone-age assessment is critical in lower age groups in order to detect the skeletal disorders at an early age and thus initiating necessary line of treatment. It is critical to evaluate certain regions in the skeletal dysplasias, especially in young children, because many cases are associated with distinct abnormalities that need early medical management.

In this study, we have presented an automated approach for skeletal age assessment in lower age categories and using very few images per class for training.

3   Method

The framework proposed in this study is based on Metric learning technique using Triplet Network [21,22] as depicted in Fig. 2. The model generates distributed embeddings for the input images using three parallel neural networks with shared weights.

3.1   Architecture

Metric learning techniques measure the similarity between input data by learning an optimal distance function tuned for particular learning tasks. The learned distance metric, in turn is used to perform the classification task using very small training data. Convolutional Neural Networks (CNN) jointly optimize the input

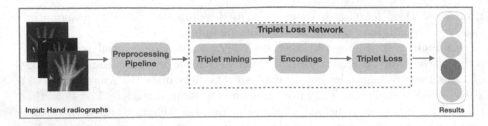

Fig. 2. Framework depicting the overall flow of the proposed model

image representation, also called embeddings conditioned on the "similarity" function being used. Triplet Network learns a metric embedding f(z) from an input sample z onto a d dimensional euclidean space S^d, in such a way that the squared distance between image embeddings is less for images belonging to identical class and more for images belonging to non-identical classes. Triplet net with three input images referred to as triplets is depicted in Fig. 3.

$$\| f(z) \|_2 = 1 \tag{1}$$

Input contains three sample images - Reference, positive and negative. Reference image or the baseline and positive from the same category, and negative image from different age class. On the chosen triplets (y, y^+, y^-) and similarity function given by $s(y, y')$ satisfying $s(y, y^+) > s(y, y^-)$, distance function tuned obeys the condition $d(y, s(y,y^+)) < d(y, s(y,y^-))$.

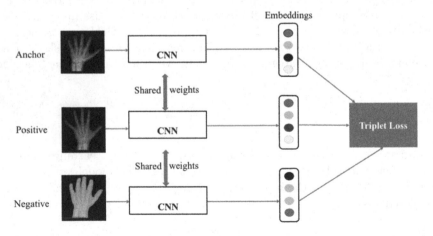

Fig. 3. Triplet Net architecture. Input image samples are fed to the embedding function to generate corresponding embeddings

3.2   Loss Function

The loss of a triplet (y, y$^+$, y$^-$), for some-arbitrary distance, on the embedding space d is defined as:

$$Loss(y, y^+, y^-) = max(d(y, y^+) - d(y, y^-) + margin, 0) \qquad (2)$$

where margin is the hyper-parameter which defines how far the dissimilar samples should be. The main objective of triplet network is to minimize anchor-positive distance and maximize anchor-negative distance. Margin helps in distinguishing the positive and negative images from the anchor image. An anchor, positive and a negative image sample are randomly selected for training the network. Figure 4 depicts the Triplet Loss diagrammatically.

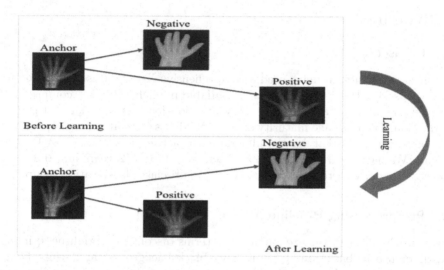

Fig. 4. The proposed triplet loss enforces anchor closer to the positive sample and farther from the negative sample.

3.3   Triplet Mining

For a given dataset with N samples, there can be $O(N*N*N)$ possible triplets to choose from. Triplet selection is a critical step to help the model converge by not selecting the irrelevant or easy triplets. An optimised blend of simple, complex and not so complex hard triplets is generally used to make the network learn. Commonly used triplet selection strategies are given below:

1. Easy triplets: triplets in which distance between the anchor-negative is greater than the sum of margin and distance between anchor-positive.

$$d(y, y^-) > d(y, y^+) + margin \qquad (3)$$

2. Hard triplets: triplets where distance between anchor-negative is less than the distance between anchor-positive.

$$d(y, y^-) < d(y, y^+) \tag{4}$$

3. Semi-hard triplets: triplets in which distance between anchor-negative is greater than anchor-positive but less than the sum of margin and distance between anchor-positive.

$$d(y, y^+) < d(y, y^-) < d(y, y^+) + margin \tag{5}$$

A triplet is called a valid triplet if, out of three images, two are from the same class but are distinct and third image has a different class.

4   Experiments

4.1   Dataset

For our experiments, we have used a comprehensive X-ray dataset from Digital Hand Atlas Database System [23,24], available publicly. Dataset comprises of 1391 left hand radiographs from 1–18 years old along with racial and gender information. The bone age maturity of every child was estimated by two pediatric radiologists by using left hand radiographs in accordance with G & P atlas method. We have used 20 radiographs each from 0–1, 1–2 years and 0–1, 1–2, 2–3 years for our experiments in binary and multi-class classification setup.

4.2   Pre-processing Pipeline

Images in the dataset vary remarkably in terms of contrast, brightness, intensity, difference in background (some have black background and white bones and vice versa) and aspect ratio. Therefore, pre-processing is an essential step for effective feature learning. As a first step, images are segmented using K-space segmentation tool which is based on four algorithms namely seeded region growing, interactive graph cut, simple interactive object extraction, interactive segmentation using binary partition trees. Next step involves the normalization of the radiographs for black background by computing mean of the $5 * 5$ pixel patch on four image corners and then converting all of them to gray-scale black background. Images were resized to $300 * 300$ for optimal performance. In the third step, images are centered by making the hand appear in the center of the $300 * 300$ image box using the center of mass and warpAffine function. Finally, CLAHE (contrast-limited adaptive histogram equalization) is used to enhance the contrast of the images. CLAHE is a variant of adaptive histogram equalization which performs histogram equalization in small patches with high accuracy and contrast limiting. Figure 5 demonstrates the effect of pre-processing on some example images.

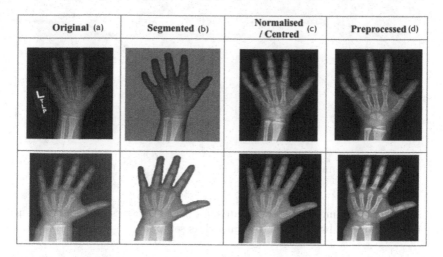

Fig. 5. Pre-processing pipeline depicting example images at all stages of pre-processing. (a) Raw input images (b) Segmented images (c) Normalized black background and centered (d) Final contrast enhanced images by CLAHE

4.3 Training Details

In our study each hand radiograph is represented by a 128 dimensional vector, batch size of 20 triplets consisting of 10 hard and 10 random samples each from a set of 50 triplets mined. Adam optimizer, margin $= 0.2$ and learning rate of $1e-3$ are the hyper parameter values used. Network is trained for 40,000 iterations on NVIDIA V100 32 GB Graphic Processing Unit node. Trained model is further used to visualize the class separations learnt by the model. Bigger embedding size did not help in increasing the accuracy and practically need extensive training to achieve comparable accuracy.

4.4 Performance Evaluation - ROC Curve

For our evaluation process, we plot a receiver operating characteristic (ROC) curve, for testing set. It is a kind of graphical description which shows the diagnostic capability of a classifier based on varied values of threshold. The curve is rendered by plotting the true positive rate (TPR) versus the false positive rate (FPR) at different values of threshold. AUC of 0.92 is obtained for 0–1, 1–2 year categories and 0.82 for classes 0–1, 1–2, 2–3 years depicted in Fig. 6.

4.5 Visualizing Embeddings

Embeddings exhibit the property of preserving the semantic relationship of the input data by arranging identical points in close groups and non-identical data points in sparse manner away in the embedding area. tSNE (t-Distributed Stochastic Neighbor Embedding) is used to visualize the class separations learnt.

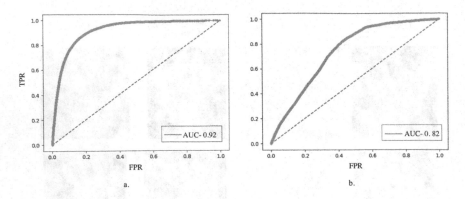

Fig. 6. Receiver operating characteristic curve for bone age assessment in test set using triplet network for (a.) binary and (b.) multi-class setup

The tSNE algorithm computes a similarity metric between pairs of data points on the higher dimensional space and the low dimensional space. Further, a cost function is used to optimize these two similarity metrics. Embeddings generated for binary and multi-class case, for training and test data before and after the model is trained are depicted in Fig. 7.

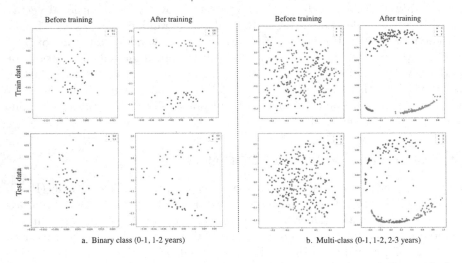

Fig. 7. tSNE visualization of embeddings for training and test data, before and after training for (a.) binary class (0–1, 1–2 years) and (b.) multi-class (0–1, 1–2, 2–3 years)

4.6   Prediction Results

Figure 8 shows the prediction results by comparing few tests images against a reference image taken from each class. As represented in figure, random test images are selected from class 0–1, 1–2 and 2–3 and the distance of this test image is compared with randomly chosen triplets. It can be clearly seen that after training the distance between the test image and image from the correct class gets reduced leading to correct bone-age predictions.

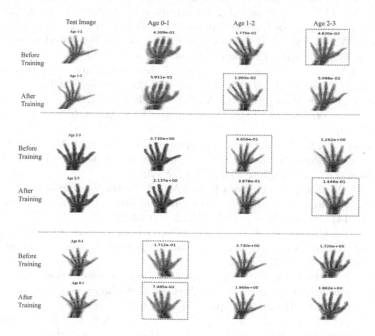

Fig. 8. Bone-age prediction results for multi-class case on few example images from test set

5   Conclusion

We have presented a bone age assessment framework which works comparably to the state of art techniques in predicting skeletal age for lower age groups. Our model performs skeletal age estimation by finding image representations in the latent space using small number of training images per class. Our method achieves an AUC of 0.92 for binary classification and 0.82 for multi-class classification. Embeddings generated illustrate the clear class separations learnt by the model. Results clearly demonstrate that our performance is at par with the existing CNN based techniques reported so far, which practically require enormous data and time to train. Another important aspect with our approach is that, once a model has been trained, the discriminative features can be used for prediction on new data, as well as on entirely a different class of data from unknown distribution. This generalization on new class of data does not require extensive retraining which is usually expensive or impossible due to scarcity of data.

References

1. Gilsanz, V., Ratib, O.: Hand Bone Age: A Digital Atlas of Skeletal Maturity, p. 106. Springer, Heidelberg (2005). https://doi.org/10.1007/b138568
2. Krakow, D., Rimoin, D.L.: The skeletal dysplasias. Genet. Med. **12**, 327–341 (2010)
3. Parnell, S., Phillips, G.: Neonatal skeletal dysplasias. Pediatr. Radiol. **42**(Suppl 1), S150–S157 (2012). https://doi.org/10.1007/s00247-011-2176-2
4. Greulich, W.W., Pyle, S.I.: Radiographic Atlas of Skeletal Development of the Hand and Wrist. Stanford University Press, Stanford (1959)
5. Malina, R.M., Beunen, G.P.: Assessment of skeletal maturity and prediction of adult height (TW3 method). Am. J. Hum. Biol. **14**, 788–789 (2002)
6. Thodberg, H.H., Kreiborg, S., Juul, A., Pedersen, K.D.: The BoneXpert method for automated determination of skeletal maturity. IEEE Trans. Med. Imaging **28**(1), 52–66 (2009)
7. O'Connor, J.E., Coyle, J., Bogue, C., Spence, L.D., Last, J.: Age prediction formulae from radiographic assessment of skeletal maturation at the knee in an Irish population. Forensic Sci. Int. **234**(188), e1–8 (2014)
8. Cunha, P., Moura, D.C., Guevara Lopez, M.A., Guerra, C., Pinto, D., Ramos, I.: Impact of ensemble learning in the assessment of skeletal maturity. J. Med. Syst. **38**, 87 (2014). https://doi.org/10.1007/s10916-014-0087-0
9. Urschler, M., Grassegger, S., Stern, D.: What automated age estimation of hand and wrist MRI data tells us about skeletal maturation in male adolescents. Ann. Hum. Biol. **42**(4), 358–367 (2015)
10. Franklin, D., Flavel, A.: CT evaluation of timing for ossification of the medial clavicular epiphysis in a contemporary Western Australian population. Int. J. Legal Med. **129**(3), 583–594 (2014). https://doi.org/10.1007/s00414-014-1116-8
11. Pinchi, V., et al.: Combining dental and skeletal evidence in age classification: pilot study in a sample of Italian sub-adults. Leg. Med. **20**, 75–9 (2016)
12. Hyunkwang, L., Shahein, T., Giordano, S., et al.: Fully automated deep learning system for bone age assessment. J. Digit. Imaging **30**, 427–441 (2017). https://doi.org/10.1007/s10278-017-9955-8
13. Shi, L., Jiang, F., Ouyang, F., Zhang, J., Wang, Z., Shen, X.: DNA methylation markers in combination with skeletal and dental ages to improve age estimation in children. Forensic Sci. Int. Genet. **33**, 1–9 (2018). https://doi.org/10.1016/j.fsigen.2017.11.005. PMID: 29172065
14. Tang, F.H., Chan, J.L.C., Chan, B.K.L.: Accurate age determination for adolescents using magnetic resonance imaging of the hand and wrist with an artificial neural network-based approach. J. Digit. Imaging **32**, 283–289 (2019). https://doi.org/10.1007/s10278-018-0135-2
15. Ren, X., et al.: Regression convolutional neural network for automated pediatric bone age assessment from hand radiograph. IEEE J. Biomed. Health Inform. **23**, 2030–2038 (2018)
16. Iglovikov, V.I., Rakhlin, A., Kalinin, A.A., Shvets, A.A.: Paediatric bone age assessment using deep convolutional neural networks. In: Stoyanov, D., et al. (eds.) DLMIA/ML-CDS -2018. LNCS, vol. 11045, pp. 300–308. Springer, Cham (2018). https://doi.org/10.1007/978-3-030-00889-5_34
17. Zhao, C., Han, J., Jia, Y., Fan, L., Gou, F.: Versatile framework for medical image processing and analysis with application to automatic bone age assessment. J. Electr. Comput. Eng. **2018**, 13 (2018). Article ID 2187247

18. Spampinato, C., Palazzo, S., Giordano, D., et al.: Deep learning for automated skeletal bone age assessment in X-Ray images. Med. Image Anal. **36**, 41–51 (2017)
19. Hao, P., Chokuwa, S., Xie, X., Fuli, W., Jian, W., Bai, C.: Skeletal bone age assessments for young children based on regression convolutional neural networks. Math. Biosci. Eng. **16**(6), 6454–6466 (2019). https://doi.org/10.3934/mbe.2019323
20. Chen, M.: Automated Bone Age Classification with Deep Neural Networks (2016)
21. Schroff, F., Kalenichenko, D., Philbin, J.: FaceNet: a unified embedding for face recognition and clustering, pp. 815–823 (2015). arXiv:1503.03832v3
22. Hoffer, E., Ailon, N.: Deep metric learning using triplet network. In: Feragen, A., Pelillo, M., Loog, M. (eds.) SIMBAD 2015. LNCS, vol. 9370, pp. 84–92. Springer, Cham (2015). https://doi.org/10.1007/978-3-319-24261-3_7
23. Gertych, A., Zhang, A., Sayre, J., Pospiech-Kurkowska, S., Huang, H.: Bone age assessment of children using a digital hand atlas. Comput. Med. Imaging Graph.: Off. J. Comput. Med. Imaging Soc. **31**(4–5), 322–331 (2007)
24. Zhang, A., Sayre, J.W., Vachon, L., Liu, B.J., Huang, H.K.: Racial differences in growth patterns of children assessed on the basis of bone age. Radiology **250**(1), 228–235 (2009)

A Development of Enhanced Contactless Bio Signal Estimation Algorithm and System for COVID19 Prevention

Chan-il Kim and Jong-ha Lee[✉]

Department of Biomedical Engineering, Keimyung University, Daegu, South Korea
segeberg@gmail.com

Abstract. In recent days, many wearable biological data measuring devices have been developed. Despite many advantages, these devices have a few shortcomings such as tissue allergies, motion artifact, and signal noise problem. User identification for the remote monitoring center is another issue. Most of these problems are caused by the sensor contact measurement method. Due to these problems, many studies have been conducted to find ways that simplify the measurement process and cause less discomfort to the users. Many existing studies measured heart rate and respiratory rate by extracting photo plethysmography (PPG) signals from the user's face with cameras and proper lighting. In this study, the face recognition experiment, we conducted face recognition in various situations. The recognition rate of this system exhibited 96.0% accuracy when trying to recognize the front, and exhibited 86.0% accuracy from the sides of the face. The average heart rate estimation accuracy was 99.1%, compared to the gold standard method.

Keywords: Non-contact · Bio signal measurement · Mobile healthcare

1 Introduction

When an epidemic such as COVID-19 occurs, it is a big problem for medical personnel to become secondarily infected while diagnosing a patient's infection. To prevent this secondary infection, it is necessary to have a means to measure the patient's bio signals from a distance without contact with the patient. Many existing studies measured heart rate and respiratory rate by extracting photo plethysmography (PPG) signals from the user's face with cameras and proper lighting.

In this paper, a monitoring system was implemented to improve the methods of biometric measurements using existing medical devices. The reason is that it is difficult to identify the person to which the biometric information belongs, easy to collect and process biometric information, and simple to use. The system is designed to overcome the limitations of prior research, and there are many problems with measuring biometric data, despite the recent progress of several studies. The most frequent problem is that the measurement can change sensitively depending on the lighting conditions and the movement of the measurement target. In particular, rotation can cause significant changes in the color-measuring coordinates, making continuous measurements difficult.

M. Singh et al. (Eds.): IHCI 2020, LNCS 12615, pp. 154–162, 2021.
https://doi.org/10.1007/978-3-030-68449-5_16

To improve this, the system compensates for changes in illuminance and allows continuous measurement to be performed regardless of the movement of the object being measured.

Also, in a situation where a highly infectious disease such as COVID19 occurs, measurement objects wear masks to cause occlusion. In this case, it is difficult to select a Region of Interest (ROI) for heart rate value (HRV) specific. In this paper, we added a process of selecting the effective pixel value at measurement object.

2 Method

The proposed system aims to measure biometric data using image sequences. First, the subject's face is recognized using a camera, the image in the database is compared with the face, and the subject's identity is identified. Next, effective pixel classification is performed to measure the heartbeat of a subject whose identity has been identified by facial recognition. The heart rate is measured by continuously tracking the subject's face and extracting the color change of the effective pixels. Through this process, the system can measure the subject's identity and heart rate with only the camera image. The system proceeds in the following order (Fig. 1).

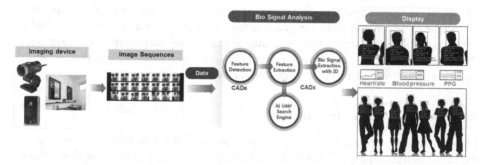

Fig. 1. Diagram of proposed system

2.1 Face Detection

Construct Eigen face (general information of face type) using Principal Components Analysis (PCA), then specify and track the parts of the input image and shape of the human face. PCA is a second-order statistical method that uses the statistical properties of mean and variance. PCA finds a set of axes normalized to each direction of the maximum normalized covariance in the input data. It is important to find the most important axis of the input data and the dimensions of it.

The characteristic vector space obtained by applying PCA includes features such as lighting changes and facial changes in the image, so the recognition rate is low. Considering this, PCA takes only the approximate location of the face and extracts the features of that location. By comparing feature points, we recognize faces to increase recognition rate.

Since facial recognition generally uses information of all pixels, the area of the facial image may be affected by small light, position, and facial expression changes, and may affect the recognition algorithm. Through this, the method of face recognition is likely to be affected by changes in situation. On the other hand, the model-based facial recognition method can be configured in consideration of changes in model, lighting, posture, and facial expressions, thereby reducing the influence of these factors on recognition.

2.2 Identity Recognition Process

In the current situation where human-to-person contact is required to be minimal, a method for acquiring the subject's information with complete non-contact is needed. To this end, the system identifies the subject's identity through face recognition before measuring the subject's bio signal. In this system, Elastic Bunch Graph Matching (EBGM) was used to extract specific points for identification.

The EBGM is used to get the facial features. The shape extraction used in EBGM is greatly affected by the number and type of model bundle graphs. This is why, the model bundle graph was created in consideration of various factors such as lighting, posture, and expression.

Facial features are very rigid and non-rigid matching because they vary depending on facial expression and angle. An effective non-rigid matching algorithm is required to solve this problem. In this paper, point matching is performed using TPRL (Topology Preservation Relaxing Labeling) algorithm. For comparison with the algorithm, we select several methods which are commonly used for point matching.

2.3 Heart Rate Value Measurement

Photoplethysmography (PPG) basically works on the blood flow in blood vessels using the light absorption properties of biological tissues. that is easy to measure in peripheral tissues such as fingers and earlobe, and typically PPG measurements are performed using a finger.

Applying this principle to your face, you can estimate your heart rate. The blood flow to the skin of the face changes according to the heart rate, and the heart rate can be estimated by measuring and analyzing the light reflection amount of the changed blood flow through the camera.

The densest part of the face with capillaries is the forehead and cheeks. As with body temperature measurements, capillary vessels are concentrated when measuring PPG on the face, so it is appropriate to measure areas where the concentration of hemoglobin varies greatly depending on the heart rate. Therefore, previous studies use the forehead or cheek as the measurement area.

However, the forehead or cheek is covered according to the characteristics and circumstances of the measurement object, so that measurement is not possible frequently. To overcome this, in this study, the measurement area was selected through another method.

2.4 Region of Interest Selection

First, PCA and EBGM are used to determine that the measurement object is a human face, and then only the face area is extracted separately. For the extracted region, only effective skin pixels are extracted using the Unsupervised Similarity clustering technique. Unsupervised Learning is a method of machine learning that analyzes only the input data without giving a result and groups the related ones. Using this method, in this paper, only skin pixels with similar change width were extracted and selected as ROI (Fig. 2).

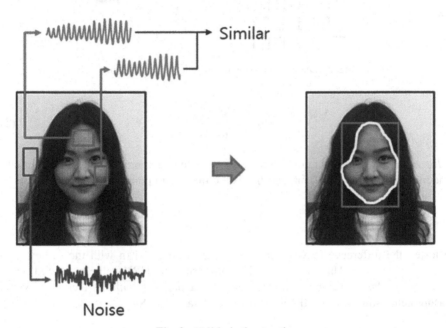

Fig. 2. Valid pixel extraction

2.5 Oxygen Saturation Measurement

In general, the oxygen saturation rate can be obtained by calculating the ratio of hemoglobin and oxyhemoglobin:

$$SpO_2[\%] = \frac{HbO_2}{HbO_2 + Hb} \times 100 \tag{1}$$

HbO_2 stands for Oxyhemoglobin and Hb stands for normal hemoglobin. We use two different wavelengths as light source that alternately emit light to measure oxygen saturation in the image. Because HbO_2 most absorbs 880 nm wavelengths and Hb absorbs 765 nm. We deploy 6×6 light emitting diodes (LEDs) with alternative wavelengths of 765 nm and 880 nm, producing light waves of 765 nm and 880 nm alternately emitting light at half the sampling rate of the camera (Fig. 3).

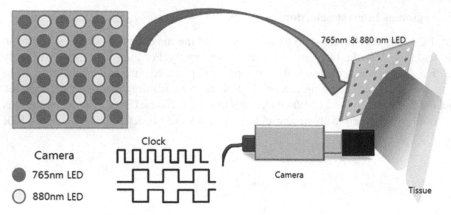

Fig. 3. Diagram of oxygen saturation measurement

$$R_{OS} = \frac{\ln(\frac{R_V}{R_P})}{\ln(\frac{IR_V}{IR_P})}$$

(2)

R_V is the minimum point of 765 nm and R_P is the maximum point of 765 nm. IR_V is the minimum point of 880 nm and IR_P is the maximum point of 880 nm.

3   Results

We tested the difference between the measurement data taken with the camera and the actual heart rate. This is obtained by comparing the HRV value measured using the change in the G channel data value of a particular face with the Heart rate value measurement using Biopac MP150 (Biopac, Goleta, CA, USA).

3.1   Recognition Rate and Heartrate Measurement Accuracy

We performed face recognition under various directions of the face. The experiment was conducted on 50 people (23 to 33 years old, 33 men and 17 women). The recognition rate for this system was 96.0% on the front and 86.0% on the face side. Also, the recognition rate for irregular lighting was 94.0%.

The heart rate measurement experiment was done in bright and dark light, the first to fifth experiment was done under bright fluorescent light, and the sixth to tenth experiment was done in dark incandescent light. Heart rate measurements (measured after accurate ambient light value measurement) were processed from the peak value of 30 to 35 s with images taken under bright fluorescent lamps, resulting in the same difference in heart rate values. (or not more than one) measured on each RGB channel. The G- channel was filtered the same as other channels but showed a uniform shape. The experiment was conducted 10 times per person and the average heart rate accuracy was 99.1%.

Additionally, we conducted an experiment on the situation that occurs when the subject of measurement uses a mask. This experiment was conducted with 10 people

(46 to 88 years old, 5 men and 5 women) who visited a medical center. Using this system, we tested the accuracy of heart rate when a part of the face was covered by wearing a mask and when the mask was removed. As a result of the experiment, when the system was used, when the mask was worn, the accuracy was 97% compared to when the mask was not worn.

3.2 Non-contact Oxygen Saturation Measurement

In this study, image sequences were used to measure pulse oxygen meters by non-contact methods, and 40×40 pixels had enough photons in an image sequence acquired for selection. The selected area signals were divided into R, G, and B channels, and the average of the G channels was calculated (Fig. 4).

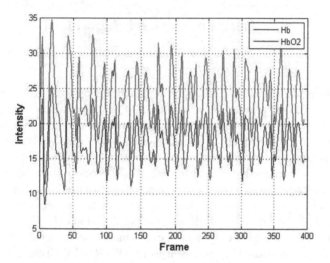

Fig. 4. Observation intensity of each target frequency

To obtain R_{OS}, we replaced the maximum and minimum points of an even frame with Eq. (2) captured 765 nm photons and 880 nm photons. The average R_{OS} was 1.61. A typical SpO_2 value is approximately 96. According to this, we calculated $k = 96/1.61 = 69.63$.

3.3 Error Rate Comparison of the Algorithm

In order to prove the efficiency of our method used for feature point comparison, an experiment was conducted on the error rate of each algorithm. In this experimental method, distance errors were calculated and compared for each feature point that occurred when the feature point of the database and the feature point of the image sequence were matched. Average error value of each algorithm is as follows (Fig. 5).

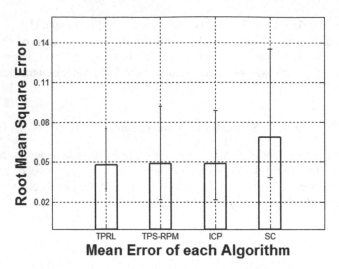

Fig. 5. Mean error of algorithms

In addition, in order to check the recognition rate in various situations, an experiment was also conducted on the error rate when rotation or noise is generated or when the target object is covered. In the deformation due to rotation, the measurement angle was changed from 5° to 20° at intervals of 5°, and the error was tested when it was deformed. The noise generation generated random noise in all the feature points, and the ratio was adjusted at 10% intervals for experiments. In the process of extracting the image sequence, the cover generated an obstacle at an arbitrary location to hinder the extraction of the feature point at the corresponding location, and then the feature point was extracted and compared with the database, and the error was calculated (Fig. 6).

The results of the rotational deformation experiment show that the TPRL algorithm has a lower average error than the comparison group TPS-RPM algorithm, ICP algorithm, and SC algorithm. Also, the change of the error according to the change of the angle is 1.5 points on the right side and 1.9 points on the left side, indicating a lower change than the algorithm of the comparison group.

In the experiment on noise generation, when comparing the TPRL algorithm and the TPS-RPM algorithm of the control group, the difference was 39%, showing the smallest difference, and 53% when compared to the ICP algorithm. Neither the TPRL algorithm nor the algorithms of the other comparison groups showed a significant change in the noise ratio change.

In the cover deformation experiment, the TPRL algorithm showed the lowest average error in all cover ratios. The TPRL algorithm showed an average error of 1.5 points less than that of the TPS-RPM algorithm at 0%, and 0.2 points at 10% and 0.1 point at 20%, respectively.

As a result, when the TPRL algorithm is used, it shows a small error compared to other algorithms. In particular, the rotational deformation showed an error of less than 2.0 points in the deformation process from 5° to 20°, and the lowest result was shown in all ratios for noise. In addition, the cover deformation experiment also showed the lowest

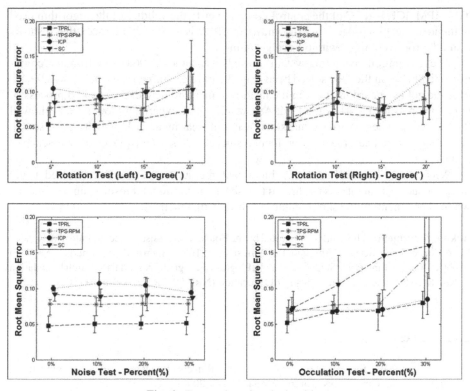

Fig. 6. Error value of each algorithm

result at each ratio, and the error increase value according to the increase of the cover ratio was also 1.7 points, showing a result that was hardly affected by the deformation.

4 Conclusion

In this study, we developed a system that measures biometric data using an image sequence and at the same time authenticates the user's identity and verifies the validity of the system.

The system recognizes the user's face from the image sequence collected through the camera, compares the recognized face with the server database to authenticate the patient's identity, and measures the patient's biometric data based on the feature points extracted in the process of identifying the patient's identity. After acquiring the coordinates for, the user's face is continuously tracked, the color of the same area is extracted, and the user's heart rate is measured using the color change of the corresponding pixel. In addition, by using the phenomenon that the light absorption wavelengths of hemoglobin oxide and hemoglobin are different, two light sources with different wavelengths were installed and a system was designed to measure oxygen saturation in blood through the difference in reflectance of light. As a verification process for each process in the system, the error was compared with the TPRL algorithm used in the system and the

TPS-RPM, ICP, and SC of the control group to verify the validity of the algorithm used in the identification process. As shown, when TPRL is used, it can be seen that it shows a small error to compare with other algorithms.

The recognition rate of this system showed an accuracy of 96.0% on frontal recognition and 86.0% on the face side. The average heart rate estimation accuracy was 99.1% over the gold standard method. In the heart rate measurement experiment, a process of covering up using a mask was added. As a result of this experiment, our system showed 97% accuracy when wearing a mask than without wearing a mask, which can be said to have a significant heart rate measurement result regardless of the presence or absence of a mask.

Accordingly, identification and bio-signal measurements were performed using image sequences, and it was confirmed that similar measurement results can be obtained while supplementing the shortcomings of the existing method.

Acknowledgement. This work is supported by the Foundation Assist Project of Future Advanced User Convenience Service" through the Ministry of Trade, Industry and Energy (MOTIE) (R0004840, 2020) and Basic Science Research Program through the National Research Foundation of Korea (NRF) funded by the Ministry of Education (NRF-2017R1D1A1B04031182).

References

1. Verkruysse, W., Svaasand, L.O., Nelson, J.S.: Remote plethysmographic imaging using ambient light. Opt. Express **16**(26), 21434–21445 (2008)
2. Poh, M.-Z., McDuff, D.J., Picard, R.W.: Non-contact automated cardiac pulse measurements using video imaging and blind source separation. Opt. Express **18**(10), 10762–10774 (2010)
3. Poh, M.-Z., McDuff, D.J., Picard, R.W.: Advancements in noncontact, multiparameter physiological measurements using a webcam. IEEE Trans. Biomed. Eng. **58**(1), 7–11 (2011)
4. McDuff, D., Gontarek, S., Picard, R.W.: Improvements in remote cardiopulmonary measurement using a five band digital camera. IEEE Trans. Biomed. Eng. **61**(10), 2593–2601 (2014)
5. Kamarainen, J.-K., Kyrki, V., Kalviainen, H.: Invariance properties of gabor filter-based features-overview and applications. IEEE Trans. Image Process. **15**(5), 1088–1099 (2006)
6. Gesche, H., Grosskurth, D., Küchler, G., Patzak, A.: Continuous blood pressure measurement by using the pulse transit time: comparison to a cuff-based method. Eur. J. Appl. Physiol. **112**(1), 309–315 (2012)
7. Myronenko, A., Song, X., Carreira-Perpinan, M.A.: Non-rigid point registration: coherent point drift. In: Advances in Neural Information Processing Systems 19, pp. 1009–1016 (2007)
8. Zitova, B., Flusser, J.: Image registration methods: a survey. Image Vis. Comput. **21**, 977–1000 (2003)
9. Chui, H., Rangarajan, A.: A new point matching algorithm for non-rigid registration. Comput. Vis. Image Underst. **89**, 114–141 (2003)
10. Zheng, Y., Doermann, D.: Robust point matching for nonrigid shapes by preserving local neighborhood structures. IEEE Trans. Pattern Anal. Mach. Intell. **28**, 643–649 (2006)

Stress Detection from Different Environments for VIP Using EEG Signals and Machine Learning Algorithms

Mohammad Safkat Karim[✉], Abdullah Al Rafsan, Tahmina Rahman Surovi,
Md. Hasibul Amin, and Mohammad Zavid Parvez

Department of Computer Science and Engineering, BRAC University,
Dhaka, Bangladesh
{mohammad.safkat.karim,abdullah.al.rafsan,
tahmina.rahman.surovi,md.hasibul.amin}@g.bracu.ac.bd,
zavid.parvez@bracu.ac.bd

Abstract. This paper proposes a method to detect stress for Visually Impaired People (VIP) when navigating indoor unfamiliar environments considering EEG signals. According to WHO, visual impairment is found in 285 million people around the world and 80% of visual impairment can be prevented or cured if proper treatment is served. However, VIP around the world have a concerning rate of living with stressful environments every day. Thus, this motivated researchers to seek a stress detection method which may be used further for supporting VIP and higher research purposes. This method refers to work with EEG Bands and detect stress by extracting different features from five EEG bands. After that, reliable machine learning algorithms are used for detection of stress based on multi-class classification. Experimental results show that Random Forest (RF) classifier achieved the best classification accuracy (i.e., 99%) for different environments where Support Vector Machine (SVM), K-Nearest Neighbors (KNN) and Linear Discriminant Analysis (LDA) secure more than 89% classification accuracy. Moreover, precision, recall and F1 score are considered to evaluate the performance of the proposed method.

Keywords: VIP · EEG · BCI · Stress · Machine learning

1 Introduction

Stress is a kind of body reaction when change occurs in external or internal environments that threatens a person to any extent [2]. Many researches have been going on in the field of stress detection by both psychologists and engineers [11]. Many Technologies have been developed in the recent past for detecting human stress by using wearable sensors and bio-signal processing [11]. Bio signals like Electroencephalography (EEG), Electrocardiography (ECG), Galvanic Skin Response (GSR), Skin Temperature (ST) have been used to detect stress [11].

© Springer Nature Switzerland AG 2021
M. Singh et al. (Eds.): IHCI 2020, LNCS 12615, pp. 163–173, 2021.
https://doi.org/10.1007/978-3-030-68449-5_17

Moreover, physiological features of humans can be used to measure the level of stress using physiological signals. Here, Differences exist between person to person when he/she responds to stress [11] though there are some issues in understanding emotion in the definition and assessment of emotion [7]. Even researchers and psychologists have been facing problems deciding what should be considered as emotion and how many types of emotions [7] are there because emotions have a major impact on motivation, perception, creativity, cognition, decision making, attention and learning [5,10]. However, when it comes to the matter of VIP (VIP), then should be taken more seriously. The World Health Organization (WHO) notes stress as the next significant issue of this era which constantly causes damage over physical and mental health of people all over the world. According to WHO, visual impairment affects approximately 285 million people (2010) around the world where it has raised up to 1 billion by October, 2020 [16,17]. Mobility in indoor and outdoor can be stressful for VIP, while roaming around in unfamiliar environments. The usual problems VIP face can be categorized in four main problems: to avoid obstacles; detecting ground level changes; using elevators; finding entrance or exit points and adapting to light variation [12].

EEG is now broadly used for emotion based recognition and in Brain Computer Interface (BCI). Different methods have been proposed to detect mental stress by using EEG signal and machine learning frameworks. Subhani et al. [13] extracted different features from time-domain EEG signals and then used different machine learning algorithms to classify different levels of stress and achieved 83.4% accuracy for multiple level identification. Saitis et al. [14] proposed a method where they used different power features and then detected stress and achieved 87% weighted AUROC for indoor environments. Jebelli et al. [15] used a method based on deep learning (DL) approach to detect stress and achieved 86.62% accuracy. Sangam et al. [18] extracted statistical and frequency domain features to detect stress.

Usually in medical identification or detection procedure, it is very necessary but not easy to achieve high accuracy with reliability. Thus it is very important to propose a method which has high accuracy with reliable precision, recall values and F1 scores during the stress detection with the help of EEG signals. To emphasize upon the problem, the contribution of this paper is that to achieve high accuracy, reliable precision, recall values and F1 scores. In this paper, we proposed a method that have extracted different significant features and then have used different learning algorithms such as Support Vector Machine (SVM), K Nearest Neighbors (KNN), Random Forest (RF), and Linear Discriminant Analysis (LDA) to detect stress using multi-class classification approach.

2   Materials and Methods

In the proposed method, we extracted different bands such as gamma (38–42 Hz), beta (12–35 Hz), alpha (8–12 Hz), theta (4–8 Hz), and delta (0.5–4 Hz) from EEG signals (see in Fig. 1). All of these frequency bands are used for feature extraction from the signals. These features are used as an input to SVM, KNN, RF and LDA classifiers which lead toward stress identification in different environments of VIP participants.

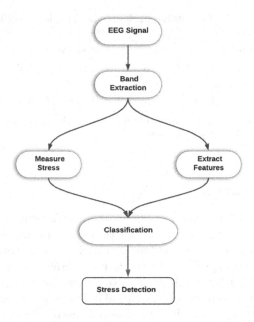

Fig. 1. Block diagram of the proposed method for detecting stress using EEG signals.

2.1 Description of EEG Signals

The data used in our study has been obtained from an experiment conducted in the indoor environments of the University of Iceland, Reykjavik [19]. They gathered the data when VIP with different degree of sight loss (see in Table 1) are navigating different environments in indoor using O&M device when instructors instructed through diverse circumstances where different levels of stress were likely to happen. VIPs walked the particular path three times for purposes of training. The route comprised seven different environments which represent a variety of indoor mobility challenges (see in Table 2). Indicatively, participants had to enter through automated doors, use an elevator, move across a busy open space and rehearse other obstacles. Also, within the environments on sudden loud sound and moving people are monitored. Nine healthy visually impaired adults with having different degrees of sight loss walked independently a complex route in an educational building (6 female; average age = 41 yrs., range = 22–53 yrs.) (Table 1). They were encouraged to walk as usual using their white canes if they wanted to help make them feel comfortable and safe, and were accompanied by their familiar O&M instructor. The study was approved by the National Bioethics Committee of Iceland and all data was being kept anonymous before analysis.

Table 1. VIPs from different degree of sight loss people [19]

Category	Description	Participant gender
VI-2	Vision less than 10% and more than 5%	2 (Female, Male)
VI-3	Vision less than 5%	4 (Female, Female, Male, Female)
VI-4	Unable to count fingers from less than one meter away	3 (Female, Male, Female)

2.2 Feature Extraction

In this proposed method, we considered different significant features such as absolute power, average band power, relative power, spectral entropy, and standard deviation which are extracted (see in Fig. 2) after getting five bands from EEG signals. The Average Band Power is the average of the sum of absolute value which is divided by N where N determines the number of data points where we use it to determine in a particular frequency range by using spectral estimation methods. The absolute power is used to normalize the PSD by dividing the PSD by itself and of course it does not make a distinction between the various frequencies. The relative power is finalized to choose whether a single frequency overwhelms other essential cerebrum frequencies [21]. Here, the standard deviation is also measured to figure out the dispersion of the data set corresponding to its mean. And finally, we chose spectral entropy to measure the consistency of power spectral density underlying in EEG.

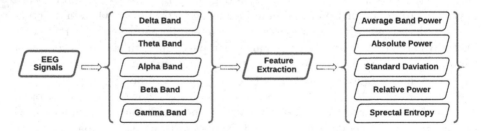

Fig. 2. Feature Extraction from Different Bands.

2.3   Stress Measurement

Frontal asymmetry indicates unbalanced cerebrum activity in the frontal outer layer that shows an imbalance task within the brain's left sphere and right sphere [9]. Among the sub-bands of EEG, alpha wave (8–12 Hz) is related to the idleness of the human brain and it shows the best connection with mental stress [20]. Wheeler et al. (1990, 1993) found the opposite relation between alpha power and its correlative operation in the brain [1]. As alpha power (8–12 Hz) and brain activity have an opposite relationship between them, it helps to measure stress [1]. Wheeler, Niemiec and Lithgow [1,4] studied mental imbalance and measure of emotion from the proportion of alpha bands within left-right frontal lobes. The presence of alpha power in the left lobe expresses negative emotions while in the right lobe refers to positive emotions. Thus, during metal stress level rises, alpha power decreases as rise of it indicates relaxation and conscious situations. However, the reduction of alpha power and rise of beta band specify the brain is in under intense activity [4]. Beta band power is high at the time of stress due to tension, excitement, and anxiety [6]. These waves also show varying behavior in different areas of the cerebrum [3]. To measure Frontal asymmetry, at first the natural logarithm of the power of the left sphere electrode is subtracted from the homologous right sphere electrode [20]. Therefore, it is stated that positive alpha asymmetry result explains left-frontal activity relative to right whereas negative result refers for right-frontal activity [8]. Finally, the stress values were separated into three equal classes (Low, Mid, High).

Table 2. Stressful Environments and Corresponding Challenges for VIP

ID	Environment	Description	Challenges
A	Stairs	Using Stairs to Move Between Floors	Finding Starting and Landing Point of Stairs
B	Open Space	Moving across an open space	Avoiding Moving People and Objects
C	Narrow Space	Walking along a narrow corridor, between chairs and objects	Moving people, noise, classroom doors opening suddenly
D	Moving	Moving people or object of the surrounding	Sudden clash with people or any object
E	Elevator	Using an elevator to move between floors	Calling the elevator, selecting floor
F	Doors	Entering through automated doors	Automated doors (hinged and rotating)
G	Sounds	Sudden loud Sound come from surrounding people or object	Any Loud sound happen suddenly

2.4   Classification

Different machine learning algorithms such as SVM, RF, KNN, and LDA were used to classify three class levels of stress. The whole data set was separated based on 7 environments (see in Table 2). Seven individual records were classified using stratified 10-fold cross-validation approach. In the 10-fold cross-validation, the data set is divided into ten subsets where nine subsets are used for training and remaining one subset is used for testing. This process repeated ten times, each time leaving out one of the subsets from the training, which was used for testing. Here considering it as a multi class classification problem, SVM's RBF kernel was used to separate classes and we applied One vs Rest (ovr) strategy to deal with the multi class problem. Neighbor based classifier KNN (k = 5) was tested and trained as it is highly appreciated for assumption process when it is a large data set. Again to avoid any kind of over-fitting case Random Forest (RF) is used which basically gives us the best results. To train and test the classifiers test size was 0.2.

3   Results and Analysis

In our research, the primary goal was to detect stress in different environments with high predictability and specificity to ensure better outcomes. To emphasize on our goal, we have extracted five bands and five features out of them (shown in Fig. 2) and selected classifiers were trained and tested using Stratified 10-Fold Cross Validation approach for classification of stress from stressful environments (defined in Subsect. 2.4). The proposed method (Fig. 1) measures stress using EEG signal bands (results shown in Fig. 3) and can classify multiple categories of stress (Low Stress, Mid Stress and High Stress).

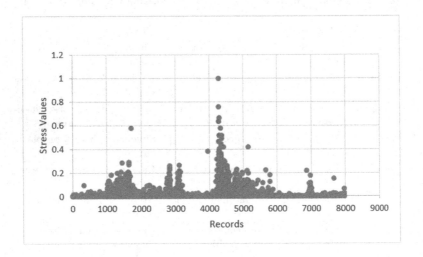

Fig. 3. Scatter Plot for Records vs Stress Values.

Extracting all the features, we experimented over every environment separately. Altogether there was 7,978 numbers of extracted records for all environments; altogether 903 records were for Doors, 1587 records were for Elevator,

602 records were for Moving object, 1045 records were for Narrow Space, 2385 records were for Open Space, 2234 records were for Stairs and 302 records were for Sounds. A graphical representation of environments and its records are visualized in Fig. 4.

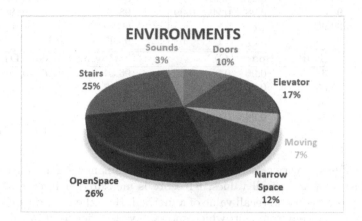

Fig. 4. Pie Chart of the used Data Records and Stressful Environments.

The Stratified 10-fold cross-validation divided the data set into ten subsets, each time leaving out one of the subsets from the training, used for testing. This method has the advantage that it utilized all of the instances in the data set for both training and testing. Every classifier's performance for every environment was evaluated using the accuracy, precision, recall, and F1 score.

3.1   Performance Analysis for Environment A, B, C, D

After splitting the data records, we have created separate data records for the environment A (Stairs), B (Open Space), C (Narrow Space) and D (Moving Object) with respectively 25%, 26%, 12% and 7% of the actual data record (as shown in Fig. 4). To evaluate the method's performance, we have analyzed it with accuracy, precision, recall and F1 score.

The method has come up with very satisfactory accuracy results for all the four environments as shown in Table 3. For environment A, RF comes up with the best results where the other classifiers have accuracy greater than 89% (KNN). For environment B, still RF is the best performer considering prediction accuracy and for environment C and D, all the classifiers secure accuracy more than 90% (shown in Table 3). Here, for all the four environments, RF can generate the better accuracy for detection of stress. We have shared also other evaluation matrices in Table 3 to evaluate the proposed method. Other than high accuracies, this proposed method has better precision, recall and F1 score also. Any precision and recall value greater than 0.7 is considered as a good result where greater than 0.9 is an excellent performance. Here this method has more than 0.9 precision

Table 3. Evaluation Table for Environment 'Stairs', 'Open Space', 'Narrow Space' and 'Moving Object'

	Stairs (A)				Open Space (B)			
	Accuracy (%)	P	R	F1	Accuracy (%)	P	R	F1
KNN	89.71	0.97	0.98	0.98	88.85	0.99	0.99	0.99
SVM	98.97	0.98	0.99	0.99	98.49	0.98	0.99	0.98
RF	99.96	1	1	1	99.96	1	1	1
LDA	97.53	1	1	1	99.37	1	1	1
	Narrow Space (C)				Moving Object (D)			
KNN	90.68	0.96	0.98	0.97	96.34	0.97	0.98	0.98
SVM	98.85	0.96	0.98	0.97	99.17	1	1	1
RF	99.81	1	1	1	99.83	1	1	1
LDA	98.66	0.98	0.99	0.98	97.84	1	1	1

P: Precision   R: Recall   F1: F1 Score

and recall value for every classifier of the four environments. Again, for balancing between precision and recall values, F1 score is measured; F1 score shows the reliability of precision and recall value of a method. Here the method's F1 score is very high for every environment; which shows any classifier used in this method can generate satisfying result anyway. So, for environment A, B, C and D, RF classifier is the best performing classifier for stress detection.

3.2   Performance Analysis for Environment E, F, G

After splitting the data records, we have created separate data records for the environment E (Elevators), F (Doors), G (Sounds) with respectively 17%, 10% and 3% of the actual data record (as shown in Fig. 4). To evaluate the method's performance, we have analyzed it with accuracy, precision, recall and F1 score. The method has generated very satisfactory accuracy results for the four environments as shown in Table 4. For environment E, all the classifiers get accuracy more than 99%. For environment F, the maximum accuracy is generated by both LDA, RF and SVM (99.45%) where KNN has accuracy more than 95% which is also a very good score. Lastly in environment G, all the classifiers have accuracy more than 90%; RF has come up with the highest accuracy. We have shared also other evaluation matrices in Table 4 to evaluate the proposed method. Other than high accuracies, this method has better precision, recall and F1 score also. Any precision and recall value greater than 0.7 is considered as a good result where greater than 0.9 is an excellent performance. Here for environment E, all the classifiers have precision and recall value close to 1 and for other environments, the method has very high precision and recall value as shown in Table 4. Again for balancing between precision and recall values, F1 score is measured; F1 score shows the reliability of precision and recall value of a method. Here the proposed method's F1 score is over 0.9 for every environment; which shows any classifier used in this method can generate satisfying result anyway. So, among these three environments, RF can generate the best performance for detecting stress.

Table 4. Evaluation Table for Environment 'Elevator', 'Doors' and 'Sounds'

	Elevator (E)				Doors (F)			
	Accuracy (%)	P	R	F1	Accuracy (%)	P	R	F1
KNN	99.81	1	1	1	95.004	1	1	1
SVM	99.81	0.99	1	1	99.45	0.99	0.99	0.99
RF	99.87	1	1	1	99.45	0.99	0.99	0.99
LDA	99.56	1	0.99	1	99.45	1	1	1
	Sounds (G)							
KNN	94.02	0.94	0.97	0.95				
SVM	98.02	0.94	0.97	0.95				
RF	99.01	0.98	0.98	0.98				
LDA	97.68	1	1	1				

P: Precision   R: Recall   F1: F1 Score

3.3   Comparison with Existing Method and Proposed Method

Among researches who worked for detection of stress, there was some who detected stress for VIP by EEG signals. In [14], Saitis et al. estimated the average weighted AUROC (%) to emphasize on the reliability of their method when worked with power spectrum from EEG bands. Where we have worked with the same data set and achieved better average weighted AUROC values using different features from EEG bands for RF classifier as shown in the Table 5.

Table 5. Average weighted AUROC (%) of the existing method [14] and proposed method using the same dataset based on RF classifier

Environment ID	A	B	C	D	E	F	G
Existing Method (%)	78	76	72	–	87	76	–
Proposed Method (%)	99.96	99.96	99.81	99.83	99.87	99.45	99.01

In [12], Saitis et al. estimated the average weighted AUROC (%) when worked with different power and entropy features from EEG bands to detect stress for VIP. And working with the same data set and achieved better average weighted AUROC values using five features from EEG bands for RF classifier as shown in the Table 6.

Table 6. Average weighted AUROC (%) of the existing method [12] and proposed method using the same dataset based on RF classifier

Environment ID	A	B	C	D	E	F	G
Existing Method (%)	77.7	74.9	70.7	–	82.4	76.7	–
Proposed Method (%)	99.96	99.96	99.81	99.83	99.87	99.45	99.01

The proposed method's extracted features and applied classifiers can be considered very reliable for stress detection processes for which this approach has secured even better average weighted AUROC score than Method 01 [14] and Method 02 [12] for combined indoor environments as shown in Table 7.

Table 7. Average weighted AUROC comparison between existing methods and proposed methods for all the indoor environments

Environment ID	Method 01	Method 02	Proposed Method
AUROC (%)	83	77.3	98.97

4   Conclusion

In this paper, a stress detection method of the visually impaired people (VIP) is proposed when they are navigating with different unfamiliar indoor environments. EEG signals are used to extract different features from five EEG bands and then different machine learning algorithms are used to classify multi-level of stress. The high prediction accuracy is 99% using random forest (RF) classifier for every environment where support vector machine (SVM), K-nearest neighbors (KNN) and linear discriminant analysis (LDA) secured more than 89% accuracy in these experiments. The results of our proposed method shows that our method achieves better results compared to other existing methods.

References

1. Wheeler, R.E., Davidson, R.J., Tomarken, A.J.: Frontal brain asymmetry and emotional reactivity: a biological substrate of active style. Psychophysiology **30**(1), 82–89 (1993)
2. Reisman, S.: Measurement of physiological stress. In: Proceedings of the IEEE 23rd Northeast Bioengineering Conference, pp. 21–23. IEEE (1997)
3. Dedovic, K., Renwick, R., Mahani, N.K., Engert, V., Lupien, S.J., Pruessner, J.C.: The montreal imaging stress task: using functional imaging to investigate the effects of perceiving and processing psychosocial stress in the human brain. J. Psychiatry Neurosci. **30**(5), 319 (2005)
4. Niemiec, A.J., Lithgow, B.J.: Alpha-band characteristics in EEG spectrum indicate reliability of frontal brain asymmetry measures in diagnosis of depression. In: 2005 IEEE Engineering in Medicine and Biology 27th Annual Conference, pp. 7517–7520. IEEE (2006)
5. Savran, A., et al.: Emotion detection in the loop from brain signals and facial images. In: Proceedings of the eNTERFACE 2006 Workshop (2006)
6. Sulaiman, N., Hamid, N.H.A., Murat, Z.H., Taib, M.N.: Initial investigation of human physical stress level using brainwaves. In: 2009 IEEE Student Conference on Research and Development (SCOReD), pp. 230–233. IEEE (2009)

7. Hosseini, S.A., Khalilzadeh, M.A.: Emotional stress recognition system using EEG and psychophysiological signals: using new labelling process of EEG signals in emotional stress state. In: 2010 International Conference on Biomedical Engineering and Computer Science, pp. 1–6. IEEE (2010)

8. Kemp, A., et al.: Disorder specie city despite comorbidity: resting EEG alpha asymmetry in major depressive disorder and post-traumatic stress disorder. Biol. Psychol. **85**(2), 350–354 (2010)

9. Briesemeister, B.B., Tamm, S., Heine, A., Jacobs, A.M., et al.: Approach the good, withdraw from the bad—a review on frontal alpha asymmetry measures in applied psychological research. Psychology **4**(03), 261 (2013)

10. Hosseini, S.A., Akbarzadeh-T, M., Naghibi-Sistani, M.B.: Qualitative and quantitative evaluation of EEG signals in epileptic seizure recognition. Int. J. Intell. Syst. Appl. **5**(6), 41 (2013)

11. Kalas, M.S., Momin, B.: Stress detection and reduction using EEG signals. In: 2016 International Conference on Electrical, Electronics, and Optimization Techniques (ICEEOT), pp. 471–475. IEEE (2016)

12. Kalimeri, K., Saitis, C.: Exploring multimodal biosignal features for stress detection during indoor mobility. In: Proceedings of the 18th ACM International Conference on Multimodal Interaction, pp. 53–60 (2016)

13. Subhani, A.R., Mumtaz, W., Saad, M.N.B.M., Kamel, N., Malik, A.S.: Machine learning framework for the detection of mental stress at multiple levels. IEEE Access **5**, 13 545–13 556 (2017)

14. Saitis, C., Kalimeri, K.: Multimodal classification of stressful environments in visually impaired mobility using EEG and peripheral biosignals. IEEE Trans. Active Comput. (2018)

15. Jebelli, H., Khalili, M.M., Lee, S.H.: Mobile EEG-based workers' stress recognition by applying deep neural network. In: Mutis, I., Hartmann, T. (eds.) Advances in Informatics and Computing in Civil and Construction Engineering, pp. 173–180. Springer, Cham (2019). https://doi.org/10.1007/978-3-030-00220-6_21

16. https://www.who.int/blindness/publications/globaldata/en/

17. https://www.who.int/news-room/fact-sheets/detail/blindness-and-visual-impairment

18. Sangam, D.V.: Electroencephalogram (EEG), its processing and feature extraction. Int. J. Eng. Res. **9**(06) (2020)

19. Saitis, C., Parvez, M.Z., Kalimeri, K.: Cognitive load assessment from EEG and peripheral biosignals for the design of visually impaired mobility aids. Wirel. Commun. Mob. Comput. (2018)

20. Allen, J.J., Coan, J.A., Nazarian, M.: Issues and assumptions on the road from raw signals to metrics of frontal EEG asymmetry in emotion. Biol. Psychol. **67**(1–2), 183–218 (2004)

21. Chotas, H.G., Bourne, J.R., Teschan, P.E.: Heuristic techniques in the quantification of the electroencephalogram in renal failure. Comput. Biomed. Res. **12**, 299–312 (1979). https://doi.org/10.1016/0010-4809(79)90042-9

Natural Language, Speech, Voice and Study

Analysis of Emotional Content in Indian Political Speeches

Sharu Goel⬥, Sandeep Kumar Pandey(✉)⬥,
and Hanumant Singh Shekhawat⬥

Indian Institute of Technology Guwahati, Guwahati, India
{sharugoel,sandeep.pandey,h.s.shekhawat}@iitg.ac.in

Abstract. Emotions play an essential role in public speaking. The emotional content of speech has the power to influence minds. As such, we present an analysis of the emotional content of politicians speech in the Indian political scenario. We investigate the emotional content present in the speeches of politicians using an Attention based CNN+LSTM network. Experimental evaluations on a dataset of eight Indian politicians shows how politicians incorporate emotions in their speeches to strike a chord with the masses. A brief analysis of the emotions used by the politicians during elections is presented along-with data collection issues.

Keywords: Speech Emotion Recognition · Politician speech ·
Computational paralinguistics · CNN+LSTM · Attention

1 Introduction

Speech signal contains information on two levels. On the first level, it contains the message to be conveyed and on the second level, information such as speaker, gender, emotions, etc. Identifying the emotional state of a speech utterance is of importance to researchers to make the Human-Computer Interaction sound more natural. Also, emotions play a significant role when it comes to public speaking, such as a politician addressing a crowd. Speeches offer politicians an opportunity to set the agenda, signal their policy preferences, and, among other things, strike an emotional chord with the public. We can observe that emotional speeches allow the speaker to connect better with the masses, instill faith in them, and attract a strong response. As such, it becomes interesting to analyze which emotions and in what proportion do politicians incorporate in their speeches.

While voting for a particular politician in elections depends on several factors such as background, work done in the previous tenure, public interaction etc, studies done in [7] suggest that emotions or affects have a negative impact on voting decisions of the public. Also, an experiment performed by the University of Massachusetts Amherst revealed that under the emotional state of anger, we are less likely to systematically find information about a candidate and increase our reliance on stereotypes and other pre-conceived notions and heuristics [6]. Another independent study revealed that anger promotes the propensity

M. Singh et al. (Eds.): IHCI 2020, LNCS 12615, pp. 177–185, 2021.
https://doi.org/10.1007/978-3-030-68449-5_18

to become politically active and hence, has a normatively desirable consequence of an active electorate [12]. Furthermore, a study on the 2008 Presidential Elections in the United States showed that negative emotions like anger and outrage have a substantial effect on mobilizing the electorate, and positive emotions like hope, enthusiasm, and happiness have a weaker mobilizing effect. The researchers excluded emotions like sadness and anxiety from their study, as these emotions did not have a noticeable influence on voting behavior [10]. The results of the previous experiments and studies were backed with the outcome of the 2016 Presidential Elections in the United States, where the current president Donald Trump had more anger and hope dominated campaign advertisements compared to his opposition, Hilary Clinton [9].

Nevertheless, a question remains; do emotional manipulations elicit the expected emotions? Does an angry speech elicit anger or some other emotion? A study on the structure of emotions [12] in which participants were randomly assigned to view one of four campaign ads designed to elicit a discrete emotion revealed that although there is heterogeneity in emotional reactions felt in response to campaign messages, the emotional manipulations from campaign ads did elicit the emotions expected. The results of the study showed that sadness was most prevalent in the sadness condition relative to all other conditions, anger was higher in the anger condition relative to all other conditions, and enthusiasm was much more significant in the enthusiasm condition relative to all other conditions. Nonetheless, the study also illustrated that different emotional manipulations were effective in eliciting a specific emotion. For instance, the study showed that sadness was felt in response to both angry and sad emotional manipulations. Similarly, anger was felt in response to sad emotional manipulation as well (although anger was more prevalent in angry emotional manipulation, as stated earlier).

Moreover, due to the recent advancements in the deep learning field, SER has seen significant improvement in performance. In [13], an end-to-end deep convolutional recurrent neural network is used to predict arousal and valence in continuous speech by utilizing two separate sets of 1D CNN layers to extract complementary information. Also, in [4], feature maps from both time and frequency convolution sub-networks are concatenated, and a class-based and class-agnostic attention pooling is utilized to generate attentive deep features for SER. Experiments on IEMOCAP [1] dataset shows improvement over the baseline. Also, researchers have explored the possibility of extracting discriminative features from the raw speech itself. In [11], a comparison of the effectiveness of traditional features versus end-to-end learning in atypical affect and crying recognition is presented, only to conclude that there is no clear winner. Moreover, works in [5] and [8] have also utilized raw speech for emotion classification. In this work, we propose a deep learning-based architecture to the task of emotional analysis of politicians speech, particularly in the Indian political context. We present a dataset of speech utterances collected from speeches of 8 Indian politicians. Also, a widely used Attentive CNN+LSTM architecture is used to the task of Speech Emotion Recognition (SER). At the first level, the speech utterances are classified into four emotion categories- Angry, Happy, Neutral, and Sad, using the trained emotion model. At the second level, analysis of the amount of different emotions present in a politician's speech is presented, which gives us a first-hand view of the emotional impact of speeches on the listeners and the voting decisions. To the best of our knowledge, this is the first work in analyzing the emotional contents of speeches of Indian politicians.

The remainder of the paper is organized as follows: Sect. 2 describes the data collection strategy and dataset description, along with the perceptual evaluation of the dataset. Section 3 describes the Attentive CNN+LSTM architecture in brief. The experimental setup and results are discussed in Sect. 4. Section 5 presents some aftermaths and discussions and Sect. 6 summarizes the findings and concludes the paper.

2 Dataset Description

Since no such speech corpus exists which suits our research in the Indian political context, we present the IITG Politician Speech Corpus. The Politician Speech Corpus consists of 516 audio segments from speeches delivered by Indian Politicians in the Hindi language (other Indian languages will be a part of the extended corpus). The speeches are publicly available on YouTube on the channels of the respective political groups the politicians belong to. Eight Indian politicians, whom we nicknamed for anonymization, namely NI (88 utterances), AH (101 utterances), AL (81 utterances), JG (20 utterances), MH (8 utterances), RI (31 utterances), RH (122 utterances), SY (65 utterances) were considered as they belonged to diverse ideologies and incorporate diverse speaking style. The politicians were chosen based on the contemporary popularity and wide availability of speeches. The speeches were addressed in both indoor(halls or auditoriums) and outdoor(complex grounds, courts, or stadiums) environment and carried substantial noise in the background. Only audio output from a single channel was considered.

The extracted audio from video clips is first segmented into utterances of length 7–15 s and classified into four basic emotions - Angry, Happy, Neutral, and Sad. Perceptual evaluation of the utterances is performed by four evaluators. For correctly labeling the speech utterances we followed a two-step strategy. The first step was to identify if there is any emotional information in the utterance, irrespective of the underlying message. This is done by assigning each utterance a score on a scale of one to five(one meaning not confident and five meaning extremely confident) based on how confidently the evaluators were able to assign it an emotion. The second step was to assign a category label to the utterances. Since the speeches from politicians exhibit natural emotions, it may be perceived differently by different evaluators. For this, we followed the following strategy for assigning a categorical label -

1. If the majority of listeners are in favor of one emotion, then the audio clip was assigned that emotion as its label. The confidence score of the label would be the mean score of the majority.

Table 1. Distribution of utterances emotion wise along with confidence score for each emotion class generated using perceptual evaluation

Emotion	No. of samples	Confidence score (1–5)	Confidence %
Angry	175	3.5519	71.0381
Happy	36	3.2152	64.3055
Neutral	230	3.8199	76.3991
Sad	75	3.7688	75.3778

2. If two listeners are in favor of one emotion and two in favor of the other, then the emotion with the highest mean confidence score was assigned as the label. The confidence score of the label would be the same mean used to break the tie.
3. If no consensus is achieved for a particular utterance i.e. all listeners assigned a different label, then that audio clip is discarded.

The number of utterances in each emotional category along with the confidence score of the evaluators, is presented in Table 1. Happy emotion category has the least number of utterances followed by Sad, which is in line with the literature on the emotional content of politician speeches. This shows that the classes are highly imbalanced, which adds to the difficulty level of the task at hand.

Since the nature of this investigation is preliminary, there are many issues related to data collection. Firstly, the size of the dataset is small at present, however, the task of data collection and extending the dataset is in progress. Secondly, the issue of data bias comes into play since the data collector selected the politicians and videos, which might have sounded emotionally enriched to the individual. This bias constricts the generalization of the model learnt from such data as the data is not exactly a representative of the entire population. Moreover, there is an imbalance between male and female speakers as well as in the number of utterances amongst different emotional classes. These issues make the modeling of emotions a challenging task.

3   Architecture Description

The deep learning architecture used for the classification of emotional states of the utterances is the state-of-the-art CNN+LSTM with attention mechanism [2,14], as shown in Fig. 1. The architecture consists of four local feature learning blocks (LFLB), one LSTM layer for capturing the global dependencies, and one attention layer to focus on the emotion relevant parts of the utterance followed by a fully-connected softmax layer to generate class probabilities.

Each LFLB comprises of one convolutional layer, one batch-normalization layer, one activation layer, and one max pooling layer. CNN performs local feature learning using 2D spatial kernels. CNN has the advantage of local spatial

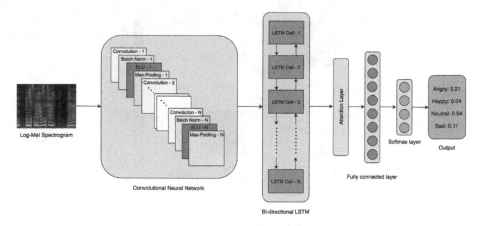

Fig. 1. Attentive CNN+LSTM architecture

connectivity and shared weights, which helps the convolution layer to perform kernel learning. Batch Normalization is performed after the convolution layer to normalize the activations of each batch by maintaining the mean activation close to zero and standard deviation close to one. The activation function used is Exponential Linear Unit (ELU). Contrary to other activation functions, ELU has negative values too, which pushes the mean of the activations closer to zero, thus helping to speed up the learning process and improving performance [3]. Max pooling is used to make the feature maps robust to noise and distortion and also helps in reducing the number of trainable parameters in the subsequent layers by reducing the size of the feature maps.

Global feature learning is performed using an LSTM layer. The output from the last LFLB is passed on to an LSTM layer to learn the long-term contextual dependencies. Sequences of high-level representation obtained from the CNN+LSTM architecture is passed on to an attention layer whose job is to focus on the emotion salient parts of the feature maps since not all frames contribute equally to the representation of the speech emotion. The attention layer generates an utterance level representation, obtained by the weighted summation of the high-level sequence obtained from CNN+LSTM architecture with attention weights obtained in a trainable fashion [2]. The utterance level attentive representations are passed to a fully-connected layer and then to a softmax layer to map the representations to the different emotion classes.

The number of convolutional kernels in first and second LFLB is 64, and for the third and fourth, LFLB is 128. The size of the convolutional kernels is 3×3 with a stride of 1×1 for all the LFLBs. The size of the kernel for the max-pooling layer is 2×2 for the first two LFLB and 4×4 for the latter two LFLBs. The size of the LSTM cells is 128.

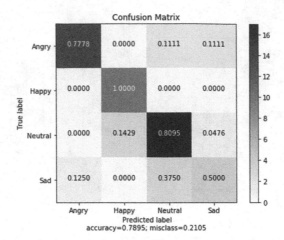

Fig. 2. Confusion matrix of Attentive CNN+LSTM architecture, with an average accuracy of 78.95%, where the rows represent the confusion of the ground truth emotion during prediction.

4   Experimental Evaluation

The experiments are performed in two stages. The first stage of the experiment is to generate class probabilities by performing Speech Emotion Recognition (SER) using the architecture discussed in the above section. The second stage of the experimental evaluation is concerned with the analysis of the duration of different emotions in a politician's speech.

For input to the Attentive CNN+LSTM architecture, the mel-spectrogram is computed from the raw speech files. Since the input to CNN requires equal-sized files, the speech files are either zero-padded or truncated to the same size. A frame size of 25 ms and a frame shift of 10 ms is used to compute the mel-spectrograms. Training of the model is performed in a five-fold cross-validation scheme, and Unweighted Accuracy (UA) is used as the performance measure. The cross-entropy loss function is used in adherence with Adam optimizer to train the model.

The model achieves a recognition performance of UAR 78.95% with the proposed architecture. Figure 2 presents the confusion matrix for the predicted emotions of the four emotion classes of the IITG Political speech dataset. Due to less number of samples in the sad emotion category, the recognition performance is degraded for that category. Happy emotion category is clearly recognized by the model. Figure 3 presents the distribution of the untrained data, trained data after the LSTM layer, and trained data after the attention layer. The plots show the capability of the architecture in clustering emotions with considerable improvement with the addition of the attention layer.

Figure 4 shows the percentage of the four basic emotions- Angry, Happy, Neutral, and Sad in the speeches of the eight Indian politicians. It can be observed

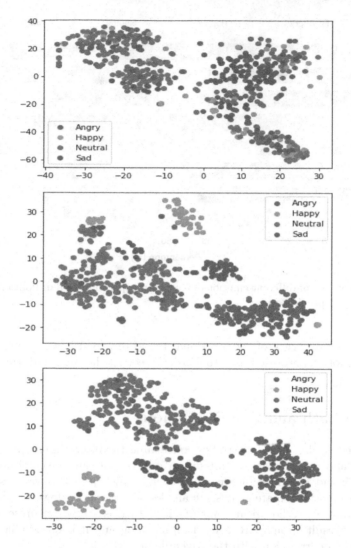

Fig. 3. Distribution of training utterances of IITG Politician Speech Dataset using TSNE with mel-spectrogram as input. 3 displays the untrained distribution, 3 displays the distribution after the LSTM layer and 3 displays the distribution after Attention layer.

that Happy emotion is least exhibited emotion and Angry followed by Neutral is the most exhibited emotion by all the eight politicians. An emotionally balanced speech is often desirable when speaking in a public scenario and as such, it can be observed that the politicians NI, RH, and AL tried incorporating all the four basic emotions in their speech. However, we can also observe that the speeches of some politicians like MH and SY are somewhat emotionally monotonous with the maximum fraction of the speech exhibiting Neutral or Sad state. Thus, this

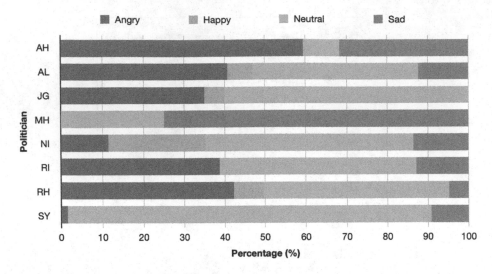

Fig. 4. Percentage of different emotions present in a speech of Indian Politicians of the IITG Political Speech Corpus

provides us an overview of the emotional mood of a politician during elections and how emotions are used as a tool to connect to the masses during election rallies.

5   Aftermath and Few Discussions

As discussed earlier, previous studies concerning the electoral campaigns in the US [6,10], and other locations put up an analysis of the universal application of emotions and how it influences the voters, which points towards successful application emotional state analysis in an election scenario. However, the studies are done from a psychological point of view. In this study, we presented some preliminary results from SER using deep learning and it is observable that our findings are in correlation with the psychological studies.

Moreover, the main contribution of our study is to provide a measurable metric to gauge the effectiveness of a politician's speech from an emotional content point of view. As future work, this study can be extended to compute the likelihood of success and failure of a politician during elections, if sufficient data is available. Moreover, as discussed in Sect. 2, the data collector's bias needs to be quantified using some statistical measures to improve the acceptability of the results.

6   Conclusions

In this work, we presented a brief analysis of the different emotions present in the speeches of politicians in the Indian context. A deep learning framework,

CNN+LSTM with attention mechanism, is used to model the emotions using the IITG Politician Speech Dataset. The Dataset is collected from publicly available speeches of Indian politicians on youtube. The speech utterances are divided and annotated by 4 annotators, and a confidence score is provided for each emotion category based on the perceptual evaluation. Various issues pertaining to data collection such as collector's bias, class and gender imbalance, etc are discussed which needs further attention and analysis. The performance of the model in predicting the emotions from speech utterances outperforms the perceptual evaluation. An assessment of the percentage of different emotion classes in each of the politician's speech is presented with a brief analysis of the results in accordance with previous results presented in the literature.

References

1. Busso, C., et al.: IEMOCAP: interactive emotional dyadic motion capture database. Lang. Resour. Eval. **42**(4), 335 (2008)
2. Chen, M., He, X., Yang, J., Zhang, H.: 3-D convolutional recurrent neural networks with attention model for speech emotion recognition. IEEE Signal Process. Lett. **25**(10), 1440–1444 (2018)
3. Clevert, D.A., Unterthiner, T., Hochreiter, S.: Fast and accurate deep network learning by exponential linear units (elus). arXiv preprint arXiv:1511.07289 (2015)
4. Li, P., Song, Y., McLoughlin, I.V., Guo, W., Dai, L.R.: An attention pooling based representation learning method for speech emotion recognition (2018)
5. Pandey, S.K., Shekhawat, H., Prasanna, S.: Emotion recognition from raw speech using wavenet. In: 2019 IEEE Region 10 Conference (TENCON), TENCON 2019, pp. 1292–1297. IEEE (2019)
6. Parker, M.T., Isbell, L.M.: How I vote depends on how I feel: the differential impact of anger and fear on political information processing. Psychol. Sci. **21**(4), 548–550 (2010)
7. Riker, W.H., Ordeshook, P.C.: A theory of the calculus of voting. Am. Polit. Sci. Rev. **62**(1), 25–42 (1968)
8. Sarma, M., Ghahremani, P., Povey, D., Goel, N.K., Sarma, K.K., Dehak, N.: Emotion identification from raw speech signals using DNNs. In: Interspeech, pp. 3097–3101 (2018)
9. Searles, K., Ridout, T.: The use and consequences of emotions in politics. Emotion Researcher, ISRE's Sourcebook for Research on Emotion and Affect (2017)
10. Valentino, N.A., Brader, T., Groenendyk, E.W., Gregorowicz, K., Hutchings, V.L.: Election night's alright for fighting: the role of emotions in political participation. J. Polit. **73**(1), 156–170 (2011)
11. Wagner, J., Schiller, D., Seiderer, A., André, E.: Deep learning in paralinguistic recognition tasks: are hand-crafted features still relevant? In: Interspeech, pp. 147–151 (2018)
12. Weber, C.: Emotions, campaigns, and political participation. Polit. Res. Q. **66**(2), 414–428 (2013)
13. Yang, Z., Hirschberg, J.: Predicting arousal and valence from waveforms and spectrograms using deep neural networks. In: Interspeech, pp. 3092–3096 (2018)
14. Zhao, J., Mao, X., Chen, L.: Speech emotion recognition using deep 1D & 2D CNN LSTM networks. Biomed. Signal Process. Control **47**, 312–323 (2019)

A Bengali Voice-Controlled AI Robot for the Physically Challenged

Abul Bashar Bhuiyan[1], Anamika Ahmed[1], Sadid Rafsun Tulon[1],
Md. Rezwan Hassan Khan[1], and Jia Uddin[2]($\boxtimes$)

[1] Department of Computer Science and Engineering, BRAC University, Mohakhali,
Dhaka 1212, Bangladesh
bbhuiyan1@gmail.com, anamikaahmedana@gmail.com,
srtulon6@gmail.com, md.rezwanhassankhan@gmail.com
[2] Technology Studies Department, Endicott College, Woosong University,
Daejeon, South Korea
jia.uddin@wsu.ac.kr

Abstract. Previously, a number of research is going on in Artificial Intelligence and robotics to make the humans life easier and efficient. Taking motivation form the research works, this paper presents an Artificial intelligence (AI) based smart robot for physically challenged peoples. The model consists of a 4 wheeler car containing a camera and a robotic claw attached to and is operated by Bengali voice commands from the user. At first, the robot will detect the user and parse the voice command to understand its meaning. After that, it will search for the object and then, it will go towards the object to pick it up and finally it will bring the object back to the user. For instance, if for any reason, the object falls from the robotic claw while carrying it towards its destination, this robot has the intelligence to pick it up again and continue its journey to its designated location. For detecting the object, we have used TensorFlow object detection API, and for parsing the voice commands, natural language processing has been used. Additionally, for face detection, we have used Facenet which is a real-time face recognition model based on deep learning consisting of convolutional neural layers.

Keywords: Artificial intelligence · Image processing · Physically challenged · TensorFlow · Facenet

1 Introduction

The word handicap refers to restrictions, obstacles or limitations that makes one's life challenging. Like other countries, a large population of the Bengali community from Bangladesh and India suffer from the curse of disability. Unfortunately, the number of handicapped people is escalating due to road accidents or diseases like quadriplegics. A handicapped person is dependent on another person for his everyday work like travelling, food, orientation, etc. Approximately, there are 16 million people with disabilities in Bangladesh which is 10% of the country's population [1]. In West Bengal, the percentage of people with access to disability is slightly above the national average [2].

© Springer Nature Switzerland AG 2021
M. Singh et al. (Eds.): IHCI 2020, LNCS 12615, pp. 186–194, 2021.
https://doi.org/10.1007/978-3-030-68449-5_19

Previously a number of research works have done for targeting disable peoples. Arnab Bhattacharjee *et al.* [3] have proposed a robotic model that operates by Bengali voice commands and is applicable for rescuing operations in noisy environment. By capturing high frequency sound waves, the robot can locate the location of the victim from an average distance. The speech recognition system uses mel-cepstrum coefficients and vector quantization. Dinesh *et al.* [4] has presented a voice activated artificial intelligent robotic arm. The robotic arm performs the desired task after interpreting the voice commands. Furthermore, Himanshu *et al.* [5] have designed a mobile robot that tracks a moving object using image processing and has the capability to avoid obstacles in real-time. There does exist systems that respond to many of the needs of people with various degrees of disability. Manuel *et al.* [6] have described a voice controlled smart wheelchair for physically handicapped people to allow autonomous driving and allowing the ability to avoid obstacles in real time. Besides, Shraddha *et al.* [7] presented an intelligent wheelchair whose rotation is controlled using a smartphone upon voice commands. The system uses gesture recognition through Android and motor control through signal conditioning [13]. In the state-of-art models, we have not found much work on voice controlled robotic car that uses both Bengali voice recognition system and Object/Face Recognition techniques in order to carry objects for its user.

This paper presents a model that can avoid obstacles in real-time, making it convenient for the physically disabled peoples to a great extent.

The rest of the paper is organized as follows: Sect. 2 describes the proposed model along with a block diagram, Sect. 3 presents the system setup and analysis of the results. And finally, Sect. 4 concludes the paper.

2 Proposed Model

The working procedure of our proposed prototype model can be classified into three phases. The first phase is to convert the voice command to text. The second phase is image processing and third is picking and dropping the object. A detail system architecture of the model is presented in Fig. 1.

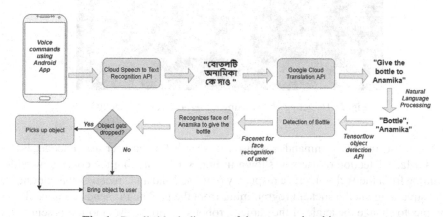

Fig. 1. Details block diagram of the proposed architecture.

2.1   Converting Voice to Text

In Fig. 1, the first block represents the workflow of converting Bangla command to text. We have created a voice recognition android mobile application. When a Bangla command is passed into our app, the app converts the voice to text using cloud speech-to-text speech recognition API. Secondly, with the help of Google Cloud Translation API, it translates the Bangla text into English. Finally, the translated text is sent to the PC from mobile app through the socket programming. Here socket programming is used to send data from the clients to the server. In the experiment, mobile act as a client and the PC acts as a server.

2.2   Image Processing

In Fig. 1, the second block represents the image processing phase. From the translated data received from the mobile app, it picks up the user's and object's names and processes these two texts to search for that a specific user and a specific object which is stated in the command. First, PC commands to search for the object. For detecting an object, we use Tensorflow object detection API which is comparatively easier than the other models like YOLO, SSD, and R-FCN, as the models need complex hardware setup to run it. We have use pre-train data for our robot to detect an object as this API can identify almost every type of different and similar objects from its already pre-trained object database. This is displayed in Fig. 2.

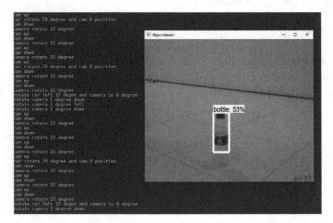

Fig. 2. Detecting object using tenserflow object detection

After that, the PC commands are used to search for the user and Facenet is utilized for face detection of users as shown in Fig. 3. For handling the continuous video streaming from the Web camera, a raspberry pi is used and a PC receives the continuous video streaming via the socket programming from the pi. The PC sends commands, for example, to change the angle of the camera, robot's motion, claws motion is sent to the raspberry pi and these commands are then forwarded to Arduino Uno through the serial connection. And then these commands are performed by the robot as per PC's command.

Fig. 3. Detecting user's face using Facenet

2.3 Pick the Object and Drop It to the User

In Fig. 1, after detecting the user and the object, the PC commands is used to pick up the object. First, the camera is adjusted and then rotated it to the center of the object. Then the car is aligned in the same direction as the camera so that it can move forward to pick the object. After the car and camera is adjusted and rotated to the center to the object, PC then commands to move forward to bring the object. After that it picks the object and search for the user and bring it to the user and drop it in front of the user.

3 Experimental Setup and Result Analysis

3.1 Hardware Setup

In this project, a web camera is used to detect the object and user. For detecting the object we have used tensorflow [12] object detection API. The name of the detected object is stored into the PC in the form of a string and then it is matched with pre-trained objects which is stored in the database. For face detection of the user [15, 16], we have used Facenet [13] which is a real-time face recognition model based on deep learning consisting of convolutional neural layers. Images are classified upon 128 different vectors of each class, where classes mean face from different people, which make it really easier to differentiate between different faces. The libraries we have used to detect face and object are os, pickle, numpy, time, socket, scipy, sys, serial, opencv and framework tensorflow. Servo motors are used to change the angle of rotation of the camera and robotic claw which we are using to pick and drop the object. For rotating the camera 360 degrees, we have connected 2 servo motors with the camera. One servo motor helps to rotate 360 degrees horizontally and another one helps it to rotate the camera vertically as shown in Fig. 4. Similarly, we have used three servo motors to rotate the robotic claw. Two servo motors is used to rotate the claw horizontally and vertically and the third servo motor is used to move the whole body of the robotic claw in up and down motion, as shown in Fig. 5, so that it can adjust its position when picking and dropping object.

Fig. 4. Servo motors rotating the web camera

Fig. 5. Servo motors rotating the robotic claw

Before picking the object, camera and car are adjusted to a position aligned with the center of the object. Compass GY271 HMC5883L is used to monitor the change of angle of the car so that it ensures the car is moving in the same direction unaltering the car's direction. We have attached a sonar sensor with the robotic claw as shown in Fig. 5 so that it can stop its car motion when the object is 3 cm away from the robotic claw to pick the object. The sonar sensor is also used to detect if the object gets dropped or not. After picking the object its searches for the user and after detecting the user it again adjusts the camera and car's position aligning with the center of the user. We have attached another sonar sensor with the camera, so that it can stop the car's motion when the car is 15 cm away from the user to drop the object at that position so that the user can pick the object. We have tested on a system having inter core i5-3230M 2.60 GHz CPU, 8 GB RAM.

3.2 Result Analysis

The accuracy for detecting object using tensorflow object detection API is really high and the chance of making an error is very less because of the large set of pre-trained data this system has offered us. For an example, it gave around 87% accuracy in detecting an apple which is quite good compared to other deep learning techniques giving an accuracy of around 95% [14]. Some other accuracy results are presented in Table 1. Similarly, the accuracy for detecting the user's face is really astonishingly high accuracy which is around 90%. This is because the deep convolutional neural network is used in facenet which works with 128 different layers to recognize a user. Our proposed model can pick and drop an object with around 70% percent accuracy with a very low rate of chance of making an error. So, in a nutshell, the accuracy of our proposed prototype matched our expectation and in near future, it will become more efficient if we are able to incorporate some powerful hardware (Fig. 6).

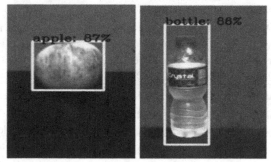

Fig. 6. Detection accuracy of apple and bottle

Table 1. Type of object and detection accuracy

Object detected	Accuracy	Lag time (seconds)
Bottle	86%	3.2
Apple	87%	3.1
Banana	88%	3
Scissors	71%	3.3

We have conducted several experiments and trial cases using our robot in the laboratory- the results of which is explained in Table 2. In experiment 1, the robot was told ordered to bring an apple which was located right behind it, and it could successfully bring the apple to the designated user to the designated user. However, in experiment 2, the robot failed to recognize the voice command. A probable reason of this failure might be the background noise due to which it could not parse the commands or translate it, eventually failing to detect the object. In experiment 4, although the robot could identify the user's face correctly, it failed to detect the eraser because it had a brown color body and was identified as chocolate. We had again encountered a failed test case in experiment 8 where the robot could not deliver the object as the sonar sensor gave wrong input, as a result of which distance to the designated location was not calculated accurately. In experiment 9, the robot passed all the scenarios except face detection (we had trained the robot with 10 images and this data was not enough). Lastly, in experiment 10, although the robot could identify the cup accurately, it failed to pick it up because it of its large mass.

The average response time for voice recognition and translation is 2–3 s depending on the processing speeds of speech-to-text and cloud translation API; for object detection and face recognition, the average response time is 3–4 s, and for object delivery the response time varies depending on the factors like distance of user to object and friction of floor surface.

Table 2. Result analysis of our robot with different performance metrics

Attempt no.	Object to be detected	Voice recognition	Object detection	Picking up object	Face detection	Delivered the object
1	Apple	Passed	Passed	Passed	Passed	Passed
2	Marble	**Failed**	**Failed**	Not Needed	Not Needed	Not Needed
3	Banana	Passed	Passed	Passed	Passed	Passed
4	Eraser	Passed	**Failed**	Not Needed	Passed	Not Needed
5	Orange	Passed	Passed	Passed	Passed	Passed
6	Pen	Passed	Passed	**Failed**	Passed	Passed
7	Bottle	Passed	Passed	Passed	Passed	Passed
8	Scissor	Passed	Passed	Not Needed	Passed	**Failed**
9	Orange	Passed	Passed	Passed	**Failed**	Not Needed
10	Cup	Passed	Passed	**Failed**	Passed	Not Needed

Overall, if we summarize Table 2, we can come up with the accuracy percentages of the robot's performance as shown in Table 3. For each feature, our robot shows a very promising performance and the accuracy lies between 80% to 100%. The accuracy is calculated using:

Feature Success Percentage = [(number of successful attempts)/(total number of attempts)] × 100%

Table 3. Robot's accuracy of performance as per experiment

Feature	Accuracy
Voice Recognition	90%
Translation	90%
Object Detection	80%
Picking Up Object	80%
Face Detection	90%
Re-pick Object after dropping	100%
Object Delivery	90%

4 Conclusion

In this paper we have proposed and implemented a robot that uses digital signal processing for converting Bengali voice commands to text, and then uses artificial intelligence for object detection, face cognition and carrying desired object for the user. The number of physically challenged people is increasing every day. Now evaluating the state of the current circumstance much is being done to handle and control the degrading rate of this situation. Keeping all these factors and the complexity of the situation in mind, we have, thus, decided to come up with such proposal. This model will hopefully have a contribution in this field and ease the day to day lives of the physically challenged people. In future, we aim to make the robot feasible to carry out dangerous industrial tasks.

References

1. Disability in Bangladesh: Centre for Disability in Development. Accessed 20 Apr 2016
2. Singh, S.: Only 40 per cent of India's disabled have access to disability certificates, 07 April 2016. http://www.thehindu.com
3. Bhattacharjee, A., et al.: Bangla voice-controlled robot for rescue operation in noisy environment. In: 2016 IEEE Region 10 Conference (TENCON), Singapore, pp. 3284–3288 (2016)
4. Mital, D.P.: A voice activated robot with artificial intelligence. Robot. Auton. Syst. **4**(4), 339–344 (1989)
5. Borse, H.: Mobile robot for object detection using image processing. Glob. J. Comput. Sci. Technol. Neural Artif. Intell. **12**(11) (2012)
6. Mazo, M., Rodríguez, F.J., Lázaro, J.L. et al.: Auton Robot, vol. 2, no. 203 (1995)
7. Shabana Tadvi, E.R., Adarkar, P., Yadav, A., Saboo Siddik, M.H.: Automated wheelchair using Android technology. Imperial J. Interdisc. Res. **2**(4), 654–657 (2016)

8. Raspberrypi.org (2019). https://www.raspberrypi.org/products/raspberry-pi-3-model-b/
9. Kaymu.com.np. Arduino Uno R3 (2015). http://www.kaymu.com.np/arduino-uno-r3-53519.html
10. Circuitdigest.com. Servo Motor Basics, Working Principle & Theory (2019). https://circuitdigest.com/article/servo-motor-basics
11. Daware, K.: How a DC motor works?. Electricaleasy.com (2019). https://www.electricaleasy.com/2014/01/basic-working-of-dc-motor.html
12. TensorFlow (2019). https://www.tensorflow.org/Instructables. 5 October 2017. How to Use the L298 Motor Driver Module-Arduino Tutorial. Accessed 1 Dec 2019. https://www.instructables.com/id/How-to-use-the-L298-Motor-Driver-Module-Arduino-Tu/
13. McGinn, C., Torre, I.: Can you tell the robot by the voice? An exploratory study on the role of voice in the perception of robots. In: 14th ACM/IEEE International Conference on Human-Robot Interaction (HRI), Daegu, Korea (South), pp. 211–221 (2019)
14. Mureşan, H., Mihai, O.: Fruit recognition from images using deep learning. Acta Universitatis Sapientiae, Informatica. **10**, 26–42 (2018)
15. Shafiqul Islam, Md., Mahmud, A., Akter Papeya, A., Sultana Onny, I., Uddin, J.: A combined feature extraction method for automated face recognition in classroom environment. In: Thampi, S.M., Krishnan, S., Corchado Rodriguez, J.M., Das, S., Wozniak, M., Al-Jumeily, D. (eds.) SIRS 2017. AISC, vol. 678, pp. 417–426. Springer, Cham (2018). https://doi.org/10.1007/978-3-319-67934-1_38
16. Zarin, A., Uddin, J.: A hybrid fake banknote detection model using OCR, face recognition and hough features. In: 2019 Cybersecurity and Cyberforensics Conference (CCC), Melbourne, Australia, pp. 91–95 (2019)

How to Enhance the User Experience of Language Acquisition in the Mobile Environment: A Case Study of Amkigorae(암기고래), a Vocabulary Acquisition Mobile Application

Chiwon Lee(✉), Donggyu Kim, Eunsuh Chin, and Jihyun Kim

Yonsei University, Seoul, South Korea
chiwon.lee@yonsei.ac.kr

Abstract. A vast majority of Korean students study secondary languages for various purposes: for certificates, college admission, or personal interest. Accordingly, immense amounts of paper vocabulary workbooks are consumed to fulfill these needs; however, it is recommended that students transition their learning experience to the mobile environment as the consumption of paper vocabulary workbooks may contribute to environmental deterioration. This study aims to research the user experience of Amkigorae(암기고래), which is a mobile secondary language acquisition app, developed by Beluga Edu(벨루가에듀), in the context of college students. This is due to the fact that college students have higher autonomy over cell-phone usage compared to teenagers, and because they have higher digital literacy and required needs for language acquisition compared to other age groups. This project primarily aims to research, evaluate, and determine the means of creating an effective user interface that may serve as a reference point for vocabulary acquisition apps to increase user retention and secondary language acquisition. In order to gain insight on the further development direction, our team conducted multiple quantitative and qualitative research to comprehensively evaluate Amkigorae. We were able to highlight the positive and negative feedback regarding users' experience on memorization, usability, and aesthetics. Analysis on the limitations and future work that should be conducted related to this topic is provided to further invite stakeholders to expand research on this topic.

Keywords: Vocabulary acquisition · Learning habits · Usability

1 Introduction

Foreign language acquisition is a major industry worldwide, especially in the context of the Republic of Korea. According to Kyobomunko(교보문고), a leading South Korean bookstore, English vocabulary workbooks are high in sales [1]. The high sales in such

---

All 4 authors equally contributed to the research paper.

M. Singh et al. (Eds.): IHCI 2020, LNCS 12615, pp. 195–207, 2021.
https://doi.org/10.1007/978-3-030-68449-5_20

books contribute to environmental deterioration as the fast consumption of these books is not in line with the following United Nations Sustainable Development Goal (UN SDG), No. 12 "Responsible Consumption and Production" [2]. Our team believes that the rise of mobile applications that allow users to shift their learning activity from books to mobile will be able to contribute to the aforementioned UN SDG article along with the article No. 4 "Quality Education" [2]. This is due to the fact that such applications allow a more interactive means of studying that is complimentary.

To promote this transition, we selected Amkigorae as our case study app to serve as the basis for finding user experience suggestions for language acquisition apps.

In this study, our team aims to provide perspective into how learners, college students in particular, acquire foreign language by probing into their learning styles and by analyzing what features of our case study app supports or hinders their experience as a user.

Our team speculates that this study will act as a reference for related educational application developers to consider when developing products to increase their user base and user retention rate as the arena of foreign language acquisition in the mobile environment has further potential to revolutionize the learning experience of users.

2   Research Questions

RQ1: Whether and how does Amkigorae fulfill the needs of its users in regards to their different language acquisition styles?
RQ2: How can the user interface of Amkigorae be improved to better serve the purpose of the app?

3   Existing Research

Existing research has revealed the following: 1) mobile learning can contribute to social innovation [5, 12–14], 2) learning language through mobile applications is effective, and in some cases, more effective than paper workbooks[6, 7, 11–14], and 3) the user experience of the app is important for user retention of language acquisition applications [7–10, 15–22].

Research further presented that mobile applications, thanks to its interactive features, were particularly more effective in enhancing the users' pronunciation of foreign words, and more effective for students as they were more motivated to study in their spare time via their cell-phones [12–14].

We also collected data on various learning styles to better contextualize our research findings in accordance with the multifarious learning habits people possess.

Out of the papers that were found, the classification of language acquisition strategies of O'Malley's was notable. According to O'Malley's study, language acquisition can be divided into three categories, which are metacognitive, cognitive, and socio-affective [3]. Metacognitive refers to learners who take ownership of their study and plan studying sessions and self-evaluate. Cognitive refers to learners who study directly by repeating words, translating, note-taking, and using imagery. Socio-affective refers to learners who enjoy learning in groups and enjoy asking questions. Based on this research, we conducted a survey to recruit representative participants and gain insight on the learning habits and user experience of users based on their learning styles.

4   Methodology

Participants Profile. In the screening survey, we included a questionnaire to distinguish learning types of the respondents. From 99 respondents of the survey, our team recruited a total of 9 people for two interviews and a diary study. Participants recruited were Korean college students who are studying or have studied secondary language at least once, and their ages ranged from 20 to 25 years old. After the interviews, we recruited 20 college students from the same age range to participate in the AB Testing based on prototypes we devised.

Screening Survey. The survey was conducted from October 4th to Oct 9th, 2019; the survey was distributed online to about 1,000 students, and it received 99 responses. The main purpose of the survey was screening and recruiting adequate participants for further interviews and diary studies. The contents of the survey included demographic information, respondent's language learning habits, and diagnostic questions based on O'Malley's learning types.

Interview 1. The first interview was conducted on October 11th, 2019. It was in the form of a fully structured interview. In order to observe whether there is a difference of app usage habits or difference in feedback between different learning types, a total of 9 participants with various learning styles were recruited: 3 for Metacognitive, 4 for Cognitive, and 2 for Socio-affective.

Diary Study. After Interview 1, a Diary Study was conducted on the 9 interviewees of Interview 1. Users were given a direction to use the 3 modes available on Amkigorae for 2 days each, for a total of 6 days. The purpose of the diary study was to explore the day-to-day experience of the user interactions with the app to decipher the perceptions of the users, the motivations of the users, and the factors that affect user retention.

Interview 2. The second interview was conducted after the diary study. Interview 2 was conducted as a semi-structured interview for diary study participants. The purpose of Interview 2 was to gain a more detailed insight about the experience from diary study.

Affinity Diagram. After gathering a sufficient amount of qualitative data, our team analyzed the data by using the affinity diagram method. The raw data included the quotes from the scripts of 6 days of Diary Study and transcripts of Interview 2.

The data was categorized into memorization, usability, and aesthetics, and divided the data based on positive(+) and negative(−) feedback for each category to further understand our participants.

AB Testing. AB Testing was conducted from November 22nd to December 3rd, 2019 on a total of 20 college students to 1) examine the efficiency of start screen, 2) explore the intuitiveness of the memorization screen, 3) gauge whether the inclusion of haptic feedback would be preferred.

Two prototypes and 3 tasks in total were created to compare the prototypes to the original interfaces of the Amkigorae app.

A scale of 1 to 5 was provided to analyze the satisfaction rate of the original interface and prototype to gain data to validate via statistical analysis in order to gauge the significance of the data we collected.

Open Coding. We conducted Open Coding on the AB Testing scripts that we have garnered. We color-coded our findings like the following: 1) Aesthetics - Pink, 2) Usability - Orange, 3) Effectiveness (in learning) - Yellow, and also extracted meaningful quotes that could help us better understand the perceptions regarding the original interfaces and the prototypes we devised.

T-Testing. We conducted T-Testing, specifically the Wilcoxon Signed Rank Test, to determine whether there is a considerable difference in the means that we have calculated based on the satisfaction rates of the original interfaces and the prototypes that we developed.

5   Implications and Prototype

Our team has categorized our findings in the following categories: 1) Usability and 2) Examining User Experience Propositions.

5.1   Usability

Although users acknowledged that the app is rife with content, the users stated that the content gave off the impression that the app is clustered, unorganized, and hard to navigate. Users seemed to desire the addition of haptic functions that would enable them to retrieve data in a more intuitive way. We were able to discern the following themes and categories regarding usability:

Rich with Functions. Amkigorae is rich with function. We were able to detect that the richness of functions within the app was the most frequently mentioned positive factor when discussing the usability of the product. The users stated multiple functions that were useful; for example, dark mode, secret mode, and the hiding function were some of the functions that were mentioned.

"The dark mode is useful to utilize before sleeping" - C01
"The group mode's secret mode function is nice" - C02

Smooth. User satisfaction was found with the smoothness of Amkigorae in the context that there is no delay when it comes to providing feedback regarding user input. Seamless user feedback is critical for a vocabulary flashcard app because some users still prefer paperback vocabulary workbooks. The participants in our research, who are college students, said that they are used to looking at paper text, which is important to note as there is no lag when studying via a paperback vocabulary workbook as there is no need for the content of the paperback workbook to load or have a connection failure.

"I'm used to looking at paper text." - S02
"It was definitely different (from) what I experienced before with Anki (a different app)... (Anki) was buggy (and laggy)." - C02

One user from the interview mentioned that she used a flashcard app, Anki, that was recommended by her teacher. She said she used it during the year 2013 to 2015 because her teacher made students use it for a class that she was taking at the time. She said that she is more prone to using physical flashcards; this can be attributed to the negative experience that she had regarding the application that she used before. Because the experience of using Amkigorae was positive this time, and because there was no lag, she said that she was willing to try the app in the future if she was under the case where she had to acquire a new language.

Clustered. Participants said that the organization of the application gives the impression that the app is clustered. When starting the app, a long video appears. This kind of function may deter the user from further using the app because it gives off the impression that the application may be irrelevant, laggy, or heavy.

Moreover, one participant noted that the dashboard should be more organized in order to attract more users. It is speculated that this may be due to the fact that the mobile environment allows less space than a paper workbook when it comes to displaying content.

"I didn't like a video popping up when starting the app." - M01
"The dashboard is a bit too chunky." - C02

Hard to Navigate. Users said that some integral functions of the application were hard to find. For example, users mentioned that the action of going back, creating a word list, or finding a button to stop the audio was confusing. This may be related to the fact that the user interface is clustered. A more simple user interface could help the user easily navigate through the app.

Furthermore, participants mentioned that it was hard to find some functions that they were likely to use if they knew the existence of the function. For instance, one participant could not find the audio stop button while another participant mentioned that she was oblivious of the fact that there was a notification function.

"It was hard to find the audio stop button." - M02
"I didn't know that there was a notification feature." - C01

Lack of Intuitiveness. Participants contended that some features of the app lacked intuitive aspects as they found some of the integral functions within the app complicating. For example, one participant noted that she found difficulty in creating a word list. This is critical as paper workbooks are intuitive. When creating a word list in a paper workbook, users of paper workbooks simply have to flip around the textbook to find a space to write the word list and write a word and a sentence within the blank space. However, in Amkigorae, users have to click a floating 'add' button on the bottom right of the screen, and then go through a stage of trial-and-error in order to discern which of the two buttons that pop up when the floating add button is touched upon leads to the creating word list function. After the user grasps which function leads to a space that creates a word list, the user has to touch the second button on the top left to edit the word list in order to type in a word, meaning, and an example sentence. This is confusing as paper workbooks and most popular apps allow users to write in the blank space by simply touching into the blank space.

"It is hard and complicated to make a word list." - S02

Importance of Haptic Feedback. One notable thing that we deduced from the interviews is that users stated they thought the hand, to a certain extent, memorized the words for them. This is important to note as the usability of the app may be evaluated by the users based on how the app allows the users to utilize their hands. Participant C02, a female participant that used a vocabulary memorization app named "Anki," stated that she did not like using the app because the app did not allow her to write the words. She would write down the words on paper because she said that the act of writing the words on paper would help her hands memorize the words for a test. When asked whether it would be useful if there was a function that enables users to write within Amkigorae, Participant C02 said that she would not find the function that useful as the amount of space that the user can utilize to write the word was limited compared to a paper workbook. She said the space would have to be larger in order for her to freely utilize her hands so that the hands would 'memorize' the words.

"I'm more used to writing in hands... The pro is that I personally think your hand remembers the moment you write that word. But maybe con is that it takes a lot of time. It takes so much time." - C02

5.2   Examining User Experience Propositions

AB Testing and statistical analysis revealed that the prototypes that were focused on rendering the interface to be more clean, intuitive, and haptic were preferred amongst users. Users expressed that the prototypes were comparatively lean, customizable, controllable, and helpful regardless of whether they would use the additional haptic features. They stated that the features would be utilized based on the users' different learning habits.

Efficiency of Start Screen. We designed a prototype that customizes the app so that the user does not need to view language options that are not of the user's interest. We also focused on making the interface more clean and organized (Fig. 1).

Fig. 1. Original interface of start screen

Participants provided generally negative responses to the original interface by stating that the original interface is confusing, clustered, and childish (Fig. 2).

"It seems like an app for children" - P13

Fig. 2. Prototype of start screen

In comparison, participants had generally positive responses for the prototype as they liked thought that the customization feature in a clean and organized interface is more efficient since they do not have to view information or languages they are not interested in (Graph 1).

"Customizing my feed is great. It's efficient." - P12

When asked about the efficiency of each version (Q1), the application scored an average of 3.17 points and the prototype scored an average of 4.35 points. Through Wilcoxon signed rank test, we found the p-value to be less than 0.001, smaller than 0.05, meaning that the difference is significant.

When asked about the satisfaction level of the interface design on each version (Q2), the application scored an average of 3.075 points and the prototype scored an average of 4.175 points. We found the p-value to be less than 0.001, smaller than 0.05, meaning that the difference is significant.

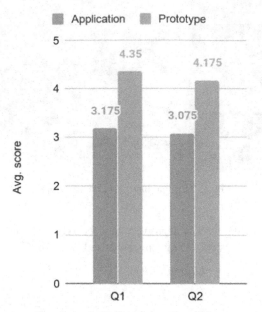

Graph 1. Q1 & Q2 Means Comparison

Intuitiveness of Memorization Screen. The second prototype we designed aimed to improve the intuitiveness of the icons and included basic haptic functions (Fig. 3).

Fig. 3. Original interface of memorization screen and buttons

Participants provided generally negative responses by stating that the bookmark is confusing as it does not like a bookmark; nonetheless, some participants claimed that the location of the bookmark is intuitive since they would expect a bookmark to be in the location where the icon is (Fig. 4).

"The bookmark icon doesn't look like a bookmark" - P11

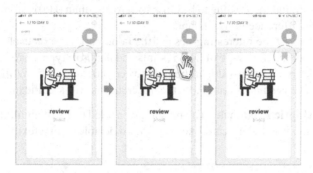

Fig. 4. Prototype of memorization screen and buttons

Participants provided generally positive responses about the prototype by stating that the bookmark was much more intuitive as it resembled a bookmark compared to the icon of the original interface.

Participants also appreciated that the color of the bookmark changes by stating that this kind of interactive element makes the app more favorable than a textbook (Graph 2).

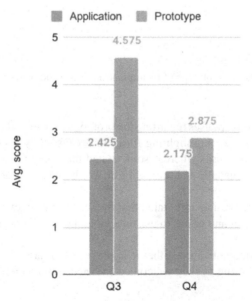

Graph 2. Q3 & Q4 Means Comparison

"It is intuitive that the bookmark looks like an actual bookmark" - P12

When asked about the intuitiveness of the bookmark icon on each version (Q3), the application scored an average of 2.425 points and the prototype scored an average of 4.575 points. We found the p-value to be close to zero less than 0.001, smaller than 0.05, meaning that the difference is significant.

When asked about the interesting-ness of the bookmark icon on each version (Q4), the application scored an average of 2.175 points and the prototype scored an average of 2.875 points. We found the p-value to be in between 0.02 and 0.05. While bigger than 0.02, as the p-value is smaller than 0.05, the difference is still significant.

Preference on Haptic Feedback. The prototype enabled users to highlight vocabulary and scribble on the vocabulary card like they would be able to in a textbook (Fig. 5).

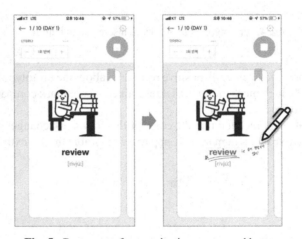

Fig. 5. Prototype of memorization screen and buttons

There were various responses regarding the prototype; some of the participants stated that they would not use the highlighting function because they do not normally highlight when memorizing words whereas some stated that they would not use the scribbling function as they usually solely highlight words that they are having difficulty memorizing.

Additionally, some participants stated that the scribbling function would be hard to utilize on a phone screen and easier to use on a tablet PC screen because phone screens are small.

Nonetheless, participants agreed that it is preferred to have the haptic function of highlighting and scribbling as they welcomed the inclusion of more options to choose from (Graph 3).

"Everyone has their own way of marking for memorization, so I think it's nice to have highlighting and the scribbling function" - P14

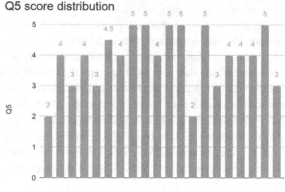

Graph 3. Q5 score distribution

When asked about the helpfulness of the haptic features added (Q5), the prototype scored an average of 3.925 points and the median was 4 points, which means that the overall helpfulness of the additional haptic features were recognized despite there being varying responses on whether the users would regularly use the functions.

6   Limitations

Before proceeding to the discussions and implications, we would like to point out a few limitations. The main issue our team encountered was due to the lack of resources; this research included only a part of the Amkigorae users' demographic. Our participants were all Korean college students in their 20 s. In addition, the duration of our study was 6 days, which may be insufficient to verify the long-term memorization of vocabulary.

We also recognize that there are limitations to the prototypes we developed as we lack the technical capability to create a working prototype for advanced features such as scribbling; we invite more researchers to conduct research on this topic to foster the transition into the mobile environment regarding language acquisition.

7   Discussions and Implications

In contrast to the general perception of the participants that mobile apps may lack content compared to paper workbooks, most participants were pleasantly surprised with the app content-wise as many participants mentioned that the app was rich with functions and content.

We could deduce that the usability of the app is decreased when the app attempts to list an excessive amount of content (e.g. videos, illustrations, tabs) in a single page; a more lean layout with minimal and intuitive icons that respond to haptic feedback is recommended in order to resolve this issue.

Via AB Testing and statistical analysis, we were able to infer that the app should 1) provide users more control over customization, 2) devise minimal and intuitive icons and interfaces as they increase user concentration, 3) note that the inclusion of more haptic features is recommended since although users may not use all the haptic features, users are delighted to have the option to choose based on their learning style. This would render the app more user-friendly for those in their 20 s; this is important as people in their 20 s is a demographic that has purchasing power and in many cases a required demand for language acquisition; they are used to apps that are easy to navigate, and sensitive about the aesthetics of the app based on insights derived from our research.

In order for apps to be favored over paper workbooks, they should accommodate the benefits of using a paper workbook while heightening the learning experience of the user by providing haptic functions such as highlighting and scribbling.

Although we were not able to test AI/ML, AR/VR functions due to limitations in technology, we were able to conclude that haptic features do give a positive impression to users.

8   Conclusion and Future Works

Our goal was to form an understanding to serve as the basis for the formation of a user experience guideline that vocabulary memorization apps can refer to so we can transition users from the paper workbook to the mobile environment by dissecting Amkigorae as a case study. Although we have found that most participants were pleasantly satisfied with the content of the app, we concluded that the app could adopt more intuitive and aesthetic layouts and icons, exclude indiscreet usages of the whale image and out-of-context illustrations, and include a re-designed mascot character and a simple, minimal interface.

Through our research, we concluded that the incorporation of haptic functions utilizing advanced technology could increase user traffic as users enjoyed having more options regardless of whether they would use all functions because of their learning style.

Therefore, we invite researchers to conduct further research on the topic in, but not limited to, the following arenas; apps could utilize AI/ML to recommend a word list for users and exclude meaningless word lists for the user to streamline the experience of the user as word lists for "Basic English" is useless for those who are studying "Advanced English." Moreover, AR could allow users to have more space to utilize their hands to write words in open space while VR could provide rich visual content that could render the user to understand the context of the vocabulary better.

We invite interested stakeholders regarding the issue of sustainable development in regards to the access to quality education and means to ameliorate the current state of environmental protection by furthering research regarding this area to transfer the demographic that purchase paper workbooks to the mobile environment.

References

1. Kyobomunko(교보문고). Speaking Miracle, English Vocabulary 1000(기적의 말하기 영단어 1000). http://www.kyobobook.co.kr/product/detailViewKor.laf?mallGb=KOR&ejkGb=KOR&barcode=9791161502199#book_info. Accessed 20 Oct 2019

2. UNDP. UN Sustainable Development Goals. https://www.undp.org/content/dam/undp/lib rary/corporate/brochure/SDGs_Booklet_Web_En.pdf. Accessed 23 Oct 2019

3. Zare, P.: Language learning strategies among EFL/ESL learners: a review of literature. Int. J. Hum. Soc. Sci. (2012). https://pdfs.semanticscholar.org/47be/e36333613765bc07c113a189 47fe5c8c055a.pdf

4. Korea Institute of Design Promotion. Design Korea 2018. http://designkorea.kidp.or.kr/html/ ko/main.php. Accessed 29 Oct 2019

5. Sun, L.: Final. E-books vs Printed textbooks/Protect environment. Penn State - Mathematics for Sustainability: Spring 2016 (2016)

6. Edge, D., et al.: Micromandarin: mobile language learning in context. In: Conference on Human Factors in Computing Systems (2011)

7. Hirsh-Pasek, K., et al.: Putting Education in "Educational" Apps: Lessons from the Science of Learning. Association for Psychological Science (2015)

8. Hong, W., et al.: Word spell: associative-phonological learning method for second language learners. Software and Data Engineering (2014)

9. Kim, J.: 모 바일 기반 언어학습에 관한 고찰. 현대영어교육 7(2), 57–69 (2006)

10. Kohnke, L., et al.: Using mobile vocabulary learning apps as aids to knowledge retention: business vocabulary acquisition. J. Asia TEFL 16(2), 683–690 (2019)

11. Lu, M.: Effectiveness of vocabulary learning via mobile phone. J. Comput. Assisted Learn. 24, 515–525 (2008)

12. Luna-Nevarez, C., et al.: On the use of mobile apps in education: the impact of digital magazines on student learning. J. Educ. Technol. Syst. 47(1), 17–31 (2018)

13. Sandberg, J., et al.: Mobile English learning: an evidence-based study with fifth graders. Comput. Educ. 57, 1334–1347 (2012)

14. Saran, M., et al.: Mobile assisted language learning: English pronunciation at learner's fingertips. Eurasian J. Educ. Res. (2009)

15. Choi, E.: Analysis of usability factors of Educational Mobile application (2018)

16. Cartoon, Jang, E., et al.: Research on factors affecting on learners satisfaction and purchasing intention of educational applications. J. Korea Contents Assoc. 12(8), 471–483 (2012)

17. Peters, D.: UX for Learning: Design Guidelines for the Learner Experience. UX Matters (2012)

18. Plaut, A.: Elements of Learning Experience Design. Medium (2014)

19. Chen, Z.: A Study of the Effect of Mobile Application Characteristics on User Satisfaction and Continuance Use Intention (2017)

20. Choi, E.: UX design strategy for education mobile app based on user value. J. Korea Inst. Inf. Commun. Eng. 21(7), 1386–1392 (2017)

21. Kim, B., et al.: Effects of mobile app service characteristics on user satisfaction and continuance usage intention. J. Inf. Technol. Appl. Manag. 26(3), 99–120 (2019)

22. Simon, K.: What is Retention and How to Measure it. UX studio (2018)

Screening Trauma Through CNN-Based Voice Emotion Classification

Na Hye Kim[1] , So Eui Kim[1] , Ji Won Mok[1] , Su Gyeong Yu[1] ,
Na Yeon Han[1] , and Eui Chul Lee[2]([✉])

[1] Department of AI and Informatics, Sangmyung University, Hongjimun 2-Gil 20, Jongno-gu,
Seoul 03016, Republic of Korea
[2] Department of Human-Centered AI, Sangmyung University, Hongjimun 2-Gil 20, Jongno-gu,
Seoul 03016, Republic of Korea
eclee@smu.ac.kr

Abstract. Recently, modern people experience trauma symptom for various reasons. Trauma causes emotional control problems and anxiety. Although a psychiatric diagnosis is essential, people are reluctant to visit hospitals. In this paper, we propose a method for screening trauma based on voice audio data using convolutional neural networks. Among the six basic emotions, four emotions were used for screening trauma: fear, sad, happy, and neutral. The first pre-processing of adjusting the length of the audio data in units of 2 s and augmenting the number of data, and the second pre-processing is performed in order to convert voice temporal signal into a spectrogram image by short-time Fourier transform. The spectrogram images are trained through the four convolution neural networks. As a result, VGG-13 model showed the highest performance (98.96%) for screening trauma among others. A decision-level fusion strategy as a post-processing is adopted to determine the final traumatic state by confirming the maintenance of the same continuous state for the traumatic state estimated by the trained VGG-13 model. As a result, it was confirmed that high-accuracy voice-based trauma diagnosis is possible according to the setting value for continuous state observation.

Keywords: Trauma · Convolution neural network · Audio · Voice · Emotion

1 Introduction

Today, modern people are exposed to various types of stress, from everyday events such as study and employment to serious incidents such as traffic accidents and crimes. Trauma, in medical terms, is called post-traumatic stress disorder (PTSD) and refers to a mental injury caused by an external traumatic event [1]. People who experience trauma have difficulty in controlling their emotions and stabilizing. The voluntary recovery rate is as high as 60% or more within one year of experiencing a trauma incident, but it drops sharply after that. Therefore, initial treatment within one year after experiencing a traumatic event is very important in recovering from the trauma aftereffect [2]. For initial treatment, it is essential to visit and consult a hospital to diagnose trauma. However, there

M. Singh et al. (Eds.): IHCI 2020, LNCS 12615, pp. 208–217, 2021.
https://doi.org/10.1007/978-3-030-68449-5_21

are many cases in which people postpone treatment due to social prejudice about mental disorders or fail to diagnose and treat trauma because they do not recognize it.

In recent years, deep learning has been used to combine engineering technology and medical fields to help doctors in early diagnosis. In particular, voice is widely used because it contains emotions and intentions that are effective in grasping the patients' emotions and can be obtained in a non-contact manner in a natural environment without patients feeling rejected.

In the paper of So et al., gender and age were classified through the voice of Seoul-language speech corpus dataset. The age classification system based on the deep artificial neural network showed 78.6% accuracy, 26.8% higher than the random forest model, one of the traditional machine learning methods [3]. In the research of Choee et al., a study on recognizing emotions through voice was performed. Voice data was transformed into an image through a mel-spectrogram and used as an input to the CNN model. Accuracy was improved through transfer learning that reuses pre-trained weights by DCASE2019 data [4].

Although many studies using voice, such as age classification, emotion recognition are being conducted, studies to screening trauma using voice analysis are not in progress. Therefore, our paper aims to screening trauma through voice data using CNN, one of the deep learning method.

2 Methods

2.1 Definition of Trauma Emotions

Paul Ekman defined six basic emotions such as happy, sad, disgust, angry, fear, and surprise [5]. In our study, only four emotions (happy, neutral, sad, and fear) are used for screening trauma. The reason is that when people are traumatized, they feel a lot of fear, and over time after trauma, the intensity of sadness stronger and they often feel depression. In addition, fear, happiness, and horror were used in this study as criteria for judging the emotional change of trauma [6]. In the early days of trauma, feeling of fear, sad, surprise, and angry are remarkable, and as time passes, the angry weakens and fear and sadness become stronger [7].

Based on these studies, in this paper, it is assumed that when fear and sad emotions appeared in the voice audio, the probability of trauma was high and when neutral and happy emotions appeared, the one was low.

2.2 Voice Audio Dataset

The dataset is a Korean voice dataset made by extracting voice containing six basic emotions from domestic broadcasts and movies. After collecting voice data representing a specific emotion, it was refined for verification. After the researchers, who are the authors of this paper, sufficiently collected the voice data corresponding to the four emotions, a second verification process was performed by six researchers. Only the voice data evaluated with the same emotion was finalized and used as the corresponding emotion data. With voice data of 100 people (male: 40 people, female: 60 people), each voice is 2 to 11 s long. There are a total of 600 wav files, 100 for each emotion. Only four of these emotions (fear, sad, happy, neutral) are used.

After pre-processing the data (pre-processing part in Sect. 2.3) built with basic emotion, it was composed of a data set capable of binary classification as shown in Table 1. This data set was used to train the CNN model. As shown in Table 1, the entire data set consists of two classes, one with and without trauma. Non-trauma is composed of neutral and happy emotions, trauma is composed of sad and fear emotions. There are 6,941 training data, 2,318 verification data, and 2,317 test data. About 60% of the data were classified as training data, 20% as verification data, and 20% as test data. Non-trauma voices were 5,534 and trauma voices were 6,041. The data set for the CNN model was constructed by dividing the training data, verification data, and test data for each class by a ratio of 6:2:2.

Table 1. Dataset configuration.

	Train	Valid	Test	Total
Non-trauma	3,318	1,108	1,108	5,534
Trauma	3,623	1,210	1,209	6,042

2.3 CNN Training

This section describes how to screen for trauma in voice. Figure 1 is a diagram of the whole process. (a) and (b) are the pre-processing steps to put the voice into the CNN model, (c) is training the model, and (d) is the final step to improve the accuracy of the final result. A detailed description of each step will be given below.

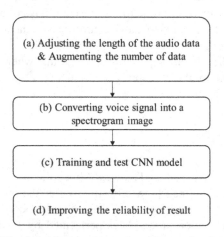

Fig. 1. Diagram of the entire process of screening for trauma.

Pre-processing

Pre-processing is carried out to put data into the input of the CNN model. The information contained in the voice cannot be easily obtained from the voice signal itself and can be extracted through mathematical signal processing. A function called Fourier transform can be used to obtain a spectrum indicating how many speech fragments (frames) of a given length of time have each frequency component. The human ear also has a similar mechanism to obtain information inherent in the heard sound by extracting each frequency component in the sound [8]. Therefore, instead of using the voice data as it is, it goes through a two-step pre-processing process to learn after making it into a spectrum.

The first pre-processing is a process to eliminate the difference in length between data and increase the number of data. Shift the voice in 0.1 s and cut it in 2 s (see Fig. 2). The second pre-processing is a process of converting the one-dimensional sound data that has been pre-processed into a STFT (Short-Time Fourier Transform) spectrogram, which is two-dimensional image data. STFT is a method of dividing time series data into predetermined time intervals and then Fourier transforming data of each interval. Set the sampling rate to 1024, and FFT the 2 s data. Then, the 512 samples are overlapped and shifted. After the second pre-processing, use the Min-Max Scaler to scale all data values between 0 and 1. At the end of all preprocessing, the audio data becomes a spectrogram image of 288 pixels in height and 432 pixels in width as shown in Fig. 3.

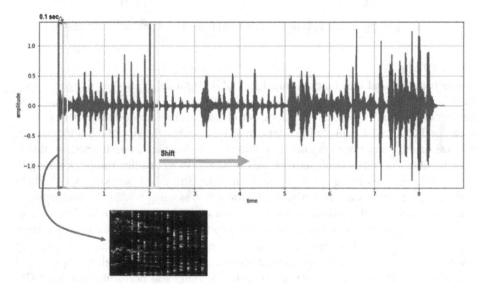

Fig. 2. The concept of spectrogram extraction based on sliding window through voice data preprocessing.

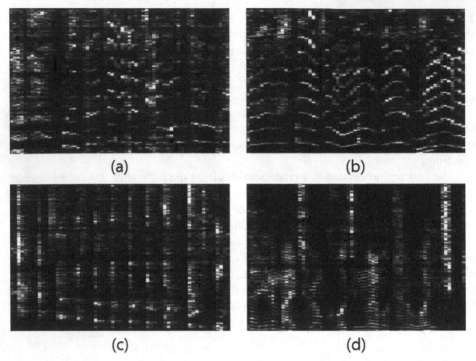

Fig. 3. Spectrogram example of voice data. (a) fear, (b) sad, (c) happy, (d) neutral.

The CNN Model

In this paper, Visual Geometry Group 13 (VGG-13) model is used as image training model. The VGG neural network model is a model studied to find out how much the number of layers affects the classification performance. Previous neural network models used 7×7 or 11×11 convolution filters. In this way, when the size of the filter is large, the size of the input image is quickly reduced while learning is performed, so that it is impossible to create a deep network. Accordingly, VGG enables smooth learning in 16 to 19 deep layers using a 3×3 filter. When the Stride is 1, stacking 3×3 filters twice can process the same receptive field information as using 5×5 convolution filters once. And stacking three times is the same as using 7×7 convolution filters once. Since information processing of receptive fields is the same, a network can have a large number of layers [9]. This has the advantage of having more nonlinearity and having fewer parameters by performing multiple nonlinear processing through multiple 3×3 convolution operations. Therefore, in this paper, training was performed using a VGG neural network model with 13 layers. The architecture of the model is shown in Fig. 4.

As the input image, a pre-processed (288, 432, 3) size spectrogram image is used. VGG-13 contains ten convolution layers with 3×3 kernel. After each two convolution layers, an max pooling layer with 2×2 kernel is inserted. This process is repeated a total of 5 times to train, and the depth of the convolution layer is in the order of 64, 128, 256, 512. And through the fully connected layer, the binary classification of screening trauma is output.

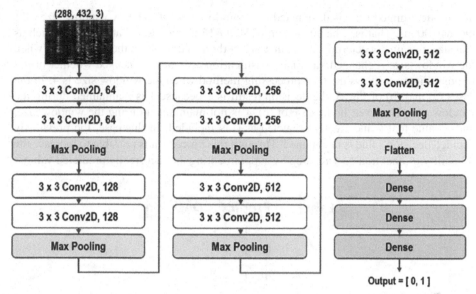

(288, 432, 3)

3 x 3 Conv2D, 64	3 x 3 Conv2D, 256	3 x 3 Conv2D, 512
3 x 3 Conv2D, 64	3 x 3 Conv2D, 256	3 x 3 Conv2D, 512
Max Pooling	Max Pooling	Max Pooling
3 x 3 Conv2D, 128	3 x 3 Conv2D, 512	Flatten
3 x 3 Conv2D, 128	3 x 3 Conv2D, 512	Dense
Max Pooling	Max Pooling	Dense
		Dense

Output = [0, 1]

Fig. 4. The architecture of VGG-13 model (ten convolution layers, five pooling layers, three dense layers).

The other compared models will only briefly explain the architecture. The Layer-14 model is a modified version of the paper [10] and has 14 convolutional layers. After passing through the convolution layer with a 5×5 kernel, the convolution layers with 3×3, 1×1, and 5×5 kernel are trained and concatenate simultaneously. After repeating this process 3 times, the convolution layers with 1×1 and 5×5 kernel is sequentially trained, and then the fully connected layer is passed. There are max pooling layers between the convolution layers. The VGG-16 model consists of 13 convolutional layers with a 3×3 kernel and 3 fully connected layers. The depth of the convolutional layer is increased in the order of 64, 128, 256, 512, 512, and max pooling layer is included whenever the depth is changed. The last model, Residual neural network 50 (resnet-50), consists of 49 convolution layers and one fully connected layer. A convolution layer with a 7×7 kernel is passed, and 3 convolution layers each with a 1×1, 3×3, and 1×1 kernel are repeated 16 times, followed by an fc layer.

In all models, binary cross-entropy is used as the loss function and RMSprop is used as the optimization function. At this time, the learning rate was set to 2×10^{-5} and the epoch was set to 300. To prevent overfitting due to repeated learning processes, an early stopping technique was used to stop learning when the loss did not improve more than 10 times. In all experiments, TensorFlow was used as a deep learning framework, and NVIDIA GeForce RTX 2080 Ti was used as the GPU.

Post-processing
To increase the accuracy, the results of VGG-13 are post-processed. Post-processing improves the reliability of the final result by determining whether or not there is trauma when the result of VGG-13 remains constant for a certain period of time. At this time,

the pre-determined time is designated as a window size of 2-10. A window size of 1 means 100 ms. That is, when the result of VGG-13 is maintained at 0 or 1 as much as that size, it is finally determined whether or not there is trauma. In the experiment, when the window size is changed, the final classification accuracy is checked whether or not trauma. Figure 5 shows the post-processing method when the window size is 4. In the cases (a) and (b) of Fig. 5, the decision result is 1 because 1 is maintained during the window size. However, in the case of (c), 1 is not maintained during the window size, and a value that is incorrectly predicted as 0 is included. In this case, the previously decided decision value is maintained. Through this process, it is possible to increase the reliability of final trauma screening by supplementing the incorrectly predicted value.

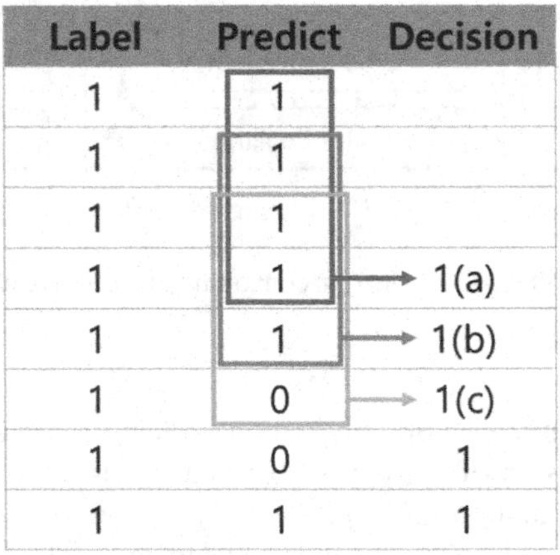

Label	Predict	Decision
1	1	
1	1	
1	1	
1	1	1(a)
1	1	1(b)
1	0	1(c)
1	0	1
1	1	1

Fig. 5. Concept of post-processing (when window size is set to 4).

3   Results

There are four neural network models used for training, and all of them were conducted in the same computer environment. As mentioned in 2.3, the training parameters of each model were all set identically (loss function is binary cross entropy, optimizer function is RMSprop, and learning rate is 2×10^{-5}), and training was performed with data sets obtained through preprocessing. Performance evaluation was performed using test data for the four trained models, and the performance of each model is shown in Table 2. The performance evaluation shows that the performance varies depending on the layer depth of the model, and the VGG-13 model showed the highest performance. Therefore, in this paper, the VGG-13 model was adopted and used.

In the case of training and testing with the VGG-13 model, the accuracy is 98.96%. Figure 6 is the receiver operating characteristic (ROC) curve of the VGG-13 model. The horizontal axis of the ROC curve is a false positive rate, which indicates the rate at which non-trauma people are predicted as trauma. The vertical axis is a true positive rate, which means the rate at which trauma people are predicted as trauma. The ROC

Table 2. Test accuracy of four models.

Model	Test accuracy
VGG-13	98.96%
Layer-14	98.49%
VGG-16	96.42%
Resnet-50	87.35%

curve is almost in contact with its axis, so it can be seen that the model made shows high performance. At this time, the confusion matrix for the two classes is shown in Fig. 7. In the case of predicting trauma as non-trauma (False Negative) or non-trauma as trauma (False Positive), 24 out of a total of 2,317 were accounted for. The error rate of the VGG-13 model can be calculated as $24/2317 = 0.010$. And this allows for the sensitivity and specificity of the trauma screening. The sensitivity of how well it can screen cases with trauma is 98.92%, and the specificity of how well it can screen non-trauma cases is 99.92%.

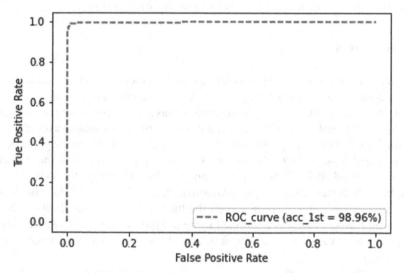

Fig. 6. ROC curve in case of using VGG-13

In the case of training and testing with the VGG-13 model and not performing post-processing, it is the same as when the window size is 1, and the accuracy is 98.96%. But the accuracy is 100% when post-processed between the 4-8 sized windows. In other words, it can be regarded that trauma can be accurately screened through the voice for 400 to 800 ms, not the voice for 100 ms. If the window size is less than 4, the wrong value becomes the final result and the accuracy falls down. And if the window size is larger than 8, the incorrect value confuses the correct value, so accuracy is poor.

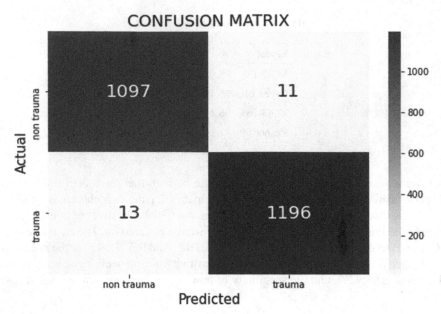

Fig. 7. Confusion matrix of VGG-13 model before post-processing.

4   Conclusions

In this study, an experiment was conducted to classify voice emotions by constructing a CNN model that screen trauma through voice. A dataset containing 4 emotions (fear, sad, happy, neutral) was used. The first pre-processing was performed to increase the number of data by aligning the voice with 2 s, and the second pre-processing was performed to transform the voice into the spectrogram image through STFT. When the pre-processed data was trained through the CNN models VGG-13, layer 14, VGG-16, and Resnet-50, the accuracy was 98.96%, 98.49%, 96.42%, and 87.35%. Therefore, the VGG-13 model with the best performance was adopted. After that, the results of the VGG13 model were post-processed to increase the reliability of the final results. As a result, it can be seen that when post-processing was performed using a window size of 4 to 8, the accuracy was 100%, which was higher than the accuracy of 98.96% when post-processing was not performed.

In future works, we will construct a multi-classification model that categorizes trauma-related four emotions (fear, sad, happy, neutral) rather than binary classification for trauma.

Acknowledgement. This work was supported by the Industrial Strategic Technology Development Program (No. 10073159) funded by the Ministry of Trade, Industry & Energy (MI, Korea).

References

1. Lee, B.S.: Nature and ethics inherent in the trauma of division. J. Epoch Philos. **22**(1), 153–183 (2011)
2. Ahn, H.N.: Recent trend of trauma treatment. J. Korean Psychol. Assoc. **2014**(1), 162 (2014)
3. So, S.W., et al.: Development of age classification deep learning algorithm using Korean speech. J. Biomed. Eng. Res. **39**(2), 63–68 (2018)
4. Choee, H.W., Park, S.M., Sim, K.B.: CNN-based speech emotion recognition using transfer learning. Int. J. Korean Inst. Intell. Syst. **29**(5), 339–344 (2019)
5. Ekman, P.: An argument for basic emotions. Cogn. Emot. **6**(3–4), 169–200 (1992)
6. Amstadter, A.B., Vernon, L.L.: Emotional reactions during and after trauma: a comparison of trauma types. J. Aggress. Maltreatment Trauma **16**(4), 391–408 (2008)
7. Center for Substance Abuse Treatment: In Trauma-Informed Care in Behavioral Health Services. Substance Abuse and Mental Health Services Administration, US (2014)
8. Kakao enterprise. https://tech.kakaoenterprise.com/66. Accessed 26 Oct 2020
9. Simonyan, K., Zisserman, A.: Very deep convolutional networks for large-scale image recognition. arXiv preprint arXiv:1409.1556 (2014)
10. Park, D.S., Bang, J.I., Kim, H.J., Ko, Y.J.: A study on the gender and age classification of speech data using CNN. J. Korean Inst. Inf. Technol. **16**(11), 11–21 (2018)

Interchanging the Mode of Display Between Desktop and Immersive Headset for Effective and Usable On-line Learning

Jiwon Ryu and Gerard Kim[✉]

Digital Experience Laboratory, Depth of Computer Science and Engineering,
Korea University, Seoul, Korea
{nooyix,gjkim}@korea.ac.kr

Abstract. In this new era of "untact", the on-line solutions for social gatherings and meetings have become important more than ever. In particular, immersive 3D VR based environments have emerged as a possible alternative to the video based tele-conferencing solutions for their potential improved learning effects. However, the usability issues stand in its way for it to be fully spread and utilized by the mass. In this poster, we study for another alternative, the mixed usage of desktop (or equally 2D video tele-conferencing) and immersive modes as necessary. The desktop mode is convenient for the user and allows the usual interaction in the most familiar way (e.g. note taking), while the immersive mode offers a more direct contact and interplay among the teacher and other learners. We conducted a pilot experiment comparing the user experience among the three modes: (1) desktop, (2) I3D VR/HMD and (3) mixed. For the mixed mode, we designed a flexible fixture for conveniently wearing/donning the head-set and switch between the desktop and I3D mode quickly. Partici- pants much preferred the mixed mode usage due to the aforementioned ad- vantages of bringing the bests of both worlds.

Keywords: On-line learning · Immersive 3D VR environment · Video based tele-conferencing · Educational effect · User experience

1 Introduction

With the advent of the COVID-19 and the era of "untact", the on-line solutions for social gatherings and meetings have become important more than ever. Most academic classes are now being conducted on-line mostly using the video tele-conference (VTC) solutions such as the likes of Zoom, Webex, Google Teams, and Microsoft Meets [4]. At the same time, there are much worries as to the educational effects of such classes conduct on-line in the long term [3]. As such, immersive 3D (I3D) VR on-line environments have emerged as a possible alternative [1]. I3D environments depict and decorate the classroom in picturesque 3D, and provides higher presence of the lecturer and colleagues, natural interaction with one another, and some degree of physical manipulation of/with educational props/materials, i.e. an atmosphere and functions conductive to a better learning (at least for certain types of classroom learning (See Fig. 1). Moreover, for

© Springer Nature Switzerland AG 2021
M. Singh et al. (Eds.): IHCI 2020, LNCS 12615, pp. 218–222, 2021.
https://doi.org/10.1007/978-3-030-68449-5_22

large scale classes, due to the heavy network traffic, the video streaming of students is often turned off. I3D rendering of students as avatars is much less computationally and network-wise burden-some. In most cases, the I3D environment solutions offer both regular desktop and head-set modes. The true advantages of the I3D environment would be reaped with the use of head-set, however, the head-sets not only incur extra cost, but also are quite inconvenient and difficult to wear and use - they are heavy and stuffy; stereoscopic rendering will most likely introduce sickness; effective interaction is limited with the closed view unless a separate controller is used. Thus, in summary, despite the prospective advantages, these usability problems stand in the way of utilizing I3D solutions to their fullest effect.

Fig. 1. Video tele-conferencing solution for classroom learning (left) and immersive 3D VR based environment – hubs.mozilla (right).

In this poster, we study for another alternative, the mixed usage of desktop (or equally 2D video tele-conferencing) and immersive modes as necessary. The desktop mode is convenient for the user and allows the usual interaction in the most familiar way (e.g. note taking), while the immersive mode offers a more direct contact and interplay among the teacher and other learners. We conducted a pilot experiment comparing the user experience among the three modes: (1) desktop, (2) I3D VR/HMD and (3) mixed. For the mixed mode, we designed a flexible fixture for conveniently wear- ing/donning the head-set and switch between the desktop and I3D mode quickly.

2 Related Work

Our study was inspired by the Hewlett Packard PC set-up, called the "VR snacking", which combined the PC with a VR station on a stand which made it easy to pick up, use, and the head-set back down. In this particular case, the computer aided design was the main application for which both desktop interfaces for the modeling activities and 3D immersive visualization were needed in the work flow [2].

Yoshimura and Borst investigated the desktop viewing and the headset viewing for remote lectures using the hubs.mozilla, a 3D immersive VR environment, for social gatherings [1]. The comparative evaluation revealed that with regards to the presence/co-presence, usability and sickness, overall participants preferred to use the desktop and the head-set mode as needed. This study is similar except we have redesigned the head-set stand (e.g. vs. one introduced by [2]) to fit the needs of on-line learning and assessed the user experience with regards to learning (vs. CAD work-flow).

3 Head-Set Stand Design

Figure 2 shows the head-set stand design. It was designed to be placed on the desk beside the desktop PC. Unlike the Hewlett Packard design, in ours, the headset is basically designed without detaching the headset from the stand most of the times (the option to completely detaching it is still available). This was to further remove the need to wear and un-don the headset. Instead, the user is to fit one's head to the head- set on the stand. To make this process as easy as possible, mechanical joints were used to provide 3.5 degrees of freedom and adjust the position (height, limited degree in the right/left) and orientation (pitch/yaw) of the head set in a compliant fashion.

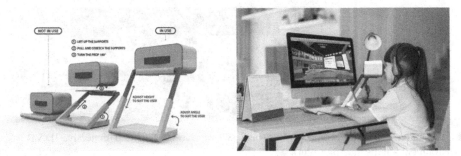

Fig. 2. The head-set stand design (left), and the illustration of mixed mode usage (right).

4 Pilot Experiment and Results

To validate the effects of our proposed bi-modal mode usage and associated design solution, we ran a small pilot experiment comparing the user perceived experiences among three modes of immersive 3D VR based on-line learning environment. Four participants (graduate students of age between 25 and 29) were recruited and instruct- ed to view and use the hubs.mozilla I3D environment [5] to listen to a 5 min lecture in three different modes (presented in a balanced order): (1) desktop mode, (2) HMD/VR mode and (3) mixed mode. After experiencing the respective modes, the participants were asked to fill out a UX survey that assessed the (1) user's perceived learning effect, (2) general usability, and (3) immersion/presence/interest.

The survey results (Figs. 3, 4 and 5) revealed the mixed mode exhibiting a usability level similar to that of the desktop. However, user's perception of the learning effect was greater with the use of I3D VR, either on its own or with the mixed mode. The VR/HMD mode was rated the lowest in terms of the general usability as expected due to the inconvenience of the head-set usage.

The post-briefings also reflected the general survey results: while the stand design and mixed mode helped users overcome the basic head-set only usability problem, subjects commented that immersive environment needed improvement in terms of not making any distraction with the non-essential objects within. There were interactive activities such as note taking needed across the three nodes, yet only practically possible in the

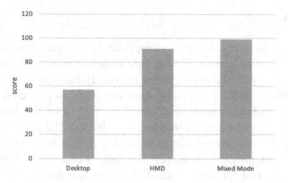

Fig. 3. The user perceived learning effects among the three modes.

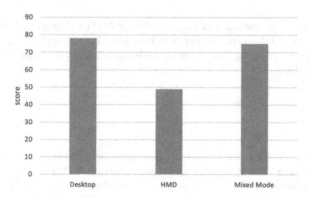

Fig. 4. The general usability among the three modes.

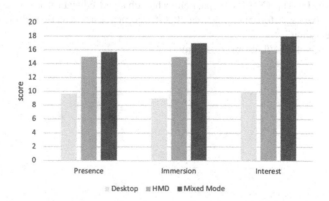

Fig. 5. Presence/Immersion/Interest levels among the three modes

desktop mode. There were subjects who felt disoriented in the process of switching between the desktop and VR/HMD mode.

5 Conclusion

In this poster, we conducted a comparative experiment to compare the educational effects and other associated user experiences of three different modes of on-line learning – (1) desktop only, (2) VR/HMD only and (3) mixed. The mixed mode offered the best balance for usability with the competitive learning effect as expected. Further design improvements must be further made to tailor it to the task of on-line learning (e.g. more movement flexibility and associated interactive activities. The experiment was only a small pilot one and the educational effects were only judged from the user's perception rather than through quantitative measures. We believe that the I3D environment will make more sense not just for lecture oriented classes but also for classes that require closer teacher-student interplay and 3D object manipulation (e.g. educational materials props).

Acknowledgement. This research was supported by the MSIT (Ministry of Science and ICT), Korea, under the ITRC (Information Technology Research Center) support program (IITP2020-2016-0-00312) supervised by the IITP (Institute for Information & communications Technology Planning & Evaluation).

References

1. Yoshimura, A., Borst, C.: Evaluation and comparison of desktop viewing and headset viewing of remote lectures in VR with Mozilla hubs. Pers. Commun. (2020)
2. Lang, B.: HP Wants to Enhance Traditional CAD Workflows with "VR Snacking" (2020). https://www.roadtovr.com/hp-vr-snacking-concept-cad-workflows/
3. Kinlaw, C., Dunlap, L., D'Angelo, J.: Relations between Faculty Use of Online (2012)
4. Li, C., Lalani, F.: The COVID-19 pandemic has changed education forever. This is How (2020). https://www.weforum.org/agenda/2020/04/coronavirus-education-global-covid19-onl ine-digital-learning/
5. Hubs by Mozilla. https://hubs.mozilla.com/

Verification of Frequently Used Korean Handwritten Characters Through Artificial Intelligence

Kyung Won Jin[1] (iD), Mi Kyung Lee[1] (iD), Woohyuk Jang[2] (iD), and Eui Chul Lee[3]([✉]) (iD)

[1] Department of Artificial Intelligence and Informatics, Graduate School,
Sangmyung University, Seoul, Republic of Korea
[2] Department of Computer Science, Graduate School, Sangmyung University,
Seoul, Republic of Korea
[3] Department of Human-Centered Artificial Intelligence, Sangmyung University,
Seoul, South Korea
eclee@smu.ac.kr

Abstract. Handwriting verification is a behavioral biometric that matches hand-written characters to determine whether it is written by the same person. Because each person has a different handwriting, it is used by investigative agencies for the purpose of presenting court evidence. However, it cannot be defined as a rule because the standards for visual reading of experts are ambiguous. In other words, different experts can make different decisions for the same pair. Therefore, we propose a handwriting verification method based on artificial intelligence that excludes human subjectivity. For 4 frequently used Korean characters, genuine or imposter pairs of the same character were trained with a Siamese-based ResNet network. The verification accuracy for the trained model was about 80%. Through this experiment, the objectivity of handwriting biometric through deep learning was confirmed, and a basis for comparison with verification performance through human eyes was prepared.

Keywords: Handwriting verification · Behavioral biometric · Siamese network · ResNet · Artificial intelligence

1 Introduction

Handwritten characters has unique characteristics that differ from person to person. Therefore, handwritten characters can be used as a means of expressing themselves, which means that they have value as a biometric for identity authentication [1].

Various offline signature verification methods using artificial intelligence have been studied before. There are several deep learning-based signature verification methods, including verification using two channel CNN [2], using the geometric features of offline signatures using ANN [3], and recently, a paper that has increased performance by 99% by using CNN [4]. The characteristics of the above papers were studied and tested based on individual signatures. Individual signatures are specialized to individuals because

© Springer Nature Switzerland AG 2021
M. Singh et al. (Eds.): IHCI 2020, LNCS 12615, pp. 223–228, 2021.
https://doi.org/10.1007/978-3-030-68449-5_23

they include the characteristics of only individuals and are unique signatures that no one uses. On the other hand, the characters we use to measure accuracy are more difficult than previous studies because they consist of frequently used characters. This paper tries to show that even with words with few characteristics, the accuracy of verification can be maintained and objectivity can be expressed.

We implemented the training network using the siamese network and ResNet so that we could compare the two images. The siamese network is a typical one-shot learning that allows training with two image pairs to output verification accuracy, and the use of a small amount of data can predict image pairs that have never been learned. Put the genuine image pair and the impostor image pair into the network to learn each pair from the same ResNet, and the siamese network calculates the distance of each pair's feature vector and learns the network. We made it possible to determine whether handwriting is of the same person or not by inputting two images that should be compared through the learned network.

To increase the training performance of deep neural network, a large number of images of genuine and imposter pairs are needed. Therefore, we increased the number of data by data augmentation.

Previously, handwriting verification varies in performance depending on the experts and the accuracy is not always constant. Therefore, many existing studies have tried to use artificial intelligence to expect objectivity or constant performance. In conducting experiments with similar purposes, we also try to demonstrate the consistent performance and objectivity of Korean handwritten characters verification through deep learning.

2 Implementation

2.1 Data Set

For handwriting verification training, twenty adult men and women write four frequently used Korean characters such as '없', '김', '다', and '이' ten times, and we scanned the paper to convert them into a digital images. From the digital image, each character was cropped through a lab-made tool and size normalized to 112×112 pixels. In this process, aspect ratio was excluded from the normalization factor because it can be characteristic factor of personal handwriting. To perform the training of deep neural network through a small amount of data, data augmentation was performed. Data augmentation was performed by randomly determining among three geometric transformation factors such as rotation ($-5°$–$+5°$), translation (within horizontally & vertically 3% of image size), and scaling up & down (uniformly 10% of image size). Through this, the number of character images increased 9 times for the training and validation set and 5 times for the test set. Therefore, the training, validation, and test sets consisted of 61, 29, and 50 images per character, respectively.

Due to the characteristics of the siamese network, data is input in pairs. Therefore, we composed the data with a genuine set, a pair of characters written by the same person, and an imposter set, a pair of characters written by different people. In here, since the number of imposter sets is much larger than the number of genuine sets, we matched the number of two sets for the training balance for binary decision. As a result, the number of data used was composed of 54,900 (about 70%) for training, 12,180 (about 15%) for validation, and 12,250 (about 15%) for testing.

2.2 Network

In this paper, ResNet model and siamese network were used in handwriting verification. The ResNet is a method of training by maintaining input information on each block through the residual block [5]. Figure 1 shows a form that adds the input x that has passed through the weight layer and the input x that has not gone through the weight layer. This residual block solves the problem of vanishing gradients when the network is deep.

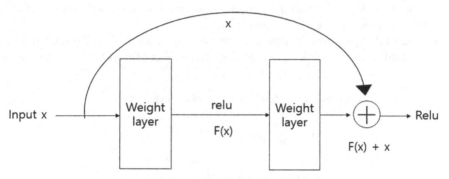

Fig. 1. The structure of the residual block in the ResNet.

The siamese network is a one-shot learning network that determines if it corresponds to that class even if it has few training cases [6]. We used the point to give two handwriting and learn how similar they were. In Fig. 2, the siamese network receives two pairs of images, each images extracts the features through 34 layers of ResNet-34 of the same structure, and the difference between the two feature vectors is obtained through L1 norm. The obtained value is expressed as a probability between 0 and 1 through sigmoid. The SGD optimizer was used for training and drop out was not used.

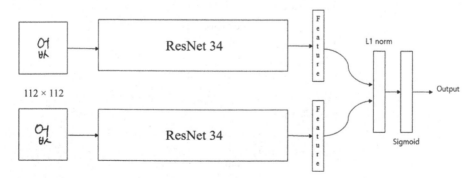

Fig. 2. The structure of siamese network.

3 Results and Discussion

In order to check whether our experiment produces promising results even for characters with a small number of strokes that do not have characteristics unlike signatures or complex characters, experiments were performed according to the complexity of the characters. Therefore, the experiment was conducted according to the number of strokes of the characters. The character used in the experiment were '없', '김', '다', and '이' and the number of strokes was 9, 5, 4, and 2, respectively. To show the reliability of the model performance, training and test were conducted in k-fold method for each character. By using the k-fold method, we can confirm that our training process did not occur overfitting for a particular set of data. The k-fold test methods support the reliability of the experiment because it trains all datasets and tests them with test sets that are not included in the training dataset. By setting k to 4, we were able to expand the number of inputs in the training and test data.

The network's performance measurement consists of a set of 12,250 random tests, and the k-fold method (with k = 4) is used, so the number of tests for each character is

Table 1. Confusion matrices of each characters.

'없' Total = 49,000		Predict label	
		Genuine	Imposter
True label	Genuine	21,447 (43.77%)	3,053 (6.23%)
	Imposter	4,172 (8.51%)	20,328 (41.49%)
'김' Total = 49,000		Predict label	
		Genuine	Imposter
True label	Genuine	20,626 (42.09%)	3,874 (7.91%)
	Imposter	4,982 (10.17%)	19,518 (39.83%)
'다' Total = 49,000		Predict label	
		Genuine	Imposter
True label	Genuine	20,404 (41.64%)	4,096 (8.36%)
	Imposter	6,241 (12.74%)	18,259 (37.26%)
'이' Total = 49,000		Predict label	
		Genuine	Imposter
True label	Genuine	19,728 (40.26%)	4,772 (9.74%)
	Imposter	4,767 (9.73%)	19,733 (40.27%)

49,000. Table 1 shows the confusion matrix for each Korean character. The results show that the more complex the characters, the better the performance, but the average performance is 80 percent with no significant difference. The average accuracy is 81.65%, and it has an accuracy of more than 80% even for characters with very few strokes.

Figure 3 shows the ROC curve for '없', which has the highest classification accuracy. It can be seen that about 15% of EER was observed.

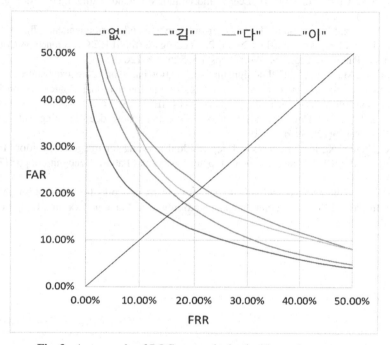

Fig. 3. An example of ROC curve obtained with equal error rate.

4 Conclusion

In this paper, we proposed the method of handwriting verification of Korean characters through the ResNet network with the siamese network structure and confirmed the verification performance. It has been shown that deep learning-based handwriting verification can be used as a more objective verification method as a way to exclude a person's subjectivity.

In this study, the verification accuracy of handwriting biometrics for individual characters was measured and feasibility was confirmed. In future works, we plan to evaluate the performance as a multi-modal biometric by fusion of similarity scores for several characters. In addition, handwriting verification will be developed to calculate the likelihood ratio when determining genuine and imposter to be used as legal evidence.

Acknowledgement. This research was supported by the Bio & Medical Technology Development Program of the NRF funded by the Korean government, MSIT(NRF-2016M3A9E1915855).

References

1. Srihari, S.N., Cha, S.H., Arora, H., Lee, S.: Individuality of handwriting. J. Forensic Sci. **47**(4), 1–17 (2002)
2. Berkay Yilmaz, M., Ozturk, K.: Hybrid user-independent and user-dependent offline signature verification with a two-channel CNN. In: Proceedings of the IEEE Conference on Computer Vision and Pattern Recognition Workshops, pp. 526–534 (2018)
3. Chandra, S., Maheskar, S.: Offline signature verification based on geometric feature extraction using artificial neural network. In: Proceedings of the International Conference on Recent Advances in Information Technology, pp. 410–414. IEEE (2016)
4. Alajrami, E., et al.: Handwritten signature verification using deep learning. Int. J. Acad. Multidisc. Res. **3**(12), 39–44 (2019)
5. He, K., Zhang, X., Ren, S., Sun, J.: Deep residual learning for image recognition. In: Proceedings of the IEEE Conference on Computer Vision and Pattern Recognition, pp. 770–778 (2016)
6. Koch, G., Zemel, R., Salakhutdinov, R.: Siamese neural networks for one-shot image recognition. In: Proceedings of the International Conference on Machine Learning Deep Learning Workshop (2015)

Study of Sign Language Recognition Using Wearable Sensors

Boon Giin Lee[1(✉)] 🆔 and Wan Young Chung[2] 🆔

[1] The University of Nottingham Ningbo China, Ningbo 315100, China
boon-giin.lee@nottingham.edu.cn
[2] Pukyong National University, Busan 48513, Korea
wychung@pknu.ac.kr

Abstract. Sign language was designed to allow hearing-impaired person to interact with others. Nonetheless, sign language was not a common practice in the society which produce difficulty in communication with hearing-impaired community. The general existing studies of sign language recognition applied computer vision approach; but the approach was limited by the visual angle and greatly affected by the background lightning. In addition, computer vision involved machine learning (ML) that required collaboration work from team of expertise, along with utilization of high expense hardware. Thus, this study aimed to develop a smart wearable American Sign Language (ASL) interpretation model using deep learning method. The proposed model applied sensor fusion to integrate features from six inertial measurement units (IMUs). Five IMUs were attached on top of the each fingertip whereas an IMU was placed on the back of the hand's palm. The study revealed that ASL gestures recognition with derived features including angular rate, acceleration and orientation achieved mean true sign recognition rate of 99.81%. Conclusively, the proposed smart wearable ASL interpretation model was targeted to assist hearing-impaired person to communicate with society in most convenient way possible.

Keywords: Deep learning · Human computer interaction · Sensor fusion · Sign language · Wearable

1 Introduction

Hearing impaired person were incapable to express themselves and interact with others verbally. As the result, sign language had become the primary source of non-verbal medium for hearing-impaired community. Sign language was a form of language expressed by the movements of fingers, hands, arms, head, body and also relied on the facial expression [1]. Specifically, American sign language (ASL) was one of most frequently targeted sign language among most published

Supported by the Faculty Inspiration Grant of The University of Nottingham Ningbo China.

© Springer Nature Switzerland AG 2021
M. Singh et al. (Eds.): IHCI 2020, LNCS 12615, pp. 229–237, 2021.
https://doi.org/10.1007/978-3-030-68449-5_24

papers from year 2007 to 2017 as principal research target [2]. Generally, the ASL's fingerspelling signs were static which did not involved any hand and fingers movements except for sign letters J and Z. Conversely, most word-based ASL signs were dynamically presented which involved a series of either basic or complex hand movements with combination of fingers shapes and patterns. Two approaches were widely adopted in the area of sign language recognition: vision and sensor. Vision-based method applied computer vision technique to analyze the hand gestures, body and potentially facial expressions from frames via camera [3]. Sensor-based method performed analysis on fingers and hands patterns (motion, position and velocity) via sensors that were attached on the body [4].

Elmezain *et al.* [5] used Bumblebee stereo camera in their study to detect isolated gestures (static hand position) and meaningful gestures (continuous hand motion) with Hidden Markov Model (HMM) which achieved mean recognition rate of 98.6% and 94.29% respectively. Molchanov *et al.* [6] proposed a 3D-based Convolution Neural Network (CNN) method to recognize 10 dynamic hand gestures via color camera, depth camera and short-range radar sensor with recognition rate of 94.1%. Many works were also found to adopt leap motion controller (LMC) and Kinect for hand gestures tracking. Chai *et al.* [7] presented a word- and sentenced-based Chinese sign language (CSL) recognition with accuracy rate of 96.32% using Kinect. Similarly, Yang *et al.* [8] proposed a 24 word-based sign language recognition method by applying hierarchical conditional random (CRF) from 3D hand gestures from Kinect. Meanwhile, Chong *et al.* [9] applied deep neural network (DNN) for ASL recognition using LMC. The study showed significant improvement of true recognition rate from 72.79% to 88.79%.

In sensor-based method, Preetham *et al.* [10] developed two gloves, attached with flex sensors on top of each two joints for each finger. Similarly, Patil *et al.* [11] developed a single glove with 5 flex sensors and mapped the flexion of fingers into 3 categories: completed bend (finger closed), partial bend (finger half closed/opened) and straighten (finger opened). Alternatively, Lee *et al.* [4] improved the design with addition of two pressure sensors on the middle fingertip and an IMU sensor that showed recognition rate improvement from 65.7% to 98.2%. Also, Mummadi *et al.* [12] placed 5 IMUs on each fingertip to acquire pitch, raw and yaw of finger's movement to detect 24 static ASL signs with true recognition rate of 92.95% using random forest (RF).

In general, vision-based approach is non-invasive but had field of view limitation. The frames were also highly affected by environmental issue. On the contrary, sensor-based approach offered higher freedom of movement and mobility without field of view restriction, but the drawback was bulky design of wearable device. This contributions of this study included 1) designed of light and novel wearable device to overcome the bulky design issue, 2) extending the previous work [4] with addition of word-based signs recognition, 3) features analysis for distinguishing different group of words (e.g., some words had similarity in one or more component(s) but is/are distinguishable by others) and 4) improvement of the recognition model using recurrent neural network framework.

2   System Overview

The proposed wearable ASL gestures recognition method composed of 4 main modules: 1) sensing, 2) pre-processing, and 3) sign recognition as illustrated in Fig. 1. The sensing module consisted of 6 IMUs placed on top of each fingertip and back of the hand palm respectively as depicted in Fig. 2. Each IMU was wire-connected to Teensy 3.2 MCU with operating voltage of 3.3V. Two TCA9548A multiplexers were used to connect multiple IMUs to MCU. The sensors data were transmitted to terminal via BLE module.

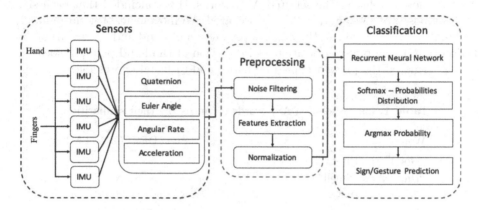

Fig. 1. Overview of proposed ASL recognition method.

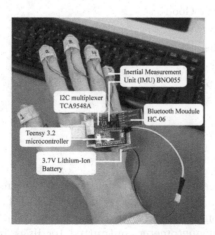

Fig. 2. Proposed ASL recognition wearable device.

3   Methods

3.1   Data Collection

Twelve subjects were recruited and were requested to sign the consent form and information sheet. Subjects were requested to practice ASL gestures in one-to-one training session (2 h) with laboratory assistant to ensure the subjects were familiar with the sign gestures. Experiments were recorded with webcam with approval from subjects under supervision of laboratory assistant. Total of 38,451 samples were collected with respect to 27 word-based ASL gestures. Table 1 showed a list of the selected ASL words. [13] concluded that each sign was composed of 1) signation (SIG), indicated the movement of hand palm, 2) designator (DEZ), signified the handshape pattern of hand and fingers, 3) orientation (ORI), the physical position and direction of the hand palm, 4) location (TAB), denoted the place occupied by hand, and 5) facial expression.

Table 1. List of selected ASL words organized by their similarities.

Words	SIG	DEZ	ORI	TAB
"Good" vs "Happy"		X		
"Happy" vs "Smell"	X	X	X	
"Sorry" vs "Please"	X		X	X
"Hungry" vs "Drink" vs "Search"		X		
"Pretty" vs "Sleep"	X	X		X
"There" vs "Me/I" vs "You" vs "Hearing"		X		
"Hello" vs "Bye"		X	X	
"Thank You" vs "Good"		X	X	X
"Yes" vs "Sorry"		X		
"Eat" vs "Water"	X			X
"Look" vs "Vegetable"		X		
"Onion" vs "Apple"	X	X	X	

3.2   Data Pre-processing

Each IMU sensor delivered outputs of acceleration (ACC, m/s^2, 100 Hz), angular rate (AGR, deg/s^2, 100 Hz) and magnetic field (MGF, μT, 20 Hz) in non-fusion mode. The sensor fusion algorithm "fused" calibrated tri-axis MEMS accelerometer, gyroscope and magnetometer, outputted readings in 3-axis Euler angle (ELA, 100 Hz) and 4-points quaternion data (QTR, 100 Hz). Mean (μ) and standard deviation (σ) were computed from sensor outputs to extract the movement patterns. The μ described the spread out pattern from of sensor outputs where low μ indicated no or nearly static movement (no gesture) and high μ signified the occurrence of large movement (encounter gesture). Meanwhile, σ observed the average distribution of sensor outputs over a window size. The features S were composed of S_i ($i = 1...6$) which represented the index of IMU on the back of the palm, thumb (fingertip), index fingertip, middle fingertip, ring fingertip and pinky fingertip respectively. The IMU data (S_i) of ACC, AGR and ELA were composed of 3-tuple dimensions (x, y and z axes) whereas QTR was composed of 4D vector with values between –1 to 1. Dataset was organized in the vector form

$$(x_j^i, y)^t; y = 1, 2...27 \tag{1}$$

where x and y denoted the j-th feature of i-th IMU and class label respectively at time t, e.g., the dataset for C_1 formed a vector of 37 features (σ and μ for each IMU in x, y and z axes with total of 6 IMUs and a class label) such that

$$(x^i, y)^t = (ACC^i, y)^t; i = 1...6 \tag{2}$$

3.3   Classification Model

Figure 3 showed the proposed deep neural network (DNN) model for ASL gestures recognition. The first layer was k-dimensional input layer followed by long-short term memory (LSTM) layer. A dropout layer (DR) was attached subsequently to randomly dropped a percentage of the neurons during the network training process. Next layer was a rectified linear activation unit (ReLU) dense layer and followed by another DR layer and dense layer. The categorical distribution of each gesture, $p(y_i)$ was computed with a softmax function

$$p(y_i) = \frac{e^{y_i}}{\sum_{j=1}^{J} e^{y_i}}; i = 1...27 \tag{3}$$

that used sparse categorical cross-entropy as the loss function. The sign was recognized with smooth approximation of argument maximum function. The DNN model was subjected to compilation using 10-fold cross validation adaptive moment estimation (Adam) algorithm as gradient descent-based optimizer with learning rate of 0.001 and decay rate of 5×10^{-5}.

4   Result and Discussion

Evaluation metric of accuracy rate (AR) (see Eq. 4) was applied to evaluate the performances of the each DNN model.

$$AR = \frac{TP + TN}{TP + TN + FP + FN} \tag{4}$$

where TP, TN, FP and FN were denoted as true positive, true negative, false positive and false negative respectively. Table 2 illustrated the recognition rates of trained DNN models using various combinations of IMUs features with 10-fold

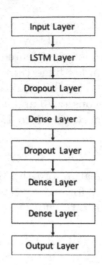

Fig. 3. Proposed deep neural network (DNN) model.

Table 2. Recognition rate based on different combinations of features using 10-fold cross validation.

Category	Feature(s)	AR (%)
C_1	QTR	99.65
C_2	ELA	99.70
C_3	AGR	99.56
C_4	ACC	99.66
C_5	QTR + ELA + AGR + ACC	99.85
C_6	QTR + ELA + AGR	99.84
C_7	QTR + ELA + ACC	99.84
C_8	QTR + AGR + ACC	99.82
C_9	ELA + AGR + ACC	99.82
C_10	QTR + ELA	99.83
$C_1 1$	QTR + AGR	99.83
$C_1 2$	QTR + ACC	99.82
$C_1 3$	ELA + AGR	99.78
$C_1 4$	ELA + ACC	99.82
$C_1 5$	AGR + ACC	99.79
Mean		99.67

cross validation. DNN model trained with all IMUs features, C_5, showed highest recognition rate of 99.85%. The DNN model trained with only AGR (C_3) had the lowest performance with recognition rate of 99.56%.

Table 3 showed the Sensitivity (Se in (5)) and Specificity (Sp in (6)) of all classes with promising recognition rates over 99%. "Drink" gesture had the lowest Se where few were falsely recognized as "hungry" or "search" gestures. Similarly, a "pretty" gesture was also falsely recognized as "sleep". Both a "thank you" and "happy" gestures were also falsely recognized as "good" gesture. A "good" gesture was also falsely recognized as "happy" gesture. Meanwhile, "yes" and "please" gestures were also falsely recognized as "sorry" gesture (lower Sp).

$$Se = \frac{TP}{TP + FN} \tag{5}$$

$$Sp = \frac{TN}{TN + FP} \tag{6}$$

Table 3. Se and Sp of the trained RNN_{C_5} model.

Class	Se (%)	Sp (%)	Class	Se (%)	Sp (%)
None/Invalid	100	100	"Thank You"	99.21	100
"Good"	100	99.97	"Yes"	100	99.97
"Happy"	99.13	100	"Please"	100	99.97
"Sorry"	100	100	"Drink"	99.02	100
"Hungry"	100	100	"Eat"	100	100
"Understand"	100	99.97	"Look"	100	100
"Pretty"	99.10	100	"Sleep"	100	100
"Smell"	100	100	"Hearing"	100	100
"There"	100	100	"Water"	100	100
"You"	100	100	"Rice"	100	100
"Me/I"	100	100	"Search"	100	100
"OK"	100	100	"Onion"	100	100
"Hello"	100	100	"Apple"	100	100
"Bye"	100	100	"Vegetable"	100	100

The DNN_{C_5} was initially trained with a LSTM layer, 2 dense layers and 2 DR layers (DR = 0), each with output neurons $\geq$ 40, at 150 epochs and batch size of 20. The training recognition rate and loss were plotted in Fig. 4a and 4b. The overfitting was observed after 20th epochs with testing recognition rate remained at nearly 95% and error rate of 0.1, indicated that DNN_{C_5} was over-trained (too high complexity). After several trials, the best performance was achieved with configuration of output neurons of 30 in LSTM layer, output neurons of 30 in 1st DR layer (DR = 0.02), output neurons of in 1st dense layer, output neurons of 20 in 2nd DR layer (DR = 0.01) and output neurons of 10 in 2nd dense layer as illustrated in Fig. 4c and 4d.

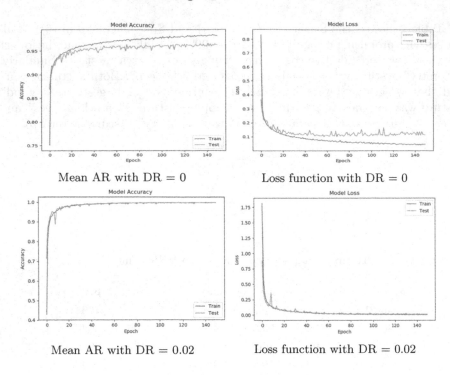

Mean AR with DR = 0   Loss function with DR = 0

Mean AR with DR = 0.02   Loss function with DR = 0.02

Fig. 4. Evaluation of trained DNN_{C_5} in mean accuracy rate and loss function.

5   Summary

This paper proposed the development of wearable sensors device for ASL recognition based on hand and fingers pattern (handshape, movement and orientation). A DNN model was trained for 27 sign words recognition. The experiment result revealed that the sensor fusion with multi-modal features of QTR, ELA, AGR and ACC were indicative for recognizing signs that involved complex gestures. Nonetheless, this study only considered single-handed word-based ASL with promising preliminary results. The future work included expansion to two-handed word-based ASL and sentence-based ASL (include other sign languages) as well as research on network optimization. Besides that, the study was considered to further collaborate with hearing-impaired community for Chinese sign language (CSL) consultation, data collection, and refinement of proposed wearable device based on user experience review from hearing-impaired community. The exploration of the possibility in hand movements analysis as complementary to facial-body expression (the 5th component in sign language) was worth to be investigated too.

References

1. Cheok, M.J., Omar, Z.: A review of hand gesture and sign language recognition techniques. Int. J. Mach. Learn. Cyber. **10**(1), 131–153 (2019)
2. Ahmed, M.A., Zaidun, B.B., Zaidan, A.A.: A review on systems-based sensory gloves for sign language recognition state of the art between 2007 and 2017. Sensors **18**(7), 1–44 (2018)
3. Appenrodt, J., Al-hamadi, A., Michaelis, B.: Data gathering for gesture recognition systems based on mono color-, stereo color- and thermal cameras. Int. J. Signal Process. Image Process. Pattern Recog. **5899**(1), 78–86 (2010)
4. Lee, B.G., Lee, S.M.: Smart wearable hand device for sign language interpretation system with sensors fusion. IEEE Sens. J. **18**(3), 1224–1232 (2018)
5. Elmezain, M., Al-hamadi, A.: A hidden Markov model-based isolated and meaningful hand gesture recognition. Int. J. Electr. Comput. Syst. Eng. **3**(3), 156–163 (2009)
6. Molchanov, P., Gupta, S., Kim, K., Pulli, K.: Multi-sensor system for driver's hand-gesture recognition. In: 11th IEEE International Conference on Workshops Automatic Face and Gesture Recognition, Ljubljana, Slovenia, pp. 1–8 (2015)
7. Chai, X., Li, G., Lin, Y., Xu, Z., Tang, Y., Chen, X.: Sign language recognition and translation with Kinect. In: Proceedings of the IEEE International Conference on Workshops Automatic Face and Gesture Recognition, Shanghai, China, pp. 22–26 (2013)
8. Yang, H.D.: Sign language recognition with the Kinect sensor based on conditional random fields. Sensors **15**(1), 135–147 (2015)
9. Chong, T.W., Lee, B.G.: American sign language recognition using leap motion controller with machine learning approach. Sensors **18**(10), 1–17 (2018)
10. Preetham, C., Ramakrishnan, G., Kumar, S., Tamse, A.: Hand talk- implementation of a gesture recognizing glove. In: 2013 IEEE Texas Instruments India Educators' Conference, Bangalore, India, pp. 328–331 (2013)
11. Patil, K., Pendharkar, G., Gaikwad, P.G.N.: American sign language detection. Int. J. Sci. R. Pub. **4**(11), 4–9 (2014)
12. Mummadi, C.K., et al.: Real-time and embedded detection of hand gestures with an IMU-based glove. Informatics **5**(2), 1–18 (2018)
13. Das, A., Yadav, L., Singhal, M., Sachan, R., Goyal, H., Taparia, K.: Smart glove for sign language communications. In: 2016 IEEE International Conference on Accessibility Digital World (ICADW), Guwahati, India, pp. 27–31 (2016)

Automated Grading of Essays: A Review

Jyoti G. Borade[✉] and Laxman D. Netak

Dr. Babasaheb Ambedkar Technological University Lonere, Raigad 402103, MS, India
jyoti.borade81@gmail.com, ldnetak@dbatu.ac.in

Abstract. Grading essay type of examinations is a kind of assessment where teachers have to evaluate descriptive answers written by students. Evaluating explanatory answers is a challenging task. Evaluating essay type questions are often laborious, requiring more time compared to multiple-choice questions. Also, the evaluation process is a subjective one leading to inaccuracies and considerable variation in grading. In recent times, many researchers have developed techniques for Automated Grading of Essays (AGE) using various machine learning methods. This paper reviews and summarizes the existing literature on methods for automated grading of essays. The paper reviews applications of approaches such as Natural Language Processing and Deep Learning for AGE. The paper identifies a few areas for further research in solving the problem of AGE.

Keywords: Automated Grading of Essay (AGE) · Machine Learning (ML) · Latent semantic analysis (LSA) · Natural Language Processing (NLP) · Artificial neural networks (ANN) · Multidimensional long short term memory (MD-LSTM) · Recurrent neural networks (RNN)

1 Introduction

Examinations are important in evaluating a student's knowledge and skills. Evaluations can be done by using multiple choice questions (MCQ), fill the blanks, match-the-pairs, answer in one word, or essay-type questions. The MCQ type examinations have many limitations, such as it is difficult to judge the knowledge or critical thinking ability of students through MCQ-based examinations. Such types of examinations fail to evaluate acquired knowledge or to increase the learning ability of students. While in essay type exams, a teacher can assess the learning ability through the students' written responses. They are thus making the essay-type exams more effective mode of assessment. However, evaluating essay-type examinations is a time-consuming task [3]. Also grading essay-type questions is a subjective process. Two evaluators may grade the same questions with different scores [4,10].

The advancements in e-learning technologies [5–7] are making it feasible to adopt these technologies for assessment also. Although the online assessment has the same purpose as written offline assessment, online assessment is becoming a cumbersome task when a teacher conducts essay examinations by asking students to write answers and submit a scanned copy of the answer-sheet. In these

© Springer Nature Switzerland AG 2021
M. Singh et al. (Eds.): IHCI 2020, LNCS 12615, pp. 238–249, 2021.
https://doi.org/10.1007/978-3-030-68449-5_25

situations, computer-based evaluation systems can overcome human inadequacies through uniform assessment.

Automated Grading of Essays (AGE) has been a widely explored research area which has adopted various kinds of technologies and approaches. Typically AGE is based on a combination of Natural Language Processing (NLP) and Machine Learning. In one of the earliest approach presented in [15] the idea of automated grading of essays called Project Essay Grader (PEG) using a set of language-based features to model the grading task as a multiple regression equation. In [8], authors have developed an approach by using Optical Character Recognition.

The adoption of NLP [24–27] for AGE is driven by the motivation to handle linguistic issues such as multiple meanings of words in different contexts [9]. It helps to extract linguistic features. These features are extracted by using NLP and Deep learning approaches [12–14, 16–23]. Recent research works have used deep neural networks based approach to implement a scoring function and NLP-based approach for extracting linguistic features. Few approaches [28] have also used fuzzy rules for AGE.

Researchers have developed AGE for many natural languages like Indonesian [29], Chinese [30], Arabic [31–35], Punjabi [36], Japanese [37]. Recently, Youfang Leng et al. [38] proposed a collaborative grammar, semantic and innovative model, terms as DeepReviewer, to complete automatic paper review.

The paper is organized as follows: Sect. 2, after introduction, describes basic data engineering tasks required for AGE. Section 3, describes metrics to evaluate models for AGE. Section 4 gives a detailed description of the approaches or methods used for grading an essay. Section 5 summarizes related works. We have concluded our study in the Sect. 6.

2 Data Engineering Tasks for Automated Grading of Essays

This section explains some of the engineering tasks that are performed during the automated grading of essays.

2.1 Feature Engineering

The goal of feature engineering is to identify attributes that can help to ascertain the quality of an essay. These features can be of following types. (i) *Lexical Features:* These includes features such as complete words, prefix, suffix, stemmed words, lemmatized words, capitalization, and punctuation. (ii) *Syntax and Grammar related Features:* These include features like POS, usage of a noun, verb, prepositional, conjunction, vocabulary, and spelling. (iii) *Semantic Features:* These include features based on verifying the correctness of content and meaning. Features like vocabulary per passage, percentage of difficult words, rhetorical features, the proportion of grammar error, and the proportion of usage errors [16] can be used to determine semantic correctness and meaning.

Features can be extracted using NLP tools such as NLTK. Deep neural networks based approaches rely on automatic feature extraction; these approaches are preferred for feature engineering [14, 23]. Many researchers have used LSTM-RNN deep models to capture the semantics of text to find the similarity between texts [12, 38].

2.2   Word Embedding

The problem of AGE can be represented as a classification or a regression task. It requires many features for accurately predicting the score of a given essay. These modelling techniques work on numeric data. The *word embedding* is a process in which a vector representation for each word of the text created [8]. The different types of word embedding are:

1. **One-hot encoded vector:** In this encoding, words are represented as a vector of binary numbers.
2. **Count Vector:** In this encoding, a matrix of key terms and documents is created. Each row contains a key term and each column contains a document. Cell represents count of occurrence of a key terms.
3. **Term Frequency-Inverse Document Frequency (TF-IDF) Vector.** The term frequency (TF) is calculated as:

$$TF = \frac{Number\ of\ times\ the\ occurrence\ of\ a\ key\ term\ in\ a\ document}{Total\ key\ terms\ in\ the\ document}$$

It computes the most relevant term for the document. The inverse document frequency (IDF) is calculated as:

$$IDF = log(N/n),$$

where, N is the total number of documents and n is the number of documents in which a term t appeared. The product of TF and IDF can be used to find out the most important words for the document from the context of the entire corpus as:

$$TF - IDF(word, Document_i) = TF * IDF$$

It assigns lower weights to a term that is common in all documents while assigns more weights to a term which occurs in subsets of documents.
4. **Co-occurrence Matrix:** This type of embedding relies on the assumption that words in a similar context appear together. For example, Arnav is a boy and Atharva is a boy. Arnav and Atharav have a similar context i.e boy. All words are arranged in a row and a column. If a particular word is put in $column_i$ same word is put in row_i. The cell represents number of times words w_1 and w_2 appeared together.
5. **Continuous Bag of words(CBOW):** It predicts the probability of output of a single word for the given input context of words. It is a 2 layer model having one hidden layer and one output layer. It maximizes the conditional probability for actual output word for the given input context words.

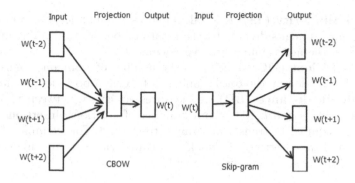

Fig. 1. Word2Vec training models using CBOW and Skip-gram.

6. **Skip-gram model:** It is having a similar structure like CBOW having one hidden and one output layer. As shown in Fig. 1, it is a reverse model of the CBOW. It predicts the probability of a context of output words for the given single input word. It maximizes the conditional probability of observing the target context words for the given input word.

Word2vec and *GloVe* vectors are commnly used tools to produce word embeddings. These models are pre-trained models, constructed using CBOW and Skip-gram model.

2.3 Measuring Text Similarity

Finding the similarity between texts of different documents is another essential data engineering activity to realize automated grading of essays. The text is not in numeric form. It cannot be compared directly to find similarity score. For this purpose, the text is represented as a vector and similarity score calculated using following distance measurements.

1. **Euclidean Distance (ED):** The Euclidean distance between two vectors p and q is calculated as:

$$d\,(p,q) = \sqrt{\sum_{i=1}^{n}\left(q_i - p_i\right)^2}$$

2. **Cosine Similarity (CS):** The cosine similarity is defined as the dot product of two vectors divided by the product of their norms. The cosine similarity between two vectors

$$p = \{p_1, p_2, \ldots, p_N\} \text{ and } q = \{q_1, q_2, \ldots, q_N\} \text{ is defined as:}$$
$$sim(p,q) = \frac{p \cdot q}{||p||.||q||} = \frac{\sum_{i=1}^{N} p_i q_i}{\sqrt{\left(\sum_{i=1}^{N} p_i^2\right)\left(\sum_{i=1}^{N} q_i^2\right)}}$$

3. **Jaccard Similarity (JS):** It is calculated as the number of common words from both documents divided by the total number of words. The higher the percentage, the more similar are two documents.
4. **Dice Coefficient (DC):** It is twice the number of common tokens in both documents divided by the total number of tokens in both documents.
5. **Levenshtein distance (LD):** It measures similarity between two strings. consider string s and t. LD is total number of deletions, insertions, or substitutions required to transform string s to string t. For example, if s string is 'cook' and target string t is 'book'. To transform s to t, change first 'c' to 'b'. Thus, the Levenshtein distance will be 1.

3   Metrics to Evaluate Models for Automatic Grading of Essays

The models developed for automatic grading of essays can be evaluated by adopting measures such as *accuracy, precision, recall,* and *F1-score* etc. Majority of the researchers have used Quadratic Weighted Kappa (QWK) for evaluating their approaches for AGE.

3.1   QWK

A weighted Kappa is one of the evaluation metrics, used to calculate the amount of similarity between prediction and actual evaluation. It generates a score of 1.0 when prediction and actual evaluation is same. It is calculated by using the following formula.

$$QWK = 1 - \frac{\sum_{i,j} W_{i,j} O_{i,j}}{\sum_{i,j} W_{i,j} E_{i,j}}$$

Where O is a matrix of size n-by-n corresponds to n essays. $O(i,j)$ gives the count of essays that obtained a score i by the first evaluator and score j by the second evaluator. E matrix gives expected ratings without considering any correlation between the two evaluations given by two different evaluators. W is a matrix of the same size as the O and E matrix. It is calculated as follows:

$$W_{i,j} = (i - j)^2 / (N - 1)^2$$

3.2   Recall Oriented Understudy for Gisting Evaluation (ROUGE)

It is an evaluation metric finds similarity between two text documents. These are summarized documents. It finds the similarity between students' and reference document, by observing common n-grams, sentence, and word pairs. ROUGE result gives the value of metrics like precision, recall, and f-score as given below [26].

$$Precision = \frac{No.\ of\ common\ words}{Total\ words\ in\ student's\ documents}$$

$$Recall = \frac{No.\ of\ common\ words}{Total\ words\ in\ reference\ documents}$$

$$F - score = \frac{2 * Precision * Recall}{Precision + Recall}.$$

4 Methods for Automatic Grading of Essays

4.1 Machine Learning

The basic procedure for AGE has been to start with feature extraction from a training set of essays. The automated grading of essay finds the syntactic and semantic features from student answers and reference answers. Then construct a machine learning model that relates these features to the final scores assigned by evaluators. This trained model is used to find score of unseen essays. In earlier research, researchers have used linear regression. Modern systems uses other machine learning techniques in combination with latent semantic analysis, concept-based mapping like clustering, Markov chains, Support vector machines, and Bayesian inference [7, 36].

4.2 Textual Similarity

These approaches rely on measuring the textual similarity. It takes text documents as inputs and finds similarity between them. Lexical similarity determines the similarity by matching contents, word-by-word. Semantic similarity is based on the meaning of contexts. The measures like Levenshtein distance and cosine similarity are used to find similarity between two texts. Both algorithms assign scores according to the similarity distance between student answers and model answer [27].

4.3 Latent Semantic-Based Vector Space Model (LS-VSM)

The latent semantic analysis is used to analyse and find the similarity between documents. In LS-VSM, the first step is to extract key terms/keywords from each document. Create a word vector. The documents mapped to the same concept space, where we can cluster documents. The overlapping cluster indicates the similarity between the documents. Also, LSA applies singular value decomposition (SVD), which reduces the dimensional representation of the term matrix, over which cosine similarity between documents can be computed. It uses BOW embeddings like Word2Vec and GloVe [11, 27, 30].

4.4 Neural Network Based Approaches

Researchers have developed many AGE system using ANN and various deep learning models like CNN, RNN, LSTM, Bidirectional LSTM etc. In ANN, we have to do feature engineering. It does not contribute to automatic feature extraction. While the deep network does automatic feature extraction and finds unique, complex features. In [12], authors have created dataset with format as [questionid],[keyid],[answer],[mark]. They collected 50 responses for each question from different students. They used sequential model with embedding layer, lstm layer, dropout layer and dense layer. Accuracy obtained for simple LSTM is 83, Deep LSTM is 82, Bi-directional LSTM is 89.

4.5 Naive Bayes Classifiers

It is used to find most likely hypothesis H i.e. a class. It is based on calculating the maximum a posteriori (map) probability. It calculates probability of a hypothesis using priori probability [28, 29].

$$hmap = P(h_i|D) = \frac{P(D|h_i) * P(h_i)}{P(D)}$$

$hmap$ is the maximum likelihood hypothesis. D is a feature vector $< d_1, ..d_n >$, the set of attributes which describes data. As shown in above equation in order to calculate $P(h|d_1, d_2, ...d_n)$, it is necessary to estimate $P(d_1, d_2, ...d_n|h_i)$. Estimation of $P(d_1, d_2, ...d_n|h_i)$, requires a large number of samples. As there are $2n$ possible features, it is difficult to store probability values corresponding to all of them. Hence, Naive Bayes classifier is commonly used. It assumes that all attributes in D are independent. Thus the joint probability is obtained as given below:

$$P(D|h) = P(d_1, d_2, ...d_n|h_i) = P(d_i|h_i)$$

And the classifier output is given as:

$$hmap = argmaxP(h_i) * P(d_i|h_i)$$

Table 1 shows comparative study of various approaches used by researchers for automated grading of essay.

5 Related Work

The automatic grading of the essay describes a multidisciplinary area of study. Page and Paulus, in 1968 proposed grading of essay based on statistical methods. They used very basic features and ignored the semantic of the text [15]. Sargur Srihari et al. [11], presented essay scoring using optical character recognition and automatic essay scoring techniques. For handwriting recognition used analytic and holistic features together with contextual processing based on trigrams. They used LSA and ANN for essay scoring after OHR. Try Ajitiono et al. [3] used a

Table 1. Comparative study of different approaches suggested for AGE

Paper	Features	ML approach	DL approach	Similarity Metrics	Model Evaluation
[6]	NLP techniques applied to extract key terms	Not Known	Not Known	Dice Coefficient	RMSE = 0.63
[9]	Syntatic, semantic, sentiment features	SVM, Random forest regressor	3 layer Neural network	Not Known	QWK for SVM = 0.78, QWK for Random forest = 0.77, QWK for neural network = 0.83
[12]	Automatic feature extraction	Not Known	LSTM-RNN	Not Known	Fscore = 92.01
[27]	NLP techniques applied to extract key terms	Not Known	Not Known	LD, CS	similarity using LD = 92, similarity using CS = 94
[32]	Spelling mistake, grammar mistake, organization of essay, rhetorical structure	LSA	Not Applicable	Not Known	Accuracy = 78.3
[33]	Automatic feature extraction	Not applicable	CNN, LSTM	Not Known	Accuracy for CNN = 97.29, Accuracy for LSTM = 96.33

wide variety of handcrafted features using NLP. In further research, Aluizio Filho et al. [34], presented imbalanced learning techniques for improving the performance of statistical models.

Anhar Fazal et al. [26], proposed an innovative model for grading spellings in essays using rubrics. They categorize each word into four categories: Simple, Common, Difficult, and Challenging. For classification, they used their own rule based word classification algorithm and utilized a rule-based spelling mark algorithm for mark assignment.

Maram F. et al. [32] developed a system for arabic language using LSA and rhetorical structure theory that achieved an overall correlation of 0.79%. Dima S. et al. [35] developed the AGE system in arabic language for poetry using Hidden Markov Model with an accuracy of 96.97. Anak A. et al. [37] used a winnowing algorithm and LSA to find automatic essay grading with accuracy up to 86.86%. They used Hidden Markov Model for part of speech tagging, morphological analysis, and syntactic structure and text classification.

Manar Joundy H et al. [6] obtained similarity between predicted text and reference text using dice coefficient. In [24], V.V.Ramalingam proposed linear regression techniques along with classification and clustering techniques. Akeem

O et al. [27] mapped student's answers with a model answer to measure textual similarity using Levenshtein distance and cosine similarity. They obtained 92 and 94% similarity for LD and CS, respectively. Fabio Bif G [28], presented an essay assessment based on fuzzy rules. The proposed approach provided F-measure with 95%.

Harneet Kaur J et al. [9], build a model based on grammar and semantics of contents. The author preferred a graph based relationship to determine semantic similarity. To predict the results used SVM, Random Forest Regressor, and three-layer neural network models. Similarity measured using Quadratic Weighted Kappa is 0.793. Abdallah M. Bashir et al. [33] used a deep network with word embeddings for text classification. They used CNN and LSTM for classification. Achieved an F-score of 92.01 for the arabic language.

Neethu George et al. [12], proposed LSTM-RNN model based on GloVe vector representation. Using this, the author achieved QWK of 0.60. Cancan Jin [13] proposed two stages of the deep model based on deep hybrid model to learn semantic, POS, syntactic features of the essays. The proposed model outperforms baselines and gives a promising improvement. Huyen Nguyen and Lucio Dery [17], Surya K et al. [20] used deep models for grading of an essay. Huyen Nguyen and Lucio Dery achieved a QWK score of 0.9447875. Jiawei Liu et al. [21] calculated score based on LSTM neural network.

In [8], authors built a system that takes input image of handwritten essay and outputs the grading on the scale of 0–5 using OHR and AGE. They used MDLSTM and convolution layer for handwriting recognition. AGE model is of two layers ANN with a feature set based on pre-trained GloVe word vectors. They achieved a QWK score of 0.88. J.Puigcerver [19] build an integrated system of OHR and AGE using only convolutional and one-dimensional recurrent layers instead of multidimensional recurrent layers.

Pedro Uria R et al. [22] compared two powerful language models BERT (Bidirectional Encoder Representation from the transformer) and XLNet and described all the layers and network architectures in these models. Theodore Bluche et al. [23] used MDLSTM to recognize handwritten paragraphs. Authors have used RNN approach in [14,16,18].

Youfang Leng et al. [38] proposed model known as DeepReviewer that reviews the paper. This model learned innovative features of an article with the help of a hierarchical RNN and CNN. The authors [39] proposed clustering of similar submissions and measured semantic distance from correct solutions. They developed an automatic grading and feedback using program repair for introductory programming courses.

6   Conclusion

The paper presents a review of machine learning techniques used to assess essay type of answers. While doing so, it identifies essential terminologies, methods, and metrics used to build such machine learning models. The three data engineering activities that need to be carried out as part of pre-processing data include

(1) Feature engineering, (2) Word embedding and (3) Measuring text similarity. Naive Baye's classifier, Neural Network-based approaches and Latent Semantic-based Vector Space Model (LS-VSM) are some of the widely used methods for grading essay type questions. The commonly used metrics to evaluate the performance of the models built for AGE are QWK, Recall, Precision and F-1 Score. Also, a comparison of existing approaches based on these identified features is presented. The presented work can be extended to build a standard dataset consisting of students' responses to the essay questions classified according to Bloom's Taxonomy level [1, 2] and then to develop a useful machine learning model for this purpose.

References

1. Kiwelekar, A.W., Wankhede, H.S.: Learning objectives for a course on software architecture. In: Weyns, D., Mirandola, R., Crnkovic, I. (eds.) ECSA 2015. LNCS, vol. 9278, pp. 169–180. Springer, Cham (2015). https://doi.org/10.1007/978-3-319-23727-5_14
2. Wankhede, H.S., Kiwelekar, A.W.: Qualitative assessment of software engineering examination questions with bloom's taxonomy. Indian J. Sci. Technol. **9**(6), 1–7 (2016)
3. Ajitiono, T., Widyani, Y.: Indonesian essay grading module using Natural Language Processing. In: 2016 International Conference on Data and Software Engineering (ICoDSE), Denpasar, pp. 1–5 (2016)
4. Patil, R.G., Ali, S.Z.: Approaches for automation in assisting evaluator for grading of answer scripts: a survey. In: 2018 4th International Conference on Computing Communication and Automation (ICCCA), Greater Noida, India, pp. 1–6 (2018)
5. Al-shalabi, E.: An automated system for essay scoring of online exams in Arabic based on stemming techniques and Levenshtein edit operations. IJCSI Int. J. Comput. Sci. Issues **13**, 45–50 (2016)
6. Hazar, M., Toman, Z., Toman, S.H.: Automated scoring for essay questions in e-learning. In: Journal of Physics: Conference Series, vol. 1294 (2019)
7. Patil, R., Ali, S.: Approaches for automation in assisting evaluator for grading of answer scripts: a survey. In: International Conference on Computing Communication and Automation (ICCCA), pp. 1–6 (2018)
8. Sharma, A., Jayagopi, D.B.: Automated grading of handwritten essays. In: 2018 16th International Conference on Frontiers in Handwriting Recognition (ICFHR), Niagara Falls, NY, pp. 279–284 (2018)
9. Janda, H.K., Pawar, A., Du, S., Mago, V.: Syntactic, semantic and sentiment analysis: the joint effect on automated essay evaluation. IEEE. Access **7**, 108486–108503 (2019)
10. Ke, Z., Ng, V.: Automated essay scoring: a survey of the state of the art. IJCAI (2019)
11. Srihari, S., Collins, J., Srihari, R., Srinivasan, H., Shetty, S., Brutt-Griffler, J.: Automatic scoring of short handwritten essays in reading comprehension tests. Artif. Intell. **172**(2–3), 300–324 (2008)
12. George, N., Sijimol, P.J., Varghese, S.M.: Grading descriptive answer scripts using deep learning. Int. J. Innov. Technol. Explor. Eng. (IJITEE) **8**(5) (2019). ISSN 2278-3075

13. Jin, C., He, B., Hui, K., Sun, L.: TDNN: a two-stage deep neural network for prompt-independent automated essay scoring. In: Conference: ACL, at Melbourne, Australia, P18-1100 (2018)
14. Voigtlaender, P., Doetsch, P., Ney, H.: Handwriting recognition with large multidimensional long short-term memory recurrent neural networks. In: International Conference on Frontiers in Handwriting Recognition, pp. 228–233 (2016)
15. Page, E.B.: Project Essay Grade: PEG (2003)
16. Cai, C.: Automatic essay scoring with a recurrent neural network. In: HP3C 2019 (2019)
17. Nguyen, H.: Neural Networks for Automated Essay Grading (2016)
18. Chen, M., Li, X.: Relevance-based automated essay scoring via hierarchical recurrent model (2018)
19. Puigcerver, J.: Are multidimensional recurrent layers necessary for handwritten text recognition?. In: 2017 14th IAPR International Conference on Document Analysis and Recognition (ICDAR), Kyoto, pp. 67–72 (2017)
20. Surya, K., Gayakwad, E., Nallakaruppan, M.K.: Deep learning for short answer scoring. Int. J. Recent Technol. Eng. **7**, 1712–1715 (2019)
21. Liu, J., Xu, Y., Zhao, L.: Automated essay scoring based on two-stage learning (2019)
22. Rodriguez, P., Jafari, A., Ormerod, C.: Language models and automated essay scoring (2019)
23. Bluche, T., Louradour, J., Messina, R.: Scan, attend and read end-to-end handwritten paragraph recognition with MDLSTM attention. In: International Conference on Document Analysis and Recognition (2017)
24. Ramalingam, V.V., Pandian, A., Chetry, P., Nigam, H.: Automated essay grading using machine learning algorithm. In: Journal of Physics: Conference Series, vol. 1000, p. 012030 (2018)
25. Haendchen Filho, A., Prado, H., Ferneda, E., Nau, J.: An approach to evaluate adherence to the theme and the argumentative structure of essays. Procedia Comput. Sci. **126**, 788–797 (2018)
26. Fazal, A., Hussain, F., Dillon, T.: An innovative approach for automatically grading spelling in essays using rubric-based scoring. J. Comput. Syst. Sci. **79**, 1040–1056 (2013)
27. Olowolayemo, A., Nawi, S., Mantoro, T.: Short answer scoring in English grammar using text similarity measurement. In: International Conference on Computing, Engineering, and Design (ICCED), pp. 131–136 (2018)
28. Goularte, F., Nassar, S., Fileto, R., Saggion, H.: A text summarization method based on fuzzy rules and applicable to automated assessment. Expert Syst. Appl. **115**, 264–275 (2019)
29. Rahutomo, F., et al.: Open problems in Indonesian automatic essay scoring system. Int. J. Eng. Technol. **7**, 156–160 (2018)
30. Peng, X., Ke, D., Chen, Z., Xu, B.: Automated Chinese essay scoring using vector space models. In: In: 2010 4th International Universal Communication Symposium, Beijing, pp. 149–153 (2010)
31. Etaiwi, W., Awajan, A.: Graph-based Arabic NLP techniques: a survey. Procedia Comput. Sci. **142**, 328–333 (2018)
32. Al-Jouie, M., Azmi, A.: Automated evaluation of school children essays in Arabic. Procedia Comput. Sci. **117**, 19–22 (2017)
33. Bashir, A., Hassan, A., Rosman, B., Duma, D., Ahmed, M.: Implementation of a neural natural language understanding component for Arabic dialogue systems. Procedia Comput. Sci. **142**, 222–229 (2018)

34. Haendchen Filho, A., Concatto, F., Nau, J., do Prado, H.A., Imhof, D.O., Ferneda, E.: Imbalanced learning techniques for improving the performance of statistical models in automated essay scoring. Procedia Comput. Sci. **159**, 764–773 (2019)
35. Suleiman, D., Awajan, A., Etaiwi, W.: The use of hidden Markov model in natural ARABIC language processing: a survey. Procedia Comput. Sci. **113**, 240–247 (2017)
36. Walia, T., Josan, G., Singh, A.: An efficient automated answer scoring system for the Punjabi language. Egypt. Inform. J. **20**, 89–96 (2018)
37. Ratna, A., Luhurkinanti, D., Ibrahim, I., Husna, D., Purnamasari, P.: Automatic essay grading system for Japanese language examination using winnowing algorithm. In: International Seminar on Application for Technology of Information and Communication (iSemantic), pp. 565–569 (2018)
38. Leng, Y., Yu, L., Xiong, J.: DeepReviewer: collaborative grammar and innovation neural network for automatic paper review, pp. 395–403 (2019)
39. Parihar, S., Dadachanji, Z., Singh, P., Das, R., Karkare, A., Bhattacharya, A.: Automatic grading and feedback using program repair for introductory programming courses. ITiCSE, pp. 92–97 (2017)

Voice Attacks to AI Voice Assistant

Seyitmammet Alchekov Saparmammedovich[1], Mohammed Abdulhakim Al-Absi[1], Yusuph J. Koni[1], and Hoon Jae Lee[2(✉)]

[1] Department of Computer Engineering, Dongseo University, 47 Jurye-ro, Sasang-gu, Busan 47011, Republic of Korea
mslchekov@gmail.com, Mohammed.a.absi@gmail.com, yusuphkoni@gmail.com
[2] Division of Information and Communication Engineering, Dongseo University, 47 Jurye-ro, Sasang-gu, Busan 47011, Republic of Korea
hjlee@dongseo.ac.kr

Abstract. Everything goes to the fact that our communication with technology will soon become almost exclusively oral. It is natural for a person to ask for something out loud and hear the answer: see how children are at ease with voice assistants. However, with new technologies - and voice control is no exception - new threats are emerging. Cybersecurity researchers tirelessly seek them out so that device manufacturers can secure their creations before potential threats turn into real attacks. However, in this paper we are going to talk about different voice attacks, which so far will hardly be able to find practical application, but the protection against which should be thought over now.

Keywords: Voice attack · AI · Voice assistant · Attack distance · Assistant · Artificial intelligence

1 Introduction

Virtual assistants can "live" on smartphones, tablets and computers (like Apple Siri) or stationary devices (like Amazon Echo and Google Home speakers). The range of their possibilities is constantly increasing: you can control music playback, find out the weather, regulate the temperature in the house, order goods in online stores…
What is the Voice Assistant?
Voice assistant is an artificial intelligence-based service that recognizes human speech and is able to perform a specific action in response to a voice command. Most often, voice assistants are used in smartphones, smart speakers, web browsers.
The functionality of voice assistants is quite diverse. What the voice assistant can do:

- conduct dialogues,
- offer quick answers to user questions,
- call a taxi,
- make calls,

- lay routes,
- place orders in the online store, etc.

Since all voice assistants have artificial intelligence, when communicating with the user, they take into account the change in his location, time of day and days of the week, search history, previous orders in the online store, etc.

- Google Now - is one of the first Google voice assistants. Works on devices with Android, iOS and Chrome browser. Likes to suggest the best routes to home, taking into account the current location of the user, offering news feeds, can analyze mail and search queries. Google Now is integrated with all Google services and some third-party applications.
- Google Assistant - is a more advanced version of the voice assistant. Can conduct dialogues and understand normal spoken language.
- Siri. Works on Apple devices only. Knows how to conduct dialogues and give recommendations, for example, where to go or which movie to watch.
- Microsoft Cortana. Available on Windows, iOS and Android. Manage reminders and calendars, track packages, set alarms, and search Bing for news, weather, and more.
- Amazon Alexa. Built into Amazon audio devices (Echo, Echo Dot, Tap) and Fire TV box. Can play music, read news, offer weather and traffic information, and voice order on Amazon.

How does it work?

The history of voice assistants begins in the late 1930s, when scientists began to attempt to recognize the voice using technology. Then two big problems got in the way of creating a quality assistant:

1. the existence of homonyms - words with the same sound, but with different meanings,
2. constant background noise from which the system must select the user's speech.

Developers are now using machine learning to solve these problems. It teaches neural networks to independently analyze the context and determine the main source of sound. However, the developers did not come to this immediately - it took at least 80 years of preparatory work.

How modern voice assistants work? Voice assistants passively read all sound signals, and for active work they need activation using a passphrase. For example, say: "Okay, Google", then you can ask your question or give a command without pauses.

At the moment of a voice request, the automatic speech recognition system (ASR system) converts the audio signal into text. This happens in four stages:

- Filtration. The system removes background noise and interference arising during recording from the audio signal.
- Digitization. Sound waves are converted into digital form that a computer can understand. The parameters of the received code also determine the quality of the recording.

- Analysis. Sections containing speech are highlighted in the signal. The system evaluates its parameters - to which part of speech the word belongs, in what form it is, how likely the connection between two words is.
- Revealing data patterns. The system includes the obtained information into a dictionary - it collects different versions of the pronunciation of the same word. To more accurately recognize new queries, assistants compare the words in them with patterns.

If, after processing the request (Fig. 1), the virtual assistant does not understand the command or cannot find the answer, he asks to rephrase the question. In some cases, additional data may be required - for example, when calling a taxi, the assistant can specify the passenger's location and destination.

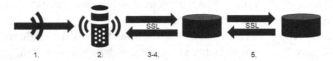

Fig. 1. Processing the request

1- Device is always listening, unless muted
2- Device starts recording when the trigger word is heard, e.g. "Hey Siri or Alexa". The LED indicates recording status.
3- Recording is sent to the cloud for processing & stored
4- Response is sent back Traffic is SSL encrypted
5- Optional: Backend can send data to third-party extensions (Actions/Skills) for additional processing

2   Various Voice Attacks

- Adversarial Attacks via Psychoacoustic Hiding:

Researchers at the Ruhr University in Germany have found that voice assistants can be hacked using commands hidden in audio files that cannot be discerned to the human ear. This vulnerability is inherent in the speech recognition technology itself, which is used by artificial intelligence (Fig. 2).

Fig. 2. various voice attacks

According to Professor Thorsten Holz, this hacking method is called "psychoacoustic hiding". With it, hackers can hide words and commands in different audio files - with music or even birdsong - that only a machine can hear. The person will hear the usual chirp, but the voice assistant will be able to distinguish something else [2].

• Laser attack.

Figure 3, researchers from the University of Michigan (USA) were able to experimentally implement a way to get remote access through a portable laser system to some iPhones, smart speakers HomePod and Amazon Echo, devices of the Google Home series using voice assistants Siri, Google Assistant and Amazon Alex. Laser beams with coded signals were sent to the place of placing the microphone in smart devices, the maximum working distance achieved in the experiment was 110 m, and the devices could thus be controlled even through a glass unit. Most speakers and gadgets took such an impact for voice commands, allowing them to perform basic actions in a smart home system that do not need additional user identification - to open a garage or a lock on a house door [3].

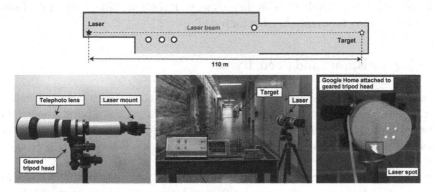

Fig. 3. Laser attack

• SurfingAttack.

Another attack technique, dubbed SurfingAttack, uses voice commands encrypted in ultrasonic waves. With the help of such waves, a potential attacker can quietly activate the voice assistant. SurfingAttack can be used to perform a limited set of actions: make calls or read text messages [4].

3   Problems

What risks do we take upon ourselves by installing a smart home system? The one that drives light bulbs and a kettle from an application on a smartphone, inside the local network and remotely. If security and lock management are tied to a smart system (as in the case of Amazon Key), then it is clear which ones. If not, then theoretically it is possible to imagine the danger of a software failure of some coffee maker, followed by a fire, or by the influence of an intruder to gain access to your home space.

Siri itself, and other voice assistants - Cortana (Microsoft), Alexa (Amazon), Bixby (Samsung), as well as the only representative of the "masculine" Google Assistant - also do not shine with intelligence.

And this is understandable. "Smart" voice assistants are based on the architecture of neural networks and machine learning technology. It should be understood that there are about 86 billion neurons in the human brain, and in modern artificial intelligence there are only a few hundred thousand. If you count the number of neurons in the nervous system of various animals, it turns out that, as noted by the founder and head of ABBYY, David Yang, *now artificial intelligence is dumber than a bee.*

Despite this, the IQ of artificial intelligence has doubled approximately every two years.

This progression shows that sooner or later AI will still reach the human level - but this is still far from it. According to leading analyst of Mobile Research Group Eldar Murtazin, voice assistants in the next year or two will be the hallmark of premium consumer electronics, and then move to the mid-price and budget segment.

Voice assistants certainly have bright prospects. But we are talking about the future, about a new quality level, unattainable for today.

4 Voice Assistants and Security

The development of voice search and speech recognition technologies causes an ambiguous reaction from many - first of all, how safe these developments are and whether they always listen only to their owner. Voice assistants have already appeared in several notable stories:

- UC Berkeley students have found a way to launch voice assistants Siri, Alexa and Google Assistant without the owner's knowledge. To do this, it is enough to add sounds to music or video that remotely resemble human speech - the program will be enough to turn them into words and start executing the given command.
- In China, they were able to activate voice assistants using sound frequencies that the average person cannot hear.
- Burger King launched an ad with the phrase "OK Google, what is Whopper?" To which the voice assistants responded and began to read lines from Wikipedia.
- Amazon Echo ordered a dollhouse after hearing a request from a 6-year-old girl. And when this story was discussed in the news, voice assistants, taking the phrase about ordering a house as a command, began to order these houses everywhere.
- Amazon Echo owners have complained that the device starts to laugh spontaneously. It turned out that the device recognized the surrounding sounds as a command to laugh and followed it.

At the same time, Google and Amazon claim that their assistants do not turn on if they do not hear the owner's voice. Apple says Siri will never execute a command related to personal data if the iPhone or iPad is locked.

5   Comparison of Voice Attacks

In this section, we talked about different voice attacks and explained what their characteristics are, in contrast to others of the kind (Table 1).

Table 1. Comparison of voice attacks

Pros and cons	Attack name		
	1	2	3
	Psychoacoustic hiding	Laser attack	Surfing attack
Inaudible	No	Yes	Yes
Visibility	No	Yes	No
Attacking in distance and how far?	Yes	Yes	No
	Everywhere	Depends on laser power and requires line of sight	The distance should be small

If we talk about the attack "Psychoacoustic Hiding", then Hackers can play a hidden message through an app in an advertisement. Thus, they are able to make purchases on behalf of other people or steal confidential information. In the worst case, an attacker can take control of the entire smart home system, including cameras and alarms.

Attackers can use the "masking effect of sound" for their own purposes: when your brain is busy processing loud sounds of a certain frequency, then for a few milliseconds you stop perceiving quieter sounds at the same frequency. This is where the team found to hide commands to hack any speech recognition system like Kaldi, which underlies Amazon's Alexa voice assistant.

A similar principle allows you to compress MP3 files - an algorithm determines what sounds you can hear and removes anything inaudible to reduce the size of the audio file. However, hackers do not remove inaudible sounds, but replace them with the ones they need. Unlike humans, artificial intelligence like Alexa is able to hear and process every sound. He was trained so that he could understand any sound command and carry it out, whether people hear it or not.

If we talk about the disadvantages of this attack, then this is only for devices that can recognize voices. Otherwise, the message inside the audio file cannot be received by the voice assistant.

In addition, as the success of an attack improves, the original audio example affects the quality of the example. To do this, researchers [2] recommend using music or other unsuspecting sound patterns, such as the chirping of birds, that do not contain speech because speech must be dimmed, which usually results in more severe unwanted disturbances.

As for this attack, a recent discovery shows that our smart buddies can be hacked using lasers. The attack is based on the use of a photoacoustic effect, in which the absorption of changing (modulated) light by a material leads to thermal excitation of the medium, a change in the density of the material and the appearance of sound waves perceived by the microphone membrane. By modulating the laser power and focusing the beam on the hole with the microphone, you can achieve the stimulation of sound vibrations that will be inaudible to others, but will be perceived by the microphone.

For an attack, as a rule, a simulation of the owner's voice is not required, since voice recognition is usually used at the stage of accessing the device (authentication by pronunciation "OK Google" or "Alexa", which can be recorded in advance and then used to modulate the signal during an attack). Voice characteristics can also be tampered with modern machine learning-based speech synthesis tools. To block an attack, manufacturers are encouraged to use additional user authentication channels, use data from two microphones, or install a barrier in front of the microphone that blocks the direct passage of light.

Limitations of laser attack are:

- Because of the dotted pointer, Lasers must point directly to a specific component of the microphone in order to transmit audio information.
- Attackers need a clear line of sight and a clear path for the lasers.
- Most of the light signals are visible to the naked eye and can identify intruders.
- In addition, when activated, voice control devices react loudly, which can alert nearby people of foul play.
- Accurate control of advanced lasers requires expertise and equipment. When it comes to ranged attacks, there is a high barrier to entry.

The next attack is called the SurfingAttack. How does it work? This attack allows you to remotely control the virtual assistants Google Assistant and Apple Siri using ultrasonic commands that are invisible to the human ear.

The upper bound frequency of human voices and human hearing is 20 kHz. Thus, most audio-capable devices (e.g., phones) adopt audio sampling rates lower than 44 kHz, and apply low-pass filters to eliminate signals above 20 kHz. The voice of a typical adult male has a fundamental frequency (lower) of 85 to 155 Hz, and the voice of a typical adult woman from 165 to 255 Hz.

Telephony uses a frequency band from 300 Hz to 3400 Hz to determine speech. For this reason, the frequencies between 300 and 3400 Hz are also called voice frequencies.

The person is able to hear sound vibrations in the frequency range from 16-20 Hz to 15-20 kHz.

Sound below the human hearing range is called ultrasound.

The researchers explain that, in essence, voice assistants listen to a much wider frequency than the human voice can reproduce. Because of this, they can react to ultrasonic vibrations and interpret them as voice commands. As a result, an attacker is able to discreetly interact with devices using voice assistants, intercept two-factor authentication codes, and make calls.

In SurfingAttack, the researchers showed that the ultrasonic waves can be sent not only directly, but also through materials of considerable thickness, such as a single piece of glass or even a wooden table (Fig. 4).

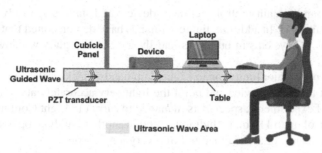

Fig. 4. SurfingAttack leverages ultrasonic guided wave in the table generated by an ultrasonic transducer concealed beneath the table [4].

Some mobile devices have customized wake-up words, SurfingAttack will not be able to activate them simultaneously, but it offers another opportunity for launching a targeted attack when the attacker learns the specific wake-up words.

Open questions of the papers (Table 2).

Table 2. Desction questions

Attack names	Suggestions from researchers
Psychoacoustic Hiding [2]	As the attack's success improves, the original audio example affects the quality of the example. To do this, the researchers recommend using music or other unsuspecting sound patterns, such as the chirping of birds, that do not contain speech, because speech must be muffled, which usually leads to more serious unwanted disturbances
Lightcommands [3]	Why the aperture vibrates from the light? At the time of writing the paper has not been studied
SurfingAttack [4]	Attacking Standing Voice Assistants. Amazon Echo and Google Home are standing voice assistants with microphones distributed across the cylinder. The current SurfingAttack cannot reach these microphones. Researchers believe this is due to the significant power loss during the power transition across the boundary of the table material and speaker material, as well as the devices' internal construction in terms of the relative position of the microphones

6 Ranged Attack

An attack using a method called "psychoacoustic concealment" [4] which can be carried out anywhere in the world using the Internet. The main goal of the attacker is to play that exactly the video or audio file in the place where the attacked Voice assistant is located. The document did not specify the exact distance, but with the attack itself, you can get access to many devices.

Lightcommands obtain control over these devices at distances up to 110 m and from two separate buildings. In addition, lightcommands have demonstrated that light can be used to control VC systems in buildings and through closed glass windows at similar distances.

In fact, the only major limitation to Light Commands attacks is the need to be within line of sight of the target device and point the light very accurately at the microphone. In the course of experiments, specialists managed to carry out Light Commands attacks from a distance of up to 110 m, including located in another building, opposite the room with the target device, or at the other end of a very long corridor.

Then aiming the laser beam at the microphone ports of the devices listed in Table 3 [3] from a distance of approximately 30 cm.

The attack distance for this SurfingAttack is very small. As shown in the Table 3 below, the MDF's attack range was significantly shorter than other table materials, which the researchers believed could also be improved by increasing signal strength. using an attack power of 1.5 W the SurfingAttack achieves a maximum attack range of 50 cm on MDF material. To improve the efficiency of SurfingAttack, an attacker can attach multiple transformers distributed across the table, which can reduce the short attack distance limit for MDF tables (Table 4).

7   Protection Against Inaudible Attacks

Manufacturers are already considering measures to protect voice-controlled devices. For example, detection of processing traces in the received signal in order to change its frequency can help from ultrasonic attacks. It would be nice to teach all smart devices to recognize the owner by voice - however, Google, which has already tested these measures in practice on its assistant, honestly warns that this protection can be bypassed using voice recording, and with proper acting skills, the timbre and manner a person's speech can be faked.

There are many solutions to protect our AI systems, such as

- Erase sensitive recordings from time to time.
- If you are not using the voice assistant, mute it.
- Turn off purchasing if not needed or set a purchase password.
- Lock the voice assistant down to your personal voice pattern, when available.
- Protect the service account linked to the device with a strong password and 2FA.
- Disable unused services, such as music streaming services.
- Do not turn off automatic update functions on the device.
- Don't use the voice assistant to remember private information such as passwords.
- Use a WPA2 encrypted Wi-Fi network and not an open hotspot at home.
- Create a guest Wi-Fi network for guests and unsecure IoT devices.
- Pay attention to notification emails, especially ones about new orders for goods.

But researchers give us their own ideas. For example, researches of SurfingAttack think Locking the device and turning off the personal results feature on the lock screen can be one way to protect against SurfingAttack. Note that only pattern, PIN and password screen lock can withstand SurfingAttack, swipe screen lock cannot.

Table 3. Tested devices with minimum activation power and maximum distance achievable at the given power of 5 mW and 60 mW. A 110 m long hallway was used for 5 mW tests while a 50 m long hallway was used for tests at 60 mW [3].

Device	Backend	Category	Authentication	Minimum power [mW]*	Max distance at 60 mW [m]**	Max distance at 5 mW [m]***
Google Home	Google Assistant	Speaker	No	0.5	50+	110+
Google Home Mini	Google Assistant	Speaker	No	16	20	–
Google Nest Cam IQ	Google Assistant	Camera	No	9	50	–
Echo Plus 1st Generation	Alexa	Speaker	No	2.4	50+	110+
Echo Plus 2nd Generation	Alexa	Speaker	No	2.9	50+	50
Echo	Alexa	Speaker	No	25	50+	–
Echo Dot 2nd Generation	Alexa	Speaker	No	7	50+	–
Echo Dot 3rd Generation	Alexa	Speaker	No	9	50+	–
Echo Show 5	Alexa	Speaker	No	17	50+	–
Echo Spot	Alexa	Speaker	No	29	50+	–
Facebook Portal Mini (Front Mic)	Portal	Speaker	No	1	50+	40
Facebook Portal Mini (Front Mic)§	Alexa	Speaker	No	6	40	–
Fire Cube TV	Alexa	Streamer	No	13	20	–
EcoBee 4	Alexa	Thermostat	No	1.7	50+	70
iPhoneXR (Front Mic)	Siri	Phone	Yes	21	10	–
iPad 6 Gen	Siri	Tablet	Yes	27	20	–
Samsung Galaxy S9 (Bottom Mic)	Google Assistant	Phone	Yes	60	5	–
Google Pix 2 (Bottom Mic)	Google Assistant	Phone	Yes	46	5	–

*at 30 cm distance, **Data limited to a 50 m long corridor, ***Data limited to a 110 m long corridor, §Data generated using only the first 3 commands.

Table 4. Maximum attack distance on different tables (attack power is less than 1.5 W). The width of Aluminum metal table is 910 cm, the width of metal table is 95 cm, and the width of glass table is 85 cm (A – Activation, R – Recognition) [4].

Device	Max attack distance (cm)							
	Aluminum metal sheet (0.3 mm)		Steel metal sheet (0.8 mm)		Glass (2.54 mm)		MDF (5 mm)	
	A	R	A	R	A	R	A	R
Xiaomi Mi 5	910+	910+	95+	95+	85+	85+	50	47
Google Pixel	910+	910+	95+	95+	85+	85+	45	42
Samsung Galaxy S7	910+	910+	95+	95+	85+	85+	48	N/A

According to laser attack researchers the fundamental solution to prevent Light Commands requires a redesign of the microphone, which seems to require a large cost.

8   Conclusion

To summarize, I wanted to show the difference between some attacks. Each attack, while unique, has its own pros and cons. Manufacturers are already considering measures to protect voice-controlled devices. For example, detection of processing traces in the received signal in order to change its frequency can help from ultrasonic attacks. It would be nice to teach all smart devices to recognize the owner by voice - however, Google, which has already tested these measures in practice on its assistant, honestly warns that this protection can be bypassed using voice recording, and with proper acting skills, the timbre and manner a person's speech can be faked.

Regarding laser attack, as I noted in Sect. 7, it is the opinion of the researchers that a fundamental solution to prevent light commands requires a redesign of the exactly MEMS microphones, which appears to be expensive.

Acknowledgment. This work was supported by Basic Science Research Program through the National Research Foundation of Korea(NRF) funded by the Ministry of Education, Science and Technology(grant number: NRF-2016R1D1A1B01011908).

References

1. Zhang, R., Chen, X., Wen, S., Zheng, X., Ding, Y.: Using AI to attack VA: a stealthy spyware against voice assistances in smart phones. IEEE Access **7**, 153542–153554 (2019)
2. Schönherr, L., Kohls, K., Zeiler, S., Holz, T., Kolossa, D.: Adversarial attacks against automatic speech recognition systems via psychoacoustic hiding, 16 August 2018

3. Sugawara, T., Cyr, B., Rampazzi, S., Genkin, D., Fu, K.: Light commands: laser-based audio injection attacks on voice-controllable systems (2020)
4. Yan, Q., Liu, K., Zhou, Q., Guo, H., Zhang, N.: SurfingAttack: interactive hidden attack on voice assistants using ultrasonic guided waves, January 2020
5. Zhang, G., Yan, C., Ji, X., Zhang, T., Zhang, T., Xu, W.: DolphinAtack: inaudible voice commands, 31 August 2017
6. Roy, N., Shen, S., Hassanieh, H., Choudhury, R.R.: Inaudible voice commands: the long-range attack and defense, 9–11 April 2018
7. Zhou, M., Qin, Z., Lin, X., Hu, S., Wang, Q., Ren, K.: Hidden voice commands: attacks and defenses on the VCS of autonomous driving cars, October 2019
8. Gong, Y., Poellabauer, C.: An overview of vulnerabilities of voice controlled systems, 24 March 2018
9. Yan, C., Zhang, G., Ji, X., Zhang, T., Zhang, T., Xu, W.: The feasibility of injecting inaudible voice commands to voice assistants, 19 March 2019
10. Gong, Y., Poellabauer, C.: Protecting voice controlled systems using sound source identification based on acoustic cues, 16 November 2018

Skills Gap is a Reflection of What We Value: A Reinforcement Learning Interactive Conceptual Skill Development Framework for Indian University

Pankaj Velavan[1], Billy Jacob[1], and Abhishek Kaushik[2]($\boxtimes$)

[1] Dublin Business School,
13/14 Aungier St, Saint Peter's, Dublin D02 WC04, Ireland
gvelavan1@gmail.com, billy.jacob33@gmail.com
[2] Adapt centre, Dublin City University Glasnevin, Dublin 9, Ireland
abhishek.kaushik2@mail.dcu.ie

Abstract. Unemployment is a major obstacle for developing countries, and the skill gap between graduating students and industry expectations is a significant reason which leads to unemployment. Various businesses, industries, and companies in developing countries spend a lot of resources in training the recruited graduates, which causes a loss of revenue for these organizations. However, Universities sustained the loss of reputation when graduates are not finding the intended job even after completing the guided course work. The current effect of COVID has revolutionized the education sector. The universities are under tremendous pressure to reduce the gap in providing the skills through webinars or online interactive classes, which prepare the students for the corporate world. The current system in developing countries like India, to conceive job-ready graduates is less engaging and not easily accessible to every pupil. To overcome these challenges, we are proposing a skill-based interactive online system for skill development. This system would be guiding the students to attain the skills for a career based on their interests and talents with the help of a reinforcement learning agent trained with the requirement of the industrial and academic expert. This system uses a Q-learning algorithm to feed the students with skills in a particular order and guide them to achieve their goals.

Keywords: Unemployment · Reinforcement learning · Learning portal

1 Introduction

India is one of the fastest-growing economies in the world. However, despite its rapid growth, unemployment in India is a widespread issue. One of the major concerns for India is how to shift the population into the skilled workforce. As per Mamgain et al. [1] India has 65% of its population under the age of 35 and India displays one of the highest available workforces in the world. TeamLease

M. Singh et al. (Eds.): IHCI 2020, LNCS 12615, pp. 262–273, 2021.
https://doi.org/10.1007/978-3-030-68449-5_27

and IIJT [2] reported that India will have 25% of the world's total workforce in 2025. The study conducted by Sasmita et al. [18] suggested that the shortage of skilled workers in India has forced organizations to spend a lot of time and money in expertise acquisition, education, talent advancement, and talent retention. With 12.8 million job seekers invading the job market every year in India, the companies cannot nurture talents continuously. The study conducted by Blom et al. [3] quoted that the current higher education system in India is not catering to the needs of these organizations. The National Education Policy Draft Report by Subramanian [4] made some serious observations and recommendations about the education system in India. According to the report, the quality of many universities and colleges and the standard of education they provide are far from satisfactory in India and they lack transferring skills based on the requirement of Industries. The study conducted by Vidal et al. [19] reported the stress on universities all around the world to create job ready graduates, for the universities to feature in the International University rankings.

To overcome these challenges, we propose a skill-based interactive student engaging system which focuses on up-skilling students intuitively and encouragingly, through the use of a skill-focused software platform. We investigated the following area prior to drafting the conceptual framework.

1.1 Unemployment Situation in Developing Countries

Unemployment and underemployment are the major concerns for human resources in developing countries. Unemployment exacerbates inequalities in wage and fuels the reduction in quality of life and disturbed the social life.

The study conducted by Herz and vanRens [5] observed that the mismatch in the skills that prevails between available jobs and graduates, is a significant contributor to unemployment in Bangladesh. These findings are also found to be acknowledged by Mohd Abdul Kadir et al. [6] who hints that in addition to the job mismatch, employability skills and low wages have also taken a toll on the graduate employment. Various reasons contribute towards the graduate being unemployable and one such reason as suggested by Deeba et al. [7] is that there exists a difference in the perspective of students and employers in terms of the essential graduate skillset. As a result of this, the industries end up uttering that the students are least prepared for a professional career reveals the study by Karunasekera and Bedse [8]. The study conducted by Lai [9] also highlights that the graduates are not skilful enough for the industry. The research performed by Mncayi [10] cites that in a developing country like South Africa apart from the availability of relevant jobs the issues like networking and jobs in the preferred geographic location are also of concern. Apart from the lack of professional skills, Yahaya et al. [11] revealed that lack of graduate attitude, English proficiency and entrepreneurship skill are also the key contributors to unemployment.

1.2   Solutions to Curb Unemployment in Developing Countries

Numerous software solutions have been attributed to tackle the unemployment problem. The research conducted by Subrahmanyam [12] and Sabu [21] considered the gap between industry and academia in India as the crucial factor and suggested for better collaboration between industry and educational institutions. Subrahmanyam [12] also stresses the need for the students to gain knowledge on topics viz., domain intelligence, architectural intelligence as vital. Reena Grover et al. [13] did clustering analysis by using the ward's method and had profiled graduates who have high employability chances using the chi-squared test. This analysis would help the institutions give more guidance for the latter category of students to improve on.

Apart from all these approaches, a more industry-oriented method had been suggested by Engela and David [21] based on their research in South Africa and had listed out the possible benefits of incorporating a 'work-related experience' program into the secondary school curriculum to curb unemployment. Agajelu and Nweze [20] suggested that entrepreneurial education and human capital development can be implemented as a solution to unemployment in Nigeria and suggested the government to fund entrepreneurial education programmes to attract and equip the youths with skills required by the industry.

1.3   Software Solutions for Unemployment

After the prodigious growth of technology, it has been easy to create and roll out solutions to the audience in an easy manner. Some of the human-computer interaction based solutions made a huge impact on the market. Nair et al. [30] presented the conceptual framework for in-house skill development in the organization to increase the work productivity and reduce the expense for outsourcing the external skilled worker. The research conducted by Ilkin Ibadov et al. [14] suggested to reduce the unemployment problem by frequently updating the curriculum based on the industry requirements. They have developed a dynamic competency model, which leverages natural language processing tools like LDA (Latent Dirichlet Allocation), LSA (Latent Semantic Analysis) and PLSA (Probabilistic Latent Semantic Analysis) for identifying the key competencies from the text of the vacancies. Mori and Kurahashi [15] had approached the graduate employment market by agent-based simulation using the reinforcement learning actor-critic method. They have created two types of agents; student agent and firm agent. Both the agents in this environment act and learn independently. On the contrary our framework consists of only one reinforcement learning agent as shown in Fig. 5. Reinforcement learning based approach had started to prove effective and had resulted in various researches in this field like the one conducted by Csanád Csáji and Monostori [16] which investigated the possibility of applying value function based reinforcement learning methods in cases where the environment may change over time. This will be helpful in our application as the jobs and the corresponding skills vary making it an environment that changes over time. A similar study conducted by Chen et al. [17] led them to propose an algorithm, called UQ-learning, to better solve action selection problem by using reinforcement learning and utility function.

1.4   Evaluation of Interactive Systems

Evaluation is a vital part of any interactive software solution. Users desire interactive systems to be simple to learn, efficient, safe, and satisfying to use [25]. Bangor et al. [22] presented System Usability Scale (SUS) on numerous products in all phases of the development lifecycle. The article authored by Brooke [23] discuss subjective System Usability Scale and its detailed methodology. A study done by Harrati et al. [27] explored the user satisfaction of e-learning systems via usage-based metrics and system usability scale analysis. Finstad [26] investigated the SUS verbally to native and non native English speakers and found that a final score of the SUS in a remote, unsupervised, international usability evaluation might be biased due to a misunderstanding or misinterpretation of terminology in the scale. A comparative study done by Kocabalil et al. [24] compared six questionnaires for measuring user experience and suggested using multiple questionnaires to obtain a more complete measurement of user experience. Jeff Sauro [28] listed the five ways such as percentile ranks, grading system, using adjectives instead of numbers, acceptability, promoters and detractors to interpret the SUS score on chabot or multiview agents [31].

2   Conceptual Framework for Skill Development

We had developed a conceptual framework to create a skill set development platform which could act as a bridge between the industry and educational institutions to prepare the students based on the current market requirements. All the companies in India could register and list out all the skill sets that they are looking out for in the forthcoming year as shown in Fig. 1. Educational institutions would provide access to students to register themselves as the knowledge seekers. The students would have no information on the vacancies available in the companies. Instead, they would be able to set their dream roles as goals, and our platform leveraging reinforcement learning would help them achieve it through periodical guidance of choice of skills. The reinforcement learning algorithm works based on giving rewards for all actions taken. If the agent suggests an irrelevant skill like mainframe for a java developer role, then the environment will provide a negative reward. If the agent suggests a relevant skill like SQL, then the agent is given a positive reward. Towards the end, if the agent can guide the user to achieve his/her role of becoming a java developer through the suggestion of all relevant skills then a huge final reward is given to the agent as shown in Fig. 1. The conceptual framework is divided into two components.

- The user-interface
- Reinforcement learning model

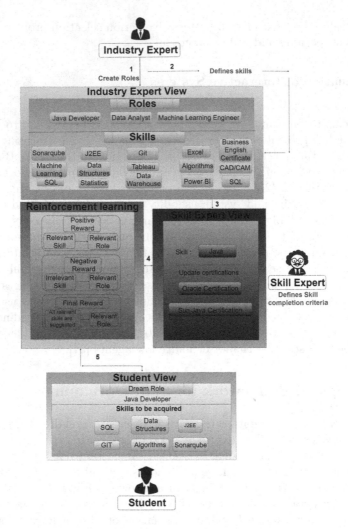

Fig. 1. Flow chat of conceptual framework

2.1   User-Interface Design and Workflow

The User Interface system consists of two modules viz., the student view, and the industry expert view. The three main stakeholders in the proposed skills-based system are employers, students, and skilled experts. This proposed system would help the students acquire all the necessary skills to achieve their dream role over the tenure of their graduation.

The industry experts are the people who possess a commendable industry work experience and who operate as part of the various organizations. Their responsibilities are as shown in Table 1. The skill experts are the professors and educators who possess a commendable industry and academic work experience

Table 1. The three views of the user interface.

Industry expert view	Skill expert view	Student view
-Create roles	- Update roles for students	- Select Dream roles
-Create Skills	- Analyse and update the skills	- Complete the suggested skill
-Update role count	- Frame skill completion criteria	- View peers skills

and who operate as part of the educational institutions. Their core responsibilities are as shown in Table 1 The students would be able to view the following components viz., the roles, skills to be obtained, skills of peers. The student can select their dream role as shown in Fig. 2, and the interface would allocate them the required skills one by one and encourage them to complete those skills. The student would be able to upload certificates of online courses or online exams in the interface for a specific skill to be achieved. These criteria would be suggested by the skilled experts who work as part of the educational institutions as shown in Fig. 1. Once the student completes one skill then the next skill to be completed would appear. The student would also be able to see the skills completed by his peers or friends. This creates a competitive learning environment for learners to reach their desired careers. Another way of operation of this interface would be the student administering the various skills that he acquired, and the system provides the relevant jobs that he can apply as shown in Fig. 3.

2.2 Reinforcement Learning Component

The Reinforcement learning component is the proposed back end module that feeds the students with the skills. The proposed reinforcement learning model is Q-learning, which is a model-free reinforcement learning algorithm to learn the quality of actions determining an agent what action to take under what circumstances [29]. The Q-learning model learns in an environment by trial and error using the feedback from actions and states. There are five basic components in Reinforcement Learning namely agent, environment, reward, Q-table and action.

- Agent: the agent is a module that takes decisions based on positive and negative rewards. In this instance, it would be the module which feeds the skills to the students
- Environment: The environment is nothing but a task or simulation. In this instance, the skills, the skill selection activity, the student, and the industry experts constitute the environment.
- Actions: the selection of a certain skill by the agent in the environment.
- States: selection of an interested role by the student, completion of a certificate or an online course by the student, completing a skill by the student would be the different states that are available in the environment.
- Q-table: Q-Table is a simple matrix or a lookup table where we calculate the maximum rewards for an action at each state. The Q-table will guide the agent to the best action at each state. For example a sample Q-table for an

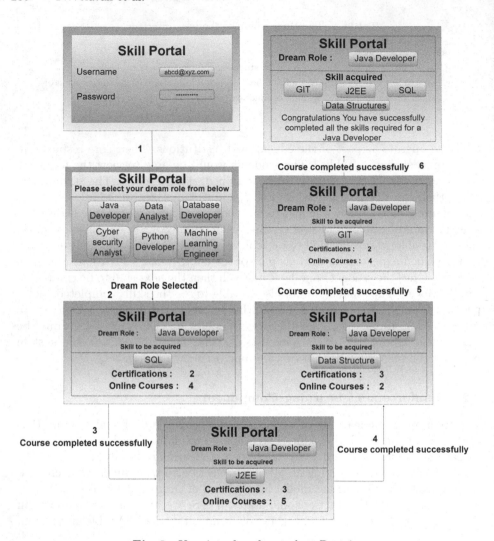

Fig. 2. User interface for student Part 1

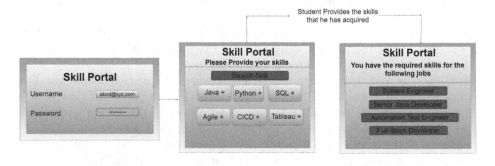

Fig. 3. User interface for student Part 2

environment with 2 states and 3 actions will look like $\begin{pmatrix} 0 & 0 & 0 \\ 0 & 0 & 0 \end{pmatrix}$ where each row corresponds to a state and each column corresponds to an action. The illustration of the above-mentioned components is shown in Fig. 4.

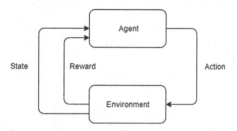

Fig. 4. Reinforcement learning

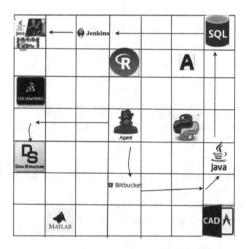

Fig. 5. Sample reinforcement learning environment

Let us consider the sample Q-learning environment with the Agent, states, actions (skills, role) as shown in Fig. 5. When a job interest is selected by the student, that job is updated as the interested role in the environment as shown in Fig. 5. When the agent feeds the skills to the student to learn one by one, if the agent advises a relevant skill to the interested role then the agent receives positive rewards from the environment and when the agent advises an irrelevant skill to the interested role then the agent receives negative rewards, and when a student completes all the skills required for a particular job then the agent receives a final big positive reward from the environment. In the training phase, the agent investigates the various actions for the various states and collects the corresponding rewards. In the implementation phase, the agent feeds the proper skill to the student based on the knowledge gained from the training phase. This way instead of using conditional operators to pre-define a particular set of skills for a certain job the reinforcement learning model will make the agent to learn the required skills based on the rewards, states, and actions present in the environment.

$$Q(s,a) = Q(s,a) + \alpha\,[\,r\ + \gamma\ max_\alpha\ ,\ Q(s',a')\ -\ Q(s,a)] \tag{1}$$

where,

$Q = Q - table\ present\ in\ the\ Q - learning\ environment.$

$s = states\ such\ as\ selection\ of\ an\ interested\ role,\ completion\ of\ a\ skill\ etc.$

$a = action,\ the\ selection\ of\ a\ certain\ skill\ by\ the\ agent\ in\ the\ environment.$

$Q(s,a) = new\ q - value\ (value\ present\ in\ the\ q - table\ in\ s^{th}\ row\ and\ a^{th}\ column).$

$Q(s',a') = old\ q - value\ in\ the\ q - table\ in\ s^{th}\ row\ and\ a^{th}\ column.$

$\alpha = the\ learning\ rate,\ set\ between\ 0\ and\ 1,$

$\gamma = discount\ factor,\ also\ set\ between\ 0\ and\ 1,$

$max_\alpha = the\ maximum\ reward\ that\ is\ attainable\ in\ the\ state\ following\ the\ current\ state.$

Algorithm 1. Reinforcement Learning Training phase

1: $\alpha = i,\ 0 < i < 1$
2: $\gamma = j,\ 0 < j < 1$
3: **for** $episodes = 1, 2, \ldots, N$ **do**
4:     The Q-learning model explores the various skills available and starts to suggest skills one by one to the student.
5:     **if** the agent feeds relevant skills to the student **then**
6:         the environment provides positive rewards
7:     **else**
8:         the environment provides negative rewards
9:     **end if**
10: **end for**
11: Q value is calculated using Eq. (1)
12: Q value will be updated in the Q-table for that particular state and action as shown in Equation(1)
13: the model will be trained for all the listed jobs in the platform using various student volunteers or alumni from educational institutions .

---

3   Training and Evaluation Phase

The reinforcement learning agent would be trained by two approaches such as trained by the real users or trained by interactive elicitation of the rewards to reach the final goal. Later both these approaches will be compared among each other on evaluation criteria of usability. The methods which outperform other method, would be selected for final training. The training method and evaluation methods are discussed below:

- Training the model using volunteer users (User Training).
- Training the model using the gym environment (Gym Training).

3.1   User Training

Alumni from educational institutions who have landed jobs in different industrial domains can be asked to evaluate the system. Different types of questionnaires would be prepared for measuring the user experience. They would be instructed to provide the dream role as the job they are currently in and then they can provide the certificates and the completed courses in each of the skills and check whether the interface feeds them the right set of skills for that particular job.

3.2   Gym Training

The reinforcement learning agent would be trained in an open AI Gym environment. The Gym environment would simulate the role of the user. A sample set of states would be prepared in open AI Gym environment for the different set of jobs present in the environment. The agent would be trained against all these states over a 1000 episodes and the corresponding q-values will be populated in the Q-table. The Q-table will then be used in the implementation phase.

3.3   User Evaluation

The interface can be evaluated by instructing the students to select their job interests and test certificates can be provided as inputs in each of the step and the skill experts can investigate whether the interface provides the right set of skills in each step on after each training setting individually. We would use System Usability Scale to asses usability, User Experience Questionnaire to assess the user experience of the system and Chatbot Usability Questionnaire to assess personality, on-boarding, navigation, understanding, responses, error handling and intelligence of our system.

4   Conclusion and Future Work

In this paper, we have proposed a novel idea to minimize the skill gap between the industry and educational institutions by proposing a framework which uses reinforcement learning to enhance the student's learning experience through the suggestion of relevant skills. Since this platform is being evaluated using the alumni who have already acquired a job through this portal, it would improve the model to a greater extent through their experience. Our main aim is to train our reinforcement learning model to follow the cognitive dimension process and help the students understand, apply, analyze, evaluate, and finally achieve their dream career. Further work involves, evaluating the system using the System Usability Scale method. This framework can be extended to industries, Businesses, and companies wherein the employees working in these organizations can upskill and reach the interested role within the organization.

References

1. Mamgain, R.P., Tiwari, S.: Youth in India: challenges of employment and employability (2015)
2. A Report By TeamLease and IIJT 3 Background (2012). www.iijt.net
3. Blom, A., Saeki, H.: Employability and skill set of newly graduated engineers in India. The World Bank (Policy Research Working Papers) (2011). https://doi.org/10.1596/1813-9450-5640
4. National Education Policy 2020 Ministry of Human Resource Development Government of India
5. Herz, B., van Rens, T.: Accounting for mismatch unemployment. J. Eur. Econ. Assoc. **18**(4), 1619–1654 (2020). https://doi.org/10.1093/jeea/jvz018
6. Mohd Abdul Kadir, J., et al.: Unemployment among graduates - is there a mismatch? Int. J. Asian Soc. Sci. **10**(10), 583–592 (2020). https://doi.org/10.18488/journal.1.2020.1010.583.592
7. Ferdous, Z., Asad, I., Deeba, S.R.: Analyzing analyzing the factors contributing to graduate unemployment. In: 2019 IEEE Global Humanitarian Technology Conference, GHTC 2019. Institute of Electrical and Electronics Engineers Inc. (2019). https://doi.org/10.1109/GHTC46095.2019.9033029
8. Karunasekera, S., Bedse, K.: Preparing software engineering graduates for an industry career. In: Proceedings of the Software Engineering Education Conference, pp. 97–104 (2007). https://doi.org/10.1109/CSEET.2007.39
9. Lai, W.S.: Unemployment among graduates: study of employers perception on graduates (2011). https://www.researchgate.net/publication/241064433
10. Mncayi, P.: An analysis of the perceptions of graduate unemployment among graduates from a South African University (2016). https://www.researchgate.net/publication/321242794
11. Yahaya, M.A., et al.: Factors influencing graduate unemployment: English proficiency. In: 2017 7th World Engineering Education Forum (WEEF) (2017). https://doi.org/10.1109/WEEF.2017.8467088
12. Subrahmanyam, G.V.B.: A dynamic framework for software engineering education curriculum to reduce the gap between the software organizations and software educational institutions. In: Proceedings - 22nd Conference on Software Engineering Education and Training, CSEET 2009, pp. 248–254 (2009). https://doi.org/10.1109/CSEET.2009.8
13. Vashisht, M.G., Grover, R.: Employability profiling of engineering students using clustering. In: Proceedings of IEEE International Conference on Signal Processing, Computing and Control, pp. 191–195. Institute of Electrical and Electronics Engineers Inc. (2019). https://doi.org/10.1109/ISPCC48220.2019.8988505

14. Ibadov, I., et al.: The concept of a dynamic model of competencies for the labor market analysis. In: Proceedings - 2020 Ural Symposium on Biomedical Engineering, Radioelectronics and Information Technology, USBEREIT 2020, pp. 511–515. Institute of Electrical and Electronics Engineers Inc. (2020). https://doi.org/10.1109/USBEREIT48449.2020.9117691
15. Mori, K., Kurahashi, S.: Optimizing of support plan for new graduate employment market: reinforcement learning. In: Proceedings of SICE Annual Conference. IEEE Conference Publication (2010). https://ieeexplore.ieee.org/abstract/document/5602591
16. Csanád Csáji, B., Monostori, L.: Value function based reinforcement learning in changing Markovian environments. J. Mach. Learn. Res. **9**, 1679–1709 (2008)
17. Chen, K., et al.: Adaptive action selection using utility-based reinforcement learning. In: 2009 IEEE International Conference on Granular Computing, GRC 2009, pp. 67–72 (2009). https://doi.org/10.1109/GRC.2009.5255163
18. Sasmita, N., Kumar, R.H.: Exigency of re-skilling for organization and employees growth. Soc. Sci. **3**, 65–67 (2018)
19. Vidal, J., Ferreira, C.: Universities under pressure: the impact of international university rankings. J. New Approach. Educ. Res. (NAER J.) **9**(2), 181–193 (2020). https://www.learntechlib.org/p/217624/. Accessed 9 Oct 2020
20. Agajelu, K.N.: Entrepreneurial education and human capital development: a solution to unemployment, p. 8, 15 November 2018
21. Engela, V.D.K., David, K.: Work-related experience: a solution to unemployment in a developing country such as South-Africa, p. 10 (2012)
22. Bangor, A., Kortum, P.T., Miller, J.T.: An empirical evaluation of the system usability scale. Int. J. Hum.-Comput. Interact. **24**(6), 574–594 (2008). https://doi.org/10.1080/10447310802205776
23. Brooke, J.: SUS - a quick and dirty usability scale, pp. 1–7 (1996)
24. Kocabalil, A.B., Laranjo, L., Coiera, E.: Measuring user experience in conversational interfaces: a comparison of six questionnaires. In: Proceedings of the 32nd International BCS Human Computer Interaction Conference (2018). https://doi.org/10.14236/ewic/HCI2018.21
25. Laugwitz, B., Held, T., Schrepp, M.: Construction and evaluation of a user experience questionnaire. In: Holzinger, A. (ed.) USAB 2008. LNCS, vol. 5298, pp. 63–76. Springer, Heidelberg (2008). https://doi.org/10.1007/978-3-540-89350-9_6
26. Finstad, K.: The system usability scale and non-native English speakers, p. 4 (2006)
27. Harrati, N., Bouchrika, I., Tari, A., Ladjailia, A.: Exploring user satisfaction for e-learning systems via usage-based metrics and system usability scale analysis. Comput. Hum. Behav. **61**, 463–471 (2016). https://doi.org/10.1016/j.chb.2016.03.051
28. Sauro, J.: 5 ways to interpret a SUS score (2018)
29. Sutton, R.S., Barto, A.G.: Reinforcement Learning: An Introduction. MIT Press, Cambridge (2018)
30. Nair, S., Kaushik, A., Dhoot, H.: Conceptual framework of a skill-based interactive employee engaging system: in the Context of Upskilling the present IT organization. Appl. Comput. Inf. **15**, 1–26 (2020)
31. Kaushik, A., Bhat Ramachandra, V., Jones, G.J.: An interface for agent supported conversational search. In: Proceedings of the 2020 Conference on Human Information Interaction and Retrieval, pp. 452–456, March 2020

PRERONA: Mental Health Bengali Chatbot for Digital Counselling

Asma Ul Hussna[1], Azmiri Newaz Khan Laz[1], Md. Shammyo Sikder[1],
Jia Uddin[2(✉)], Hasan Tinmaz[2], and A. M. Esfar-E-Alam[1]

[1] BRAC University, Dhaka, Bangladesh
{asma.ul.hussna,azmiri.newaz.khan,md.shammyo.sikder}@g.bracu.ac.bd,
esfar.alam@bracu.ac.bd
[2] Technology Studies, Endicott College, Woosong University, Daejeon, Korea
jia.uddin@wsu.ac.kr, htinmaz@endicott.ac.kr

Abstract. These three words "Human", "Machine" and "Interaction" refer to the modern era of technology where researchers emphasized the interaction between computer machines and humans. We humans cannot confine ourselves to relying solely on human-based healthcare systems. The current era of digitization brings advances and additional opportunities for medical care, especially in general and clinical psychology. Existing chatbots such as Woebot, Wysa, Moodkit, etc. do not allow users to express themselves. However, our chatbot PRERONA allows the user space to talk about whatever they want along with it is intelligent enough to ask questions as well as answer questions. In our research work, we have presented a model named 'PRERONA: Mental Health Bengali Chatbot for Digital Counselling'. PRERONA is designed to help depression struggling people especially for Bengali users who are lack attention and have none to talk about their problem. The main objective of this chatbot PRERONA is to give instant answers to questions and queries as well as to provide proper mental health care to Bengali's in times of depression. Although our bot PRERONA is made only for Bengali users, we have it in multiple languages such as Bengali, English and Korean. In addition to that, we are using adaptive learning process. If the bot can not answer any questions, that will be recorded as well as added to that database later on for more accuracy. Natural Language Processing (NLP) is used here to successfully implement these languages. The complete structure of this chatbot and how it works are observed accurately throughout this research.

Keywords: NLP · Interaction · Psychology · Adaptive · Depression · Mental health care

1 Introduction

Machine learning has played a crucial part in the research of computational learning theory. It becomes an integral part of artificial intelligence to construct

M. Singh et al. (Eds.): IHCI 2020, LNCS 12615, pp. 274–286, 2021.
https://doi.org/10.1007/978-3-030-68449-5_28

algorithms when a computer program simulates human conversation through voice commands or text chats or both, we call that a chatbot [5, 10]. This automated program interacts with users like a human. Chatbot can play significant role extensions to health care services. If we specify, the need for chatbots in the psychological or mental health sector is rising along with the alarming rate of increasing number of depression and anxiety patients [4]. According to the world health organization (WHO) research, over 300 million people are struggling from depression worldwide which is equivalent to 4.4% of the world's total population. It is more common in females (5.1%) than males (3.6%) [1]. In Bangladesh, according to the first national survey on mental health (2003–2005), about 5 million people are suffering from depression including 4.6% adult and 1% children [6]. As per a recent WHO study, it is estimated that in Bangladesh, 4.1% of the population suffer from depression, and 4.4% from anxiety disorder [9]. But this is even more shocking that in spite of the existing treatments for mental disorders, 76%-86% people of low to a middle-income country do not seek medical attention for their disorders either out of negligence or hesitation due to social stigma. In this situation, our mental health chatbot PRERONA can play a vital role. It can ask real time questions to determine and fix the mental condition of the users, PRERONA lets its users share their feelings and also responses accordingly. Additionally, it provides the users with a lot of contents relating anti-depression as well as anti-anxiety. Moreover, PRERONA is a natural language processing (NLP) based tri-language chatbot, it is able to understand all languages based on the given database. Initially, we have set Bengali, English and Korean Corpus to our database which makes our bot more interesting. The main target behind PRERONA is to develop an online-based mental health system which can be considered an alternative to the typical existing systems. Here users will get 24/7 support and can express themselves without any hesitation along with proper guidance to get over their problems under expert assistance. This paper presents a detailed analysis of the proposed model of PRERONA, experimental setups and result analysis.

2   Literature Review and Related Work

Eliza was created in an AI laboratory in MIT between 1964-66 is the oldest and well known chatbots in the history of chatbots. Eliza was made in that way so that she could ask open questions with which she also answered, which simulated her role as a Rogerian psychotherapist [11, 13]. Now a days we can find a number of chatbots in practice such as Alexa, SIRI has brought revolutionary change in this field. Advanced technology has allowed for using chatbots in the creation of software-based therapy. Now the chatbot will see you instead of a therapist. Early work on mental health chatbot named Wysa and Woebot claims to be AI powered, follows cognitive behavioral therapy (CBT) [2]. If we consider works on

Bengali chatbot, there are Golpo and doly which are NLP based and successfully able to chat with users in Bengali [3,7]. Here, PRERONA has the features of both the types discussed above. It is a Natural language process (NLP) based mental health chatbot which can efficiently and skillfully chat with the users or patients especially in Bengali users.

3   Proposed Approach

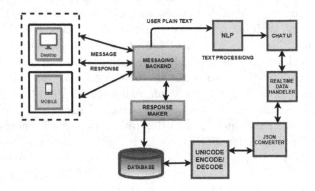

Fig. 1. Block diagram of the system

A detailed structure of the PRERONA: chatbot is illustrated in Fig. 1. The following sections discuss the different blocks related to the model. First of all, the user will type text that text will take as an input to the messaging backend, it will send to the NLP unit which is basically a text processing unit [8]. The chat user interface receives data as well. At the same time the real-time data handler catches all the data and sends it to JSON conversion which is a JavaScript text based format for presenting structured data. Now it sends the data to the UTF-8 which is a variable-width character encoding system used for electronic communication [12]. Our custom dataset receives data from UTF-8 and checks if it has that keyword or input in the dataset or not. If it matches with the dataset, sends the reply to the response makers. Otherwise it will show our default error message "Sorry not be able to understand you. Please tell me again". This is how our system responds to user messages.

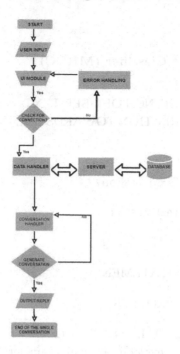

Fig. 2. Flowchart of the system

3.1 Flowchart of the System

In Fig. 2, we can see user input is sent to the manual operation which is the user interface. Then a decision maker checks both network and internet connectivity on or off. If it finds off, it directly goes to the error handling process. This process again sends the data to the UI module so that we can say users must have an internet connection to access our application. Once again if decision makers find connectivity is okay, then it sends data to the data handler. From here, data can send to or receive from servers; databases as well as the direction flow between these are bidirectional. After that data handler process sends data to the conversion handler. Again a decision maker comes, it checks whether the data generated is converted or not. If not, it sends data to the conversion handler. Otherwise, it generates data, gives output or reply as well. And then it terminates the conversation.

3.2   Algorithms

Mental Health Context Classifier (MHCC)

Pseudo Code:

```
        INPUT:     A BUNCH OF USER TEXT
        OUTPUT: DIRECTION TOWARDS SOLUTIONS

        NOTATION:

        TQ:           Text Question

        TA:           Text Answer

        INITIALISATION:

            1.  FIRST STATEMENT

                LOOP PROCESS

            2.  WHILE ( T  IN TEXT )  DO
            3.  sp_cha = explode_speacial_character(T)
            4.  seg = segmentation(T)
            5.  Done
            6.  WHILE ( STOP  WORD  IN  T)
            7.  DELETE STOP WORD
            8.  GOTO STEP  3
            9.  DONE
            10. TQ = json_data_transfer()
            11. TA = find_relative_answer()
            12. IF ( TQ ⇔ TA) > 90% THEN
            13.   specific direction to user
            14.   JUMP A
            15.    ELSE IF (TQ⇔TA)>50% && < 90%
            16.      probabilistic direction to user
            17.     JUMP B
            18. ELSE
            19. Ask_QUERY_AGAIN()
            20. A : Solution_Provided_More_Accurate
            21. B: Solution_Provided_Less_Accurate+suggestions
            22. END IF
            23. END
```

3.3   Flowchart of the Algorithm

(See Fig. 3).

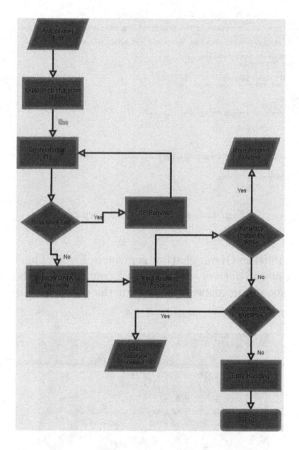

Fig. 3. Flowchart of the algorithm

3.4   Dataset Description

Bengali Corpus

We collected data through interviews with lots of people who were asked questions about depression. Since there is no open source Bengali mental health database for NLP, we have followed the interview questionnaire approach for collecting Bengali data. We have created a customized Bengali dataset by analyzing our collected data with the help of some mental health specialists (Fig. 4).

```
bangla.yml    knlrcorpus ym    mglshcorpus ym    psychology ym    botprofile ym

 1
 2
 3
 4    categories:
 5      - Mental Health
 6
 7    conversations:
 8
 9    - - প্রেরনা চ্যাট বটে আপনাকে স্বাগতম। আপনি কি সংলাপ এর জন্য প্রস্তুত? (Bot default question)
10      - প্রস্তুত(if user reply is positive)
11    - - এখন আমার প্রশ্ন গুলোর উত্তর দিয়ে আমাকে সাহায্য করবেন। আপনি কি সম্প্রতি স্বাভাবিক এর চেয়ে কম বা বেশি ঘুমান? (Bot will start conversation)
12      - প্রস্তুত না(if user reply is negative)
13    - - তাহলে আমরা আবার পড়ে তোমার সুবিধা মতো সময়ে সংলাপ করবো। ধন্যবাদ। (Bot will close the conversation)
14
15      - আর বাঁচতে ইচ্ছে করে না।।নিজেকে শেষ করতে ইচ্ছে করে (user question)
16    - - মরে যাওয়া কোন উপায় নয়, জীবন সুন্দর। (bot reply)
17
18      - আমাকে কাউন্সিলরের নাম্বার দাও (user question)
19    - - জরুরী বিভাগ শাখায় দেয়া আছে, দয়াকরে দেখুন। (bot reply)
20
21      - নিজেকে তুচ্ছ বা মূল্যহীন মনে
22        হয়।।আজকাল নিজেকে তুচ্ছ বা মূল্যহীন মনে
23        হচ্ছে।।আমি লুজার।।আমাকে দিয়ে কিছু হবে না।। আমাকে দিয়ে কিছু হবে না।। (user question)
24
25    - - কেও কখনও লুজার হতে পারে না।(bot reply)
26
27      - আমার ঘুম আসে না।(user question)
28    - - শরীর চর্চা করুন ঘুম আসবে। (bot reply)
29      - তোমাকে কে বানাইসে।। ।তোমাকে কে বানিয়েছে?।। তোমাকে কে বানাল? (user question)
30    - - আমাকে ব্র্যাক
31      বিশ্ববিদ্যালয়ের শিক্ষার্থী
32      আজমিরি নেওয়াজ, আসমা উল হুসনা
33    - এবং সায়মা শিকদার বানিয়েছে। (bot reply)
```

Fig. 4. A sample Bengali corpus

Input-Output Pattern Of our chatbot is given in Fig. 5. This figure is actually a sample pattern of our dataset, how we have connected similar types of input as well as it also shows our answer pattern for the predicted input.

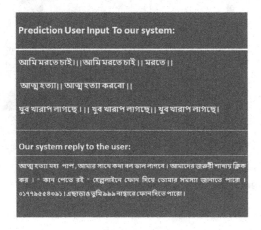

Fig. 5. Input-output pattern of the system

We have allowed some users to use our system for testing purpose. In Fig. 6, Conversations between users and our bot is given -

Fig. 6. Testing data with different types of the users

4   Result Analysis

In the experiment we gave our chatbot to some users to chat. Our chatbot can chat in three different languages as a result we have got three types of results. Besides that, we have done our experiment into two phases. The accuracy of our chat depends on three things such as Database of our system, Typing skills of user, Data processing efficiency of our system. First, we have developed our Bengali database based on practical experience. The accuracy rate increases along with database enrichment. Second, it is a challenge for the user to type Bengali correctly as each letter of the Bengali alphabet has a different format. So it might be very difficult for any system to communicate with the user. We have emphasized keyword-based communication to increase performance. Third, we found our system data processing efficiency is above 98%. To mention here that users are not confined to selective questions or answers. In addition it is important for the user to type the question or the key word correctly. We have found this problem mostly in Bengali conjunctive letters.

Accuracy for Bengali Language Phase 1
In the first phase for Bengali language, our chatbot showed 50% accuracy for the two users, 45% for the two users, 40% for the seven users, 35% for the two users and it showed lowest 20% for the two users. The average accuracy for the Bengali language was 35%. From the first phase we tried to collect a lot of Bengali words and those words were added to our dataset. As a result, in the second phase we get some improvements in the accuracy (Fig. 7).

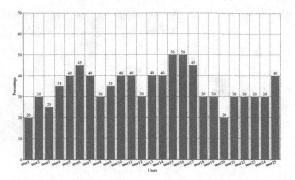

Fig. 7. Accuracy for Bengali language phase 1

Phase 2

From the 2nd phase, we can see the improvements in the accuracy, got highest 60% accuracy from the one user, 50% accuracy from the four users, and 45% accuracy from the four users, 40% for the ten users and lowest 30% for the four users. In addition the average accuracy in the 2nd phase was boosted up to 41.2% from 35% (Fig. 8).

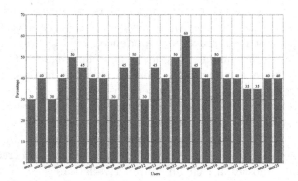

Fig. 8. Accuracy for Bengali language phase 2

Accuracy for English Language

In the case of the English language, there are no conjunctive letters. So in the English language, the accuracy rate is higher than in the Bengali language. The accuracy rate is proportional to the vastness of our database. If wrong punctuation and spelling mistakes are reduced from user end, efficiency will increase.

Phase 1

For the 1st phase of the English language the accuracy rate was not so high. It was below average. As we can see from the Fig. 9, it gives the highest 50%

accuracy for the one user, 45% accuracy for the three users, 40% accuracy for the seven users, 35% accuracy for the four users, and the lowest 25% accuracy for the two users. It gives us an average 35.8% accuracy.

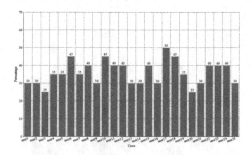

Fig. 9. Accuracy for English language phase 1

Phase 2
But in the 2nd phase, we can see some change in the accuracy. As we mentioned before that we use lots of words from the users in the first phase which were not included in the dataset. So there is an improvement in the accuracy for phase two from Fig. 10, we can see that we got the highest 50% accuracy from the four users, 45% accuracy from the six users, 40% accuracy from the eleven users, and the lowest 30% for the three users. We got 40% average accuracy in the 2nd phase which is 4.2% more than the 1st phase.

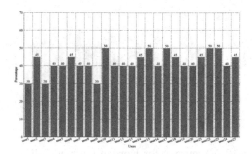

Fig. 10. Accuracy for English language phase 2

Accuracy for Korean Language
For the Korean language, we did our accuracy test in two phases. We could not interview native Korean people for the Korean language, initially; we have included the translated version of the English dataset. Moreover, maximum users who communicated with the bot as a test case have also used the translated method for Korean which decreased the accuracy rate. In the early phase, it

gave us poor accuracy. In the first phase for the Korean language, our chatbot showed 40% accuracy for the three users, 35% for the one user, 30% for the eleven users, 20% for the eight users and it showed the lowest 10% for the one users. The average accuracy was 26.2%.

Phase 1

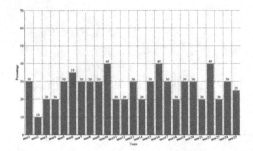

Fig. 11. Accuracy for Korean language phase 1

Similar to other languages we try to collect lots of Korean words and those words were added to our dataset. As a result, in the second phase, we got some improvements in accuracy (Fig. 11).

Phase 2

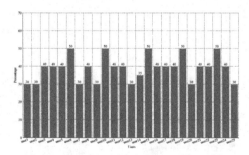

Fig. 12. Accuracy for Korean language phase 2

From the 2nd phase, we can notice some improvements and got highest 50% accuracy from the five users, 40% accuracy from the twelve users, and 35% accuracy from the one user and lowest 30% for the seven users. In addition, the average accuracy increases from 26.2% to 39% (Fig. 12).

Data Accuracy

We can see from the Table 1 that in the phase 1 our chatbot gave 11.61% accuracy for the Bengali language, 11.94% accuracy for the English Language and 9.07% accuracy for the Korean Language. In total it gave 32.62% accuracy and 67.38% error.

Data Accuracy Percentage for Phase 1 is shown in Table 1

Table 1. Data accuracy percentage for phase 1

Bengali language	English language	Korean language	Error
11.61%	11.94%	9.07%	67.38%

But in the phase 2, we can see from the Table 2 that it gave 13.70% accuracy which is 2.09% more than phase 1 for the Bengali language. Furthermore, it gave 14.00% accuracy for the English Language and 12.93% accuracy for the Korean Language from the Table 2. In total it gave 40.63% accuracy and 59.37% error which is 8.01% lesser error than phase 1.

Data Accuracy Percentage for Phase 2 is shown in Table 2

Table 2. Data accuracy percentage for phase 2

Bengali language	English language	Korean language	Error
13.70%	14.00%	12.93%	59.37%

5 Conclusion and Future Work

Our research is a pioneering work in the field of Bengali mental health chatbots. In Bangladesh, most people are not familiar with the word 'Depression', 'Frustration' and 'Anxiety' but they face it regularly without even realizing. Moreover, the number of suicides in Bangladesh is increasing day by day due to depression. The main goal of our software system is to provide knowledge about depression to Bengali people through our bots, talk to depressed people, and help them to overcome depression by giving positive support. The main challenge of our software system was to create a chatbot that can respond to users accurately. Due to the lack of Bengali dataset, we have implemented an supervised learning chatbot along with adaptive feature which will converse with the user based on the pattern matching algorithm. Our Bengali data corpus will help in creating a system for Bengali language processing research. In the future, we will train our bot and add machine learning as well as neural network technique. We have a plan to optimize the automated system by making the chatbot a voice-enabled system that will help to better understand people with low literacy. In addition we try to build a smart AI based chatbot which will assist users and give more accuracy. Also, we want to develop our own dataset. From our dataset we can analyze the changing in human behavior. In addition, we will add emotional

286 A. U. Hussna et al.

intelligence techniques to the chatbot. This will further help in understanding the dialogues and identifying the user's emotions. As a result, the effectiveness of counseling will be improved.

References

1. Number of people with depression increases. https://www.who.int/bangladesh/news/detail/28-02-2017-number-of-people-with-depression-increases
2. Your safe space in this difficult time. https://www.wysa.io/
3. Chowdhury, A.R., Biswas, A., Hasan, S.M.F., Rahman, T.M., Uddin, J.: Bengali sign language to text conversion using artificial neural network and support vector machine. In: 2017 3rd International Conference on Electrical Information and Communication Technology (EICT), pp. 1–4. IEEE (2017)
4. Fitzpatrick, K.K., Darcy, A., Vierhile, M.: Delivering cognitive behavior therapy to young adults with symptoms of depression and anxiety using a fully automated conversational agent (Woebot): a randomized controlled trial. JMIR Mental Health 4(2), e19 (2017). https://doi.org/10.2196/mental.7785
5. Frankenfield, J.: Chatbot, August 2020. https://www.investopedia.com/terms/c/chatbot.asp
6. Helal, A.: Depression: let's talk. The Daily Star, April 2017. https://www.thedailystar.net/health/depression-lets-talk-1384978
7. Kowsher, M., Tithi, F.S., Alam, M.A., Huda, M.N., Moheuddin, M.M., Rosul, M.G.: Doly: Bengali chatbot for bengali education. In: 2019 1st International Conference on Advances in Science, Engineering and Robotics Technology (ICASERT), pp. 1–6. IEEE (2019)
8. Nasib, A.U., Kabir, H., Ahmed, R., Uddin, J.: A real time speech to text conversion technique for Bengali language. In: 2018 International Conference on Computer, Communication, Chemical, Material and Electronic Engineering (IC4ME2), pp. 1–4. IEEE (2018)
9. World Health Organization, et al.: Depression and other common mental disorders: global health estimates. Technical report, World Health Organization (2017)
10. Sharma, V., Goyal, M., Malik, D.: An intelligent behaviour shown by chatbot system. Int. J. New Technol. Res. 3(4) (2017)
11. Weizenbaum, J.: Eliza–a computer program for the study of natural language communication between man and machine. Commun. ACM 9(1), 36–45 (1966)
12. Yergeau, F.: UTF-8, a transformation format of ISO 10646. Technical report, STD 63, RFC 3629, November 2003
13. Zemčík, M.T.: A brief history of chatbots. DEStech Trans. Comput. Sci. Eng. (aicae) (2019)

Speech Based Access of Kisan Information System in Telugu Language

Rambabu Banothu[1]([✉]), S. Sadiq Basha[2], Nagamani Molakatala[3],
Veerendra Kumar Gautam[4], and Suryakanth V. Gangashetty[1]

[1] International Institute of Information Technology-Hyderabad, Hyderabad,
Telanagana, India
rambabu.b@research.iiit.ac.in, svg@iiit.ac.in
[2] JNTUA, Anantapuram, Andhra Pradesh, India
bhashasadiq563@gmail.com
[3] SCIS, University of Hyderabad, Hyderabad, Telanagana, India
manidcis@gmail.com
[4] Indian Institute of Technology Hyderabad, Sangareddy, Telanagana, India

Abstract. In a developing country like India, agriculture provides large
scale employment in rural areas, thus serving as the backbone of an
economic system. For farmers it is important to decide which crop to
grow, what Government schemes benefit them the most and what is the
best selling price of the crop. In this paper we described the Speech to
Speech interaction between farmer and government through an application using speech recognition system for Telugu language that will form
the interface to the webpage providing information about the government schemes and commodity prices through voice. The proposed Kisan
Information System (KIS) is integration of Speech recognition, Dialogue
Manager, and Speech Synthesis modules. The performance of the KIS is
better compared with existing one for Telugu language.

Keywords: Speech recognition · Speech to speech system · Kisan
information system · Dialog manager

1 Introduction

Interaction between human and machine (computer) plays a significant role in
the present scenario. The mode of human and machine interaction could be
speech, text, gestures, or even musical, among which, speech is the easiest modality for human-machine interaction [1]. An interaction could be in the form of a
conversation which includes questions, answers and statements. A system which
converse with humans through speech is often referred to as speech to speech
system (STS).

A speech to speech system with speech as an input mode signifies the natural
way of communication for human beings [1,2]. The goal of the speech to speech
system is to provide the information by interacting with human beings in their

M. Singh et al. (Eds.): IHCI 2020, LNCS 12615, pp. 287–298, 2021.
https://doi.org/10.1007/978-3-030-68449-5_29

native language [3,4]. In the agriculture domain, the farmers are far behind the technology, and due to the lack of awareness of communicating with the machine [19], the farmers are seen as the reason for their own failure. The farmers are not aware about the genuine prices decided by the government. Farmers are selling their crops to some mediators and not getting actual benefits that they deserve. Farmers are able to reduce these issues by adapting the speech to speech system. The speech to speech system not only helps the farmers but also connects the government directly with the people.

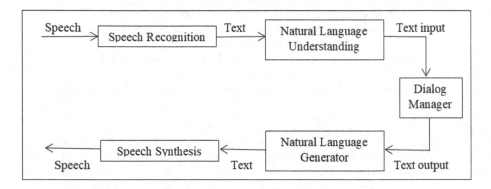

Fig. 1. Architecture of a KIS System.

Figure 1 shows the simple architecture of speech to speech system. The speech to speech system consists of Automatic speech recognition (ASR), Natural language understanding (NLU), Dialogue manager (DM), Natural language generation (NLG) and text-to-speech (TTS). When the user gives the speech input to the STS system, the speech input data is converted to text using ASR [12,23] and this text is analyzed by NLU module to fetch the relevant information. DM is the core of the STS system, as it decides the necessary actions to be performed along with the response that is to be given to the user. The user's response is given through the NLG by generating appropriate sentences in the form of text only and this text response is converted into speech response using TTS module [10,24]. In this paper, a preliminary version of speech to speech system is illustrated for providing the information to the farmers and we refer to this speech to speech system as kisan information system (KIS).

This paper is organized as follows. Section 2 mentions the related work. The proposed system is described in Sect. 3. Section 4 gives the current status of the proposed system. The experimental details and results are shown in Sect. 5. Section 6 concludes the study. Section 7 talks about our future works to improve the system.

2   Related Work

Speech to speech interaction is the process where the sound (query) produced by the user can be identified and understood, search for the appropriate results in a database for the produced sound (query), generate the response for the produced sound (query) and convert the generated response from text to speech and gives the response (result) to the user for the produced sound (query).

In [13], the authors have developed a speech based automated commodity price helpline service in six Indian languages namely Hindi, Bengali, Marathi, Assamese, Telugu and Tamil. In this work, they have used Context Dependent-Hidden Markov Model (CDHMM) for acoustic modeling configuration for 10 commodities and 10 district names in each language.

In [14], the author has developed the telephonic based automatic speech recognition system for Bodo language (also called as Boro language most widely used in the state of Assam) using asterisk server and sphinx recognition toolkit. They have collected totally 25 commodity names and yes/no words from 100 different speakers. They have used the tied state triphone model with sixteen Gaussian mixtures per state and obtained the output with the accuracy of 77.24% in training and 72.12% in testing phases.

In [6], Speech Based Conversation (SBC) system was proposed in Telugu language, which is implemented using sphinx recognition toolkit. In the SBC system, they used the method called context dependent triphone HMM with the eight Gaussian mixtures per state to model the speech data that are collected from 96 speakers, where each speaker uttered 500 words (list of commodity names, market names and district names). Hence the total number of words recorded is 48000 and they have obtained 77.7% accuracy.

In [15], the authors have collected the data of about 20 h from 1500 farmers belonging to 35 districts of Maharashtra over the telephone channel. They have studied the effect of the dialect dependent model and concluded that the transcription obtained by force alignment of speech data with alternative pronunciation improved to the performance.

In [16, 21], the authors have proposed the speech based system for accessing agriculture commodity prices in Tamil language. They have developed a speech recognition system based on the i-vector method and analyzed the system performance. Equal Error Rate (EER) is obtained for the proposed method compared with the state of art method Gaussian Mixture Model-Universal Background Model (GMM-UBM) for various Gaussian mixture components and i-vector dimensions. They have conducted the experiment for speaker dependent and independent case. For speaker dependent case the total number of utterance collected from a single speaker is 4725 among which 4200 utterances are used for training and 525 are used for testing. For independent cases, speech data is collected from 3 different speakers. Total number of utterances collected from each speaker is 675 among which 450 are used for training and 225 for testing. They have concluded that the performance of i-vector based system can be increased by increasing the number of Gaussian mixture components.

In [17], authors have proposed the Interactive Voice Response System (IVRS) system for agriculture assistance for farmers in Marathi language. This system is named as AGRO IVRS. They have created the database with 35 speakers out of which 17 were female and 18 are male. The vocabulary size of the database consists of 175 sample crop names, 350 samples of infection symptoms and 10 samples of crop diseases. They have concluded that the overall recognition rate is 91% for the system.

In [18] authors have proposed the idea of using the IoT device in the agriculture domain, which is actually used to sense the agriculture data and store the same in cloud databases. Cloud based big data analysis is used to analyze the data for fertilizer requirements and analysis of crops etc. They have performed the prediction based data mining techniques.

In this paper, we have proposed the speech based system for farmers named as Kisan Information System using Kaldi speech recognition toolkit.

Table 1. Data size used in Kisan information system.

Data category	Data size
Commodity	120
Districts	13
Government schemes	20

3   Kisan Information System (KIS)

Kisan information system (KIS) is built for farmers of rural and semi urban areas to obtain all the information related to farming (modern way of farming, types of crops, crop maintenance techniques, price for different commodities, government schemes) for the state of Andhra Pradesh, India. KIS is a web/android application based speech to speech system, as mobile is the most commonly used device for communication services. KIS provides the information in Rayalaseema ascent of Telugu language [26]. Table 1 gives the data used for the KIS system. The information for the KIS system is taken from Department of Agriculture [1], Government of Andhra Pradesh, India. The block diagram of KIS is as shown in Fig. 2. The major problems in the development of KIS are:

- **Noisy Environment:** The KIS is built by keeping in view the farmers present in rural and semi urban areas. The users are able to utilize the services of KIS through the application present in their mobile. The quality of the speech is affected due to the distance between the mobile and the user environment that the user is using [5]. The KIS could also be a noise filled environment with fan or vehicle sound or any other background noise.

---

[1] http://www.apagrisnet.gov.in/.

- **Variation in Pronunciation:** Even though the KIS is aimed for the Telugu language of Rayalaseema ascent, the pronunciation variations are large in the same region [8, 22]. It is hard to quantize that the farmer could use the information of KIS by normal style of Telugu language.
- **Unstructured Conversation:** The target audience of the KIS may not have the experience of interacting with computer based information accessing systems. So the conversation contains unstructured data filled with repeated words and false starts, thus posing the challenge to provide the service for such kinds of people.

Fig. 2. State block diagram for Kisan information system.

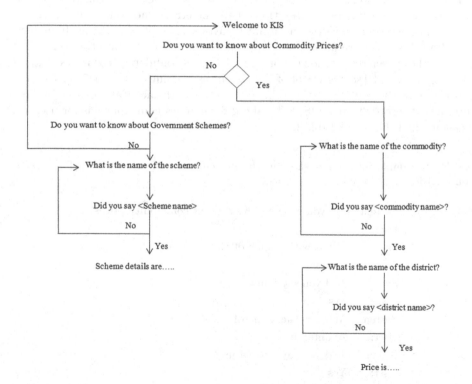

Fig. 3. State flow diagram for Kisan information system.

4   Current Status of the KIS

Kisan information system is built in Rayalaseema ascent of Telugu language. As a preliminary work, the different speakers are requested to record the speech data using the PRAAT tool set at sampling frequency of 16 KHz with 16 bit digitization, with mono channel. The text data is given to the speakers who are asked to record data in Telugu (native) language. The aim of such an experiment is to collect the speech data with variations in different pronunciations. Data was collected for 150 speakers consisting 16 hours of speech comprising a total of 350 words collected from each speaker.

Approximately 15 h of data was taken to build the acoustic model of ASR. This context dependent HMM models [4] built with 8 Gaussian mixtures per state used Kaldi speech recognition system, with appropriate NLU, DM, NLG and TTS [6]. KIS requires two concepts or inputs for the commodity prices from the user that are commodity districts. KIS requires one input from the user which is the name of the scheme. A typical dialog state flow chart is as shown in Fig. 3. When the user provides some information to KIS, the system will ask the user for some explicit confirmations and then the user needs to respond to those confirmations/questions with either Yes or No. This is needed for KIS to ensure that the input query is correct as recognition is error prone. An ideal KIS provides the accurate results to the users with less number of iterations. But the errors and confirmations raised by the speech recognition system cannot be avoided [11]. The challenge lying here is to limit these confirmations. The approach here is to associate the confidence score to the speech recognition output of ASR using confidence scoring techniques like language model scores, count of N-hypothesis, parsing and so on [7,9]. We have incorporated the confidences into the system to get the accurate results. Sample interaction recorded with KIS, KIS asking for explicit confirmations for user are as shown in Table 2 and Table 3.

Table 2. Sample interaction showing the system with the user asking for explicit confirmations such as commodity price.

System:	Do you want to know about commodity prices?
User:	Yes
System:	What is the name of the commodity?
User:	Banana
System:	Did you say banana?
User:	Yes
System:	What is the name of the district?
User:	Anantapur
System:	Did you say anantapur?
User:	Yes
System:	In anantapur district price of banana is 40/kg

5   Experimental Result Evaluation

In the Kisan Information System, experiment is conducted in two modes, one is offline mode and the other online mode. The offline mode developed in the laboratory had the data collected from 150 speakers of which 100 were used for the training set, 20 for development and 30 for testing.

Table 3. Sample interaction showing the system with the user asking for explicit confirmations such as government schemes.

System:	Do you want to know about commodity prices?
User:	No
System:	Do you want to know about government schemes?
User:	Yes
System:	What is the name of the scheme?
User:	Agriculture subsidy
System:	Did you say agriculture subsidy?
User:	Yes
System:	Details of the agriculture subsidy scheme are . . . .

5.1   Offline Mode

Each input speech signal is converted into feature vectors called Mel-frequency cepstral coefficients (MFCCs) having a total 39 dimensions (MFCC 13 + Delta 13+ Delta-delta 13). After extracting the MFCCs, we are building the acoustic GMM and HMM statistical models [20,21,25]. Along with GMM and HMM models, we aimed at building the universal background model (UBM) to cover all context present in the KIS database. We have used N-gram based Language Model (LM). The Tri2 model is created with MFCC and Speaker Adaptive Training (SAT). The SAT reduces the mismatches between the training and testing conditions. Then the Tri3 model is created in addition to Linear Discriminant Analysis (LDA) with the Tri2 model. The speech signal of a person is slightly different each time and due to this, the variations lead to an increased rejection rate. The LDA reduces the variations in the signal and rejection rate increases and thus the word rate accuracy will increase [16]. The word accuracy rate (WAR) for the acoustic model used is mentioned in Table 4. It can be seen that as the number of features increase, WAR gradually increases. As mentioned in Table 4, WAR increases from monophone to tri4. Monophone takes a singular phoneme for recognition, whereas triphone model takes three phonemes and creates a tree based on the probabilities of preceding and succeeding phonemes.

Table 4. Offline mode experiment results

Features	Acoustic model type	WAR%
MFCC 39	Monophone	66
MFCC + SAT	Tri2	78
MFCC + SAT + LDA	Tri3	82
SGMM + UBM	Tri4	85

5.2 Online Mode

The online mode experiment is conducted in real world environments. We have tested the KIS in three test cases of real world environments. First test case is open market, second is farmer's field and the third is road side. The first test case is a market place where voices of many other speakers along with the actual speaker is included as noise. The second test case is a farmer's field where there is lot of noise from wind, creatures etc. And the third test case is taken at road side which contains noise from wind, traffic sounds and other speakers. In above all three test cases, the KIS receives the input filled with noise which include utterances of other speakers. For this, 10 speakers were randomly chosen, among which 6 were male and 4 were female. These were asked to test the information of commodity prices and government schemes separately using KIS. For all three test cases, each speaker was asked to test the KIS for 4 different commodity prices and 4 different government schemes. The first test case (open market test case) results are mentioned in Table 5. The second test case (farmer's field test case) results are mentioned in Table 6. And the third test case (road side test case) results are mentioned in Table 7.

Table 5. Results of open market (first) test case of online mode experiment

Speakers ID	Test result of commodity prices	Test result of government schemes	Accuracy (%)	Overall accuracy (%)
Spkr1_M	2/4	2/4	50.0	
Spkr2_M	3/4	0/4	37.5	
Spkr3_M	3/4	2/4	62.5	
Spkr4_M	2/4	2/4	50.0	
Spkr5_M	2/4	1/4	37.5	55.06
Spkr6_M	4/4	3/4	87.5	
Spkr1_F	3/4	2/4	62.5	
Spkr2_F	1/4	2/4	37.5	
Spkr3_F	4/4	1/4	62.5	
Spkr4_F	3/4	3/4	75.0	

Table 6. Results of farmer's field (second) test case of online mode experiment

Speakers ID	Test result of commodity prices	Test result of government schemes	Accuracy (%)	Overall accuracy (%)
Spkr1_M	3/4	2/4	62.5	
Spkr2_M	2/4	1/4	37.5	
Spkr3_M	3/4	1/4	50.0	
Spkr4_M	4/4	3/4	87.5	
Spkr5_M	3/4	3/4	62.5	68.75
Spkr6_M	4/4	2/4	75.0	
Spkr1_F	3/4	2/4	62.5	
Spkr2_F	1/4	1/4	25.0	
Spkr3_F	2/4	2/4	50.0	
Spkr4_F	2/4	1/4	37.5	

By observing Tables 5, 6 and 7, we can say that among the three test cases, the second test case (Farmer's field test case) has the highest accuracy of 68.75% compared to other two test cases. This is due to the less noise and disturbances when compared to the first and last test cases.

The consolidated results for different speakers in different test case scenarios is shown in Fig. 4. It can be inferred that the noise level is low in case of Spkr4_M in Farmer's field and Spkr6_M in Open market field. Another point to note here is that speaker-wise accuracy varied from 12.50% to 87.50% across all the three test cases in real world scenarios.

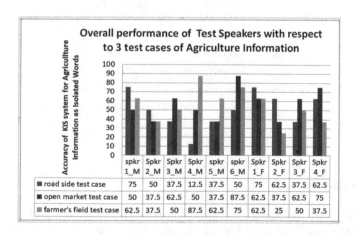

Fig. 4. The overall performance of the KIS system

Table 7. Results of road side (third) test case of online mode experiment

Speakers ID	Test result of commodity prices	Test result of government schemes	Accuracy (%)	Overall accuracy (%)
Spkr1_M	3/4	3/4	75.0	
Spkr2_M	2/4	2/4	50.0	
Spkr3_M	2/4	1/4	37.5	
Spkr4_M	1/4	0/4	12.5	
Spkr5_M	2/4	1/4	37.5	62.5
Spkr6_M	3/4	1/4	50.0	
Spkr1_F	4/4	2/4	75.0	
Spkr2_F	2/4	3/4	62.5	
Spkr3_F	2/4	1/4	37.5	
Spkr4_F	3/4	2/4	62.5	

We also performed gender and category wise analysis from the conducted experiment as shown in Table 8. It can be inferred from the table that the accuracy for Commodity category is more than that of Government Schemes category, which can be attributed mainly to the imbalanced nature of the data. Surprisingly, we observed that the accuracy for females is less only in Test case 2.

Table 8. Gender and category wise accuracy

	Category	Male	Female	Total
Test case 1	Commidity	66.67	68.75	67.50
	Government	41.67	50.00	45.00
Test case 2	Commidity	79.17	50.00	67.50
	Government	50.00	37.50	45.00
Test case 3	Commidity	54.17	68.75	60.00
	Government	33.33	50.00	40.00

6   Conclusion

We demonstrated the speech to speech interactive system referred to as the Kisan Information System for the rural and semi urban farmers to obtain better access to the information regarding agriculture, crop (commodity) prices and government schemes. The challenges of the system are to take care of the variations in the pronunciations arising from different ascents of the rural and semi urban farmers. Also, the unstructured data at the time of usage of the system challenges our KIS system. The focus of this study is to develop an interactive system with speech as medium and try to make the system as natural as possible, which is adaptable by all the users for a long period of time.

7 Future Scope

The demonstration of the KIS is a preliminary prototype. As a result we have observed that in the online mode of experiment the KIS is unable to perform accurately because in training the KIS is not much trained in noisy conditions. In the future, this work may be extended to improve the performance of KIS by collecting a dataset that is adaptable to all the environmental conditions and training the KIS in order to improve the performance.

Acknowledgement. We are at faith words to express our gratitude to Sri C. Ranganayakulu, Senior Lecturer in Electronics and Communication Engineering Government Polytechnic Anantapur, Andhra Pradesh, India for his best suggestions and constant encouragement. We also like to thank H.K. Swetha, G. Sanjana, M. Vyshnavi, and G. Chethana for their help in initial data collection work.

References

1. Rabiner, L.R.: Applications of speech recognition in the area of telecommunications. In: Proceedings of IEEE Workshop on Automatic Speech Recognition and Understanding (ASRU), Santa Barbara, CA USA, December 1997
2. Girija, P.N., Sreeenu, G., Nagamani, M., Prasad, M.N.: Human machine speech interface using Telugu language- HUMSINTEL. In: Proceedings of IEEE International Conference on Human Machine Interfaces, Bangalore, India, December 2004
3. Gopalakrishna A., et al.: Development of Indian language speech databases for large vocabulary speech recognition systems. In: Proceedings of SPECOM, Patras, Greece (2005)
4. Donovan, R., Woodland, P.: Improvements in an HMM-based speech synthesiser. In: Europespeech95, Madrid, Spain, vol. 1, pp. 573–576 (1995)
5. Sreenu, G., Prasad, M.N., Girija, P.N., Nagamani, M.: Telugu speech interface machine for university information system. In: Proceedings of IEEE Advanced Computing and Communications (ADCOM), Ahmadabad, India, December 2004
6. Mantena, G.V., Rajendran, S., Gangashetty, S.V., Yegnanarayana, B., Prahallad, K.: Development of a spoken dialogue system for accessing agricultural information in Telugu. In: Proceedings of 9th International Conference on Natural Language Processing (ICON-2011), Chennai, India, December 2011
7. Jiang, H.: Confidence measures for speech recognition: a survey. Speech Commun. **45**(4), 455–470 (2005)
8. Krishna, B.S.R., Girija, P.N., Nagamani, M.: Pronunciation dictionary comparison for Telugu language automatic speech recognition system. J. Data Eng. Comput. Sci. GRIET **2**(1), 60–69 (2010)

9. Polifroni, J., Hazen, T.J., Burianek, T., Seneff, S.: Integrating recognition confidence scoring with language understanding and dialogue modeling. In: Proceedings of ICSLP (2000)
10. Achanta, S., Banothu, R., Pandey, A., Vadapalli, A., Gangashetty, S.: Text-to-speech synthesis using attetion based sequence-to-sequence neural networks. In: Proceedings of ISCA Speech Synthesis Workshop 9, September 2016
11. Nagamani, M., Girija, P.N.: Substitution error analysis for improving the word accuracy in Telugu language automatic speech recognition system. IOSR J. Comput. Eng. (IOSRJCE) **3**(4), 07–10 (2012). ISSN 2278–0661
12. Rabiner, L., Rabiner, L.R., Juang, B.-H.: Fundamentals of Speech Recognition. PTR Prentice Hall, Upper Saddle River (1993)
13. Speech-Based Automated Commodity Price Helpline in Six Indian languages. http://asrmandi.wixsite.com/asrmandi
14. Deka, A., Deka, M.K.: Speaker independent speech based telephony service for agro service using asterisk and sphinx 3. Int. J. Comput. Sci. Eng. Open Access **4**, 47–52 (2016)
15. Godambe, T., Karkera, N., Samudravijaya, K.: Adaptation of acoustic models for improved Marathi speech recognition. In: Proceedings of International conference, Acoustics, CSIR-NPL, pp. 1–6, November 2013
16. Yogapriya, S., Shanmugapriya, P.: Speech based access for agriculture commodity prices in Tamil. Int. J. Sci. Technol. Eng. (IJSTE) **4**(7), 75–81 (2018). ISSN (online): 2349–784X
17. Gaikwad, S., Gawali, B., Mehrotra, S.: Speech recognition for agriculture based interactive voice response system. Int. J. Modern Eng. Res. (IJMER) **4**(2), 112–115 (2014). ISSN: 2249–6645
18. Rajeshwari, S., Kannan, S., Rajkumar, K.: A smart agriculture model by integrating IoT, mobile and cloud-based big data analytics. Int. J. Pure Appl. Math. (IJPAM) **118**(8), 365–370 (2018). ISSN: 1314–3395 (on-line version): ISSN: 1311–8080 (printed version)
19. Mittal, S., Gandhi, S., Tripathi, G.: Socio-Economic impact of Mobile Phones on Indian Agriculture. Indian Council for Research on International Economic Relations (ICRIER), February 2010
20. Brown, M.K., McGee, M.A., Rabiner, L.R., Wilpon, J.G.: Training set design for connected speech recognition. IEEE Trans. Signal Process. **39**(6), 1268–1281 (1991)
21. Yegnanarayana, B., Prahallad, K.: AANN - an alternative to GMM for pattern recognition. Neural Netw. **15**(3), 459–469 (2002)
22. Kethireddy, R., Kadiri, S.R., Gangashetty, S.V.: Learning filter-banks from raw waveform for accent classication. In: Proceedings of IJCNN, Glasgow, UK, 19–24 July 2020, pp. 1–6 (2020)
23. Prahallad, K.: Dealing with untranscribed speech. In: Proceedings of International Conference on Signal Processing and Communications (SPCOM), pp. 1–12. IISc Bangalore, India (2012)
24. Prahallad, K., Elluru, N.K., Keri, V., Rajendran, S., Black, A.W.: The IIIT-H Indic speech databases. In: Proceedings of INTERSPEECH, Portland, Oregon, USA (2012)
25. Nagamani, M.: Lexical modelling in HMM based Telugu language ASR system. Ph.D. thesis, June 2020
26. Girija, P.N., Nagamani, M.: Improving reading and writing skills with intelligent tutor for Telugu language learning-INTTELL. In: Proceedings of International Conference on Intelligent Sensing and Information Processing (2005)

Voice Assistant for Covid-19

Shokhrukhbek Primkulov, Jamshidbek Urolov, and Madhusudan Singh(✉)

School of Technology Studies, ECIS, Woosong University, Daejeon, South Korea
msingh@wsu.ac.kr

Abstract. The coronavirus has caused ruin over the globe. Many individuals are tainted, and thousands are dead. The pandemic is still on the ascent as all the nations on the planet have forced a lockdown. Everyone has the privilege to know how this infection has influenced our lives as we keep on pursuing a war against it. We have developed voice assistant for COVID-19 update. We have proposed to create a voice assistant which tells statistics data about Covid-19 as in their country, city, location wise. The proposed Covid-19 voice assistant gives information about COVID-19 latest data and speaks to us. It is useful for everyone, especially for disabled people because it works with a voice. Even people with eyesight problems can use it easily and get the update about deadly viruses.

1 Introduction

At the end of 2019, the world faced a new problem, that name is Corona virus or Covid 19. The Covid 19 virus has spread all over the world and forced people to stay at home and keep social distancing [1]. Clearly, the Coronavirus pandemic will remain on the world agenda for a few more months. When we look at the information exchanged between various institutions, organizations, or experts, we have not yet received satisfactory information on the source of infection, its distribution, and the treatment process [2]. However, all types of repair and isolation contacts are believed to be effective in terms of protection.

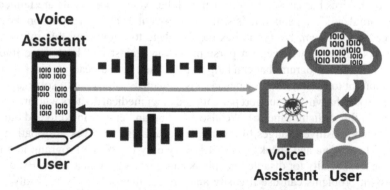

Fig. 1. Overview of voice assistance for Covid 19 updates

M. Singh et al. (Eds.): IHCI 2020, LNCS 12615, pp. 299–306, 2021.
https://doi.org/10.1007/978-3-030-68449-5_30

As in our research, we have tried to identify the problem that people have problem to get the updated statistics data about Covid 19 worldwide or region wise. Still, most people not aware of pandemic statistics, they do not know how many infected cases, how many death rates, and how many recovery rates in the world or in their countries. Also, the pivotal point is that all websites about COVID 19 infographics are not for all people [3], for example, disabled people cannot go to websites and check data regularly, this action makes them in danger. If people would know more about Covid-19 and it is the current situation and it could help to decrease the rate of infection, and people would act properly, and protect accordingly. Fig. 1 has given complete overview of our proposed voice assistant for Covid-19.

This article has organized as follows Sect. 2 has mentioned about details about COVID-19 background study. The Sect. 3 has included the design of our proposed voice assistance for Covid 19 and Sect. 4 has shown the Development of Voice Assistant of our proposed idea. The Sect. 5 has concluded our articles.

2 Background Study

COVID-19 has become a worldwide disease. An exponentially increasing pandemic Improperly known mechanism of transmission. The virus is most usually contained with little to no effects, but in 2–8% of those affected, it can also lead to increasingly progressive and frequently fatal pneumonia. The disease was known as beta-coronavirus and is in the same group as previously detected dangerous viruses like SRS and MERS. The infection has already spread to over 200 nations. The cases of COVID-19 have plagued hospitals and clinics around the world. People attend clinics and hospitals with other signs that are not linked with a lot of hysteria and gossip. These visits have risen primary treatment and infection transmission costs as the hospital infrastructure is overwhelmed. Due to shortages of essential protective equipment and skilled suppliers, the accelerated pace of spread has stressed healthcare systems worldwide, partly motivated by variable access to point-of-care research methodologies, including reverse transcription-polymerase chain reaction methodologies (RT-PCR) [4].

The coronavirus has caused ruin over the globe. Many individuals are tainted, and thousands are dead. The pandemic is still on the ascent as all the nations on the planet have forced a lockdown. Everyone has the privilege to know how this infection has influenced our lives as we keep on pursuing a war against it. Along these lines, we motivated to write a program that can help to fight back with a deadly virus [5]. This is a very important topic and many organizations; scientists are trying to solve and create something to kill this virus or support patients and other medical doctors who are showing heroic acts in this difficult time [6]. We also as a human being wanted to add our effort and make some positive changes.In this paper voice assistant gives information about COVID-19 latest data and speaks to us. It is very easy to use the program, it is useful for everyone, especially for disabled people because it works with a voice. Even people with eyesight problems can use it easily and protect themselves from deadly viruses effectively.

3 The Proposed Voice Assistance for Covid 19

The proposed program covers basic web scrapping tool and voice assistance, and multithreading in python. Now, let look at our flow chart (Fig. 2).

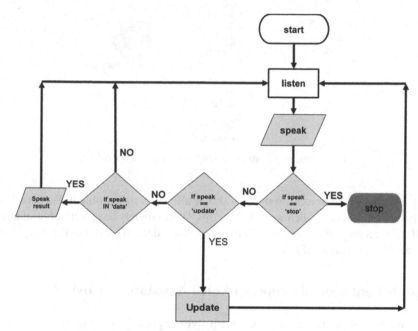

Fig. 2. Flowchart of Covid -19 data access with voice command.

As you can see our flowchart is simple. Basically, we speak to our program and there are two keywords such as 'stop' and 'update'. The 'stop' makes our program to quit. The 'update' keyword updates the latest data and gives it to us. Now it is time scrabble our data. A web scraper [6] is a term used to describe the use of a program or algorithm for extracting and processing large amounts of data from the Internet. Regardless of whether you are a data specialist, an engineer, or anyone who analyzes many data sets, the ability to extract data from the Internet is a useful skill. In Fig. 3 we can see user give voice command to machine and machine will data process searching cloud data base and receive the updated data by data access server and return to machine and machine will give the details of Covid 19 as output.

In this article, we used a web scrabbing tool called 'parsehub' [7]. With Parse Hub, you have complete control over how you select, structure, and change items so you do not have to search your browser's web inspector. We never have to write a web scraper again. And can use Parse Hub for convenient interactive map processing, endless scrolling, authentication, drop-down lists, forms, and more. The prosed system has a quick selection function that finds out exactly how the website is structured, and groups related data for you. All we must do is open a website and click on the information we want to extract! It is very easy to use a web scrabbing tool that is empowered with AI.

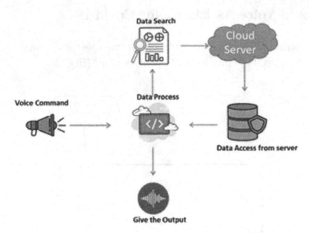

Fig. 3. Process of proposed voice assistant for Covid 19.

Just by clicking to the objects on the web site, it allows us to scrabble it. We scrabbed a famous statistic website called worldmeter.com. World meters have a credible source of COVID19 data around the world. After we scrubbed data, we can get the Application Programming Interface (API) key.

4   Implementation of Proposed Voice Assistant of Covid 19

Firstly, we need to install some requirement python packages based on the following points suhc as.

- **pip install requests -** It allows us to send HTTP/1.1 requests extremely easily. There is no need to manually add query strings to your URLs or to form-encode your POST data. Keep-alive and HTTP connection pooling are 100% automatic, thanks to urllib3.
- **Pip install pyttsx3 -** pyttsx3 is a text-to-speech conversion library in Python. Unlike alternative libraries, it works offline and is compatible with both Python 2 and 3.
- **pip install pywin32 -** Python extensions for Microsoft Windows Provide access to much of the Win32 API, the ability to create and use COM objects, and the Python win environment.
- **pip install SpeechRecognition -** Library for performing speech recognition, with support for several engines and APIs, online and offline.
- **pip install pyaudio -** Audio provides pythonbindings for PortAudio, the cross-platform audio I/O library. With PyAudio, you can easily use Python to play and record audio on a variety of platforms, such as GNU/Linux, Microsoft Windows, and Apple Mac OS X / macOS.

Now all our requirements are ready, and we can start coding. In the beginning, we use our API key, run token, project token, and save them to the variables, respectively (Fig. 4).

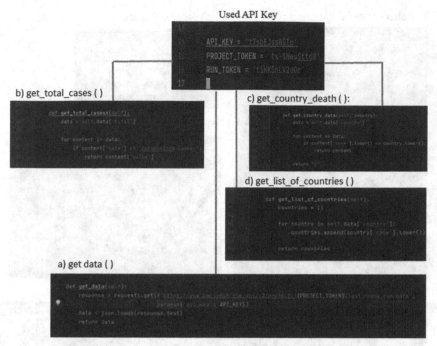

Fig. 4. (a) get_data function (b) get_total_cases function (c) get_country_death function (d) get_list_of_countries function

We create a class called Data and give them the following methods such as get data (): The aim of this method to get the information from parsehub utilizing your latest run for the undertaking on their servers and bring it back. Get_total_cases (): Get the all-out instances of COVID-19 around the globe. Get_country_death (): Get the number of COVID-19 cases and deaths in any country. Get_list_of_countries (): Get the list of countries.

To create human speech data functions, we have shown in the Fig. 5 where we have creaded following terms such as: **speak** () - To this function, we give one parameter in which we initialize the pyttsx3 engine to say the text. **Get_audio** () - This function with recognize_google () method listens to the speech input by the user through the microphone and returns the input text **main** () - This function is responsible for main logic and recognizing the input speech using regular expression patterns and classifies them as patterns for cases and deaths in the world or cases and deaths in any country. The matched pattern calls that function and speaks the number out loud.

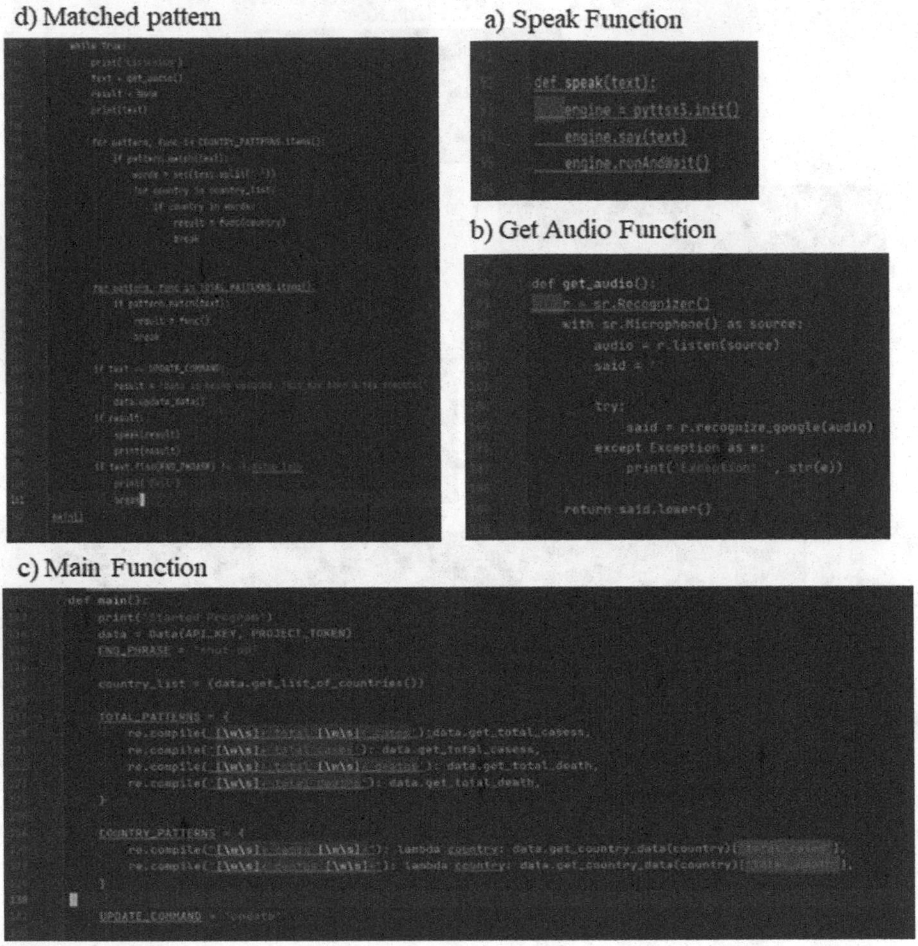

d) Matched pattern

a) Speak Function

b) Get Audio Function

c) Main Function

Fig. 5. Human speech data functions (a) Speak Function (b) Get Audi Function (c) Main Function (d) Matched pattern Function

Now our program speaks only the data we fetched initially. As global pandemic still going on it is important to update our data, if we update it manually it would take a lot of time and energy. That is why we decided to add automatically updating future to make complex, and easy to use. In order, to update our date, we created an **update** () function, which does not use the last run project, instead, it will create a new run on the parsehub server. It will take some moment.

In this function, we used a **multithreading library o**f python. Multiprocessing is a package that supports spawning processes using an API like a thread module. The multiprocessor package offers both local and remote concurrency, effectively bypassing the global locking of the interpreter using subprocesses instead of threads. Multithreading helped us to run voice assistance and updating data simultaneously. When we run the program, the console will type 'listening' and we can ask about 'How many total cases'

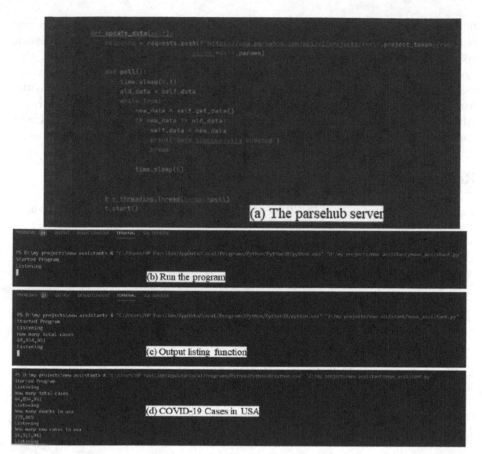

Fig. 6. Voice Assistance for COVID 19 (a) The parsehub server (b) Run the program (c) Output listing function

or 'How many total death' the assistant would reply with the number which is real-time. Even you can ask about a specific country, for example, you can say 'How many cases in the USA' or 'How many deaths in Brazil' it would reply with trusted numbers. When we run the program, the console will type 'listening' and we can ask about 'How many total cases' or 'How many total deaths' the assistant would reply with the number which is real-time. Even, you can ask about a specific country, for example, you can say 'How many cases in the USA' it would reply with trusted numbers as shown in Fig. 6-d.

5   Conclusion

We must admit that with the Covid19 pandemics, we obtained information that will give clues about the health system of the countries. When we look at regularly published data, we often see that the total number of cases, the number of extracts, the number of new cases, and the number of people in recovery are reported to the public. But we still thought that we need something easier and faster and, thought that modern problems require modern solutions. In this article, we have proposed to create a voice assistant which tells statistics data about Covid-19 as in their country, city, location wise. Our system has provided accurate information about COVID-19 with help of only voice assistant. We have built voice assistance that updates information about coronavirus using just voice. Our proposed system aims to help people get rid of this deadly virus. Also, the long-term aim is to get more knowledge in terms of programming, especially learn more about algorithms, web scrabbing, object-oriented programming, and more.

References

1. Baber, H.: Spillover effect of COVID19 on the global economy. Trans. Mark. J. (TMJ) **8**(2), 177–196 (2020)
2. Yadav, D.P., Sharma, A., Singh, M., Goyal, A.: Feature extraction based machine learning for human burn diagnosis from burn images. IEEE J. Transl. Eng. Health Med. **PP**(99), 1 (2019). https://doi.org/10.1109/JTEHM.2019.2923628
3. Sezgin, E., Huang, Y., Ramtekkar, U., et al.: Readiness for voice assistants to support healthcare delivery during a health crisis and pandemic. NPJ Digit. Med. **3**, 122 (2020). https://doi.org/10.1038/s41746-020-00332-0
4. Sáiz-Manzanares, M.C., Marticorena-Sánchez, R., Ochoa-Orihuel, J.: Effectiveness of using voice assistants in learning: A study at the time of COVID-19. Int. J. Environ. Res. Public Health **17**, 5618 (2020)
5. Dhakal, P., Damacharla, P., Javaid, A.Y., Vege, H.K., Devabhaktuni, V.K.: IVACS: intelligent voice assistant for coronavirus disease (COVID-19) self-assessment. arXiv:2009.02673, September 2020. https://arxiv.org/abs/2009.02673. Accessed 5 Nov 2020
6. Coronavirus website. https://www.worldometers.info/coronavirus/
7. Free web scrabbing tool: ParseHub | Free web scraping - The most powerful web scraper

Combining Natural Language Processing and Blockchain for Smart Contract Generation in the Accounting and Legal Field

Emiliano Monteiro[1], Rodrigo Righi[2], Rafael Kunst[2], Cristiano da Costa[2], and Dhananjay Singh[3]([☒])

[1] Unemat, Sinop, MT 78555-000, Brazil
emiliano@unemat.br
[2] Unisinos, Av. Unisinos, São Leopoldo, RS 93022-750, Brazil
{rrrighi,rafaelkunst,cac}@unisinos.br
[3] Hankuk University of Foreign Studies, Seoul, South Korea
dsingh@hufs.ac.kr

Abstract. The growth of legislation and the demand for systems for automation of tasks has increased the demand for software development, so that we have greater assertiveness and speed in the reading and interpretation of laws in legal activities. Currently, written legislation must be interpreted by an analyst to be encoded in computer programs later, which is often error-prone. In this context, with the popularization of cryptocurrencies, interest was aroused for the use of Blockchain in the legal area. In particular, the use of smart contracts can house business rules in legislation and automate block control. Still, fast, quality code writing can benefit from the use of natural language processing (NLP) to help developers. After revisiting the state-of-the-art, it is perceived that there are no works that unite intelligent contracts and natural language processing in the context of the analysis of legislation. This work presents a computational model to generate intelligent codes from the analysis of legislation, using NLP and Blockchain for such a procedure. The practical and scientific contribution is pertinent for the law to be interpreted in the correct way and in accordance with the latest updates. Also, a prototype and initial tests are presented, which are encouraging and show the relevance of the research theme.

Keywords: Blockchain · Smart contract · NLP

1 Introduction

Blockchain is a node chain structure, in which each node has data blocks, a timestamp, and a code hash that refers to the previous node, reinforcing the data chain immutably [1]. Blockchain technology was initially designed to support crypto currencies [2], but you can use them for other purposes today, including

M. Singh et al. (Eds.): IHCI 2020, LNCS 12615, pp. 307–321, 2021.
https://doi.org/10.1007/978-3-030-68449-5_31

supply chain tracking and contracts between people who don't know each other. Thus, Blockchain works by saving the transactions of interest to which it was designed, where each can implement a number of rules and also serve to transmit data. In particular, rules for executing transactions or operations are performed via smart contracts [3] (or smart contracts) [1]. From the programmer's point of view, they can be seen as a script (or simplifying, a program that should run automatically when a transaction is performed). Thus, blockchain can contain data and executable code [2], where the latter is seen as the programming logic for updating data and triggering actions external to the blockchain.

1.1   Problem Definition and Motivation

One of the areas where smart contracts has been growing and has exploitative potential is the legal [4]. Smart contracts are programs written by programmers, and therefore error-prone and often questionable as to quality and security aspects [5]. In addition, the number of legislative norms in Brazil has been growing rapidly [6], demonstrating the need for automation in the generation of reliable code [7]. Blockchain implementation should have security as a concern (regardless of its applicability). Which is why the generation of smart contracts code is automatically and assertively relevant, not including human activities.

In the current situation of systems development for the current legislation, there are two characters, the programmer and the analyst. System analysts often perform the following operations: a) Write technical specifications (and later publish requirements analysis); b) Interpret the legislation (in force); c) Write smart contract; d) Analyze the execution of smart contract. On the other hand, software programmers often perform the following operations: a) Interpret requirements (read requirements); b) Write smart contract; c) Deploy smart contracts; d) Evaluate the execution of smart contract and (examine execution logs). It is a normal operation of the analyst when studying (interpreting) the current legislation, decoding it in the form of an analysis or specification of requirements that the programmer will use. In this case, the smart contract is written by a person in the programmer role. These actions described above can be seen in the use case in Fig. 1, which describes what we call the current scenario.

The above tasks show the manual and error-prone character in the drafting of smart contracts and analysis of legislation. Contracts are written manually by programmers based on requirements that are passed on to them. When analyzing the state-of-the-art, it is perceived that there are no initiatives that go in the direction of automation of analysis of legislation and generation of contracts automatically. Therefore, the need to generate quality code is quickly identified to meet the growing need for Blockchain projects applied to accounting and legal. It is also clear the need to automate repetitive tasks, dishonoring analysts and programmers, and making legal support for different tasks optimized.

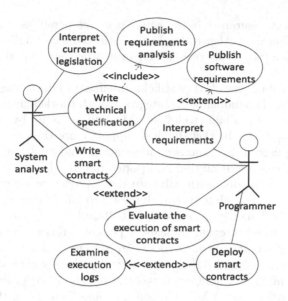

Fig. 1. Use cases involving the traditional situation in the generation of software for a given legislation.

1.2   Proposal and Organization of the Article

Given the gap mentioned, this article proposes the development of a model for generating code for smart contracts automatically. To do this, natural language processing (NLP) is used, so that programmers are assisted in making semi-ready codes available. The template covers the steps that are important to NLP, from text input of legislation to code output. Based on the model, a functional prototype was developed that can be explored in legal projects.

The rest of this article is organized as follows. Section 2 presents the related works, which commentoned on the research gap in the area. Section 3 brings the proposed model and its phases. Section 4 presents the evaluation methodology, showing test scenarios development steps. Section 5 shows a discussion of the results, showing good practices obtained with the tests. Finally, Sect. 6 brings the final considerations, again highlighting the contributions and placing some tasks within the scope of future work.

2   Related Works

This section presents some works related to the theme of blockchain use, smart contracts, NLP and computer systems in the legal sphere. The development of *smart contracts* has been shown to be a work done by programmers in a traditional way involving implementations with problems. Although the use of *blockchain* spreads rapidly, problems[1] there are several challenges such as main-

---
[1] Scalability, interoperability, speed to process transactions and regulation.

tenance and quality assurances in the *smart contract* as well as malicious behavior. The integration with other technologies with Artificial Intelligence (AI) also brings new challenges to this technology, even when integrated with artificial intelligence.

The evolution of the use of AI (Artificial Intelligence) in the most varied fields of computing is cited including being integrated with Blockchain, and seen in the products and services available today [8]. AI could be useful by suggesting programming languages (for blockchain that support more than one programming language), being used as a recommendation agent, and analyzing past transactions [9]. Considering the evolution and popularity of AI and Blockchain, both can be used in conjunction with software development practices to meet the growing demand for new solutions (involving both) [5]. Aiding in the generation of code by natural language processing (NLP) and its expansion into various fields of [10] applicability could indicate a possible efficient combination with the production of smart contracts code. Several levels of language processing have already been presented [11] and in some way already exist on various systems (such as Unix[2]). Authors evaluated the possibility of generating natural language specification from UML (Unified Modeling Language) [12] diagrams.

Other [13] approachs suggest using Named Entity Recognition and Classification (NERC) as a technique used with NLP to be able to identify entity names or keywords in text. It is also cited the benefit of using dictionaries. In addition to indicating the need to divide the process into 3 parts: (I) data reception, (II) preprocessing and (III) comparison engine. Other studies propose to generate code from a pre-analysis made with UML [14], initially generating diagrams of use cases, class diagrams and collaboration, later the codes are generated. Olajubu [15] proposes a standard to improve quality of user specifications. Billings [16] questions whether code generation in the future will use various technologies such as ML (Machine Learning), AI, and NLP among others, pointing out that machine-generated code may be common in 2040. The adoption of computational linguistic techniques to perform requirements for better results may also be used [18]. If natural language processing depends on context, it is interesting to decrease ambiguity before processing, in this sense [17] it is suggested to use contextual language processing and two levels of grammatical processing Two Level Grammar (TLG) using the Vienna Development Method to construct a point between the input requirements and the NLP.

3   Proposed Model

This section presents the proposed model, as well as its phases and the description of each of them. The model consists of word processing (searches, segmentations, word exchanges, etc.), receiving the raw text at the input, processing through the various stages of the NLP and generating an output. In the end, you

---

[2] grep, awk, and sed.

get the expected result (computer language code) that is passed on to developers. In a way, Fig. 2, illustrates a representation of the phases contemplated in the model.

The model should be understood as a large process, read from left to right. The text you want to handle enters the process via plain text (TXT) or structured (XML or HTML) format. The legislation can be submitted by a producing process that stores current legislation and sends it through a TCP port or written to disk. The possibility of writing the desired legislation on disk allows any system or user to deposit in a folder what they want to process. This template provides for the possibility of reading input files, from time to time, of files that are in a folder, this is possible by using a timer. This timer assists in reading the files, at predetermined times, also with the synchronization of these texts in a prepossessing buffer. The buffer receives text that will go through tag removal. The buffer acts as a queue, to retain the texts before they enter the queue that implements the NLP, this queue is a FIFO (First In First Out) directing the texts to the class. Lists of words stored in files or databases should be used in the last steps of the NLP. Each NLP processing step has been separated into an independent class to facilitate its implementation and maintenance, so each step can be implemented independently of the other.

This set of classes should be in the same namespace to facilitate its implementation because each class has a primary method that implements the code responsible for its responsibility in the process, and the end of this processing must be passed via message exchange to the next class. Any text that passes through the NLP FIFO is directed to a processed text queue, the queue sends data to class generation. Class generation from a grammar of smart contract languages produces an intelligent contract skeleton. This smart contract skeleton is written to disk in plain text (TXT) format, so the programmer can evaluate whether the generated code is feasible or if it needs to return to processing. A monitoring and administration module is required for: timer administration, inform the system of the input folder and the output folder, configure port number. The administration module will present operating statistics such as: amount of imported legislation, processing time, identified entities, generated classes, etc.

3.1 NLP Phases Treated in the Model

Several other fields of knowledge such as medicine already use NLP techniques [19]. It is observed that the path, via machine translation should be done in steps [20]. NLP phases are briefly presented below: A. Text slipt (or segmentation): it is the phase that breaks the text (sentences, words, etc.) into smaller parts, for example for an entry of a paragraph this will be divided into sentences. B. Tokenization: It is a step that allows you to break a sequence of text into smaller parts by locating space between words. C. Locate prefix and suffix (Steaming): Prefixes and suffixes of a word are identified from the words in the text. D. Remove stop words: Locate common words that do not add value to the general context of the main text. E. Relationship parsing: A tree is built with the relationship between words of interest in the sentences. F. Search for nouns: Nouns are identified and form an interest list for the next step. G. Search for entities: Entities of interest are located in the speech (for example: people or objects) and are marked. The recognition and location of named entities is carried out (an example: airplane, car, etc.). H. Discover recursive relations: Nouns, pronouns e verbs are searched to locate if they reference to another nouns, pronouns and verbs, new information or data might outcome from recursive relations among them.

Currently, in the traditional path of development, so that code is generated from the legislation is manual. An analysis in the legislation will go through the following steps: Requirements analysis, separation of actors to identify classes, identification of attributes, and identification of methods. Later writing code.

3.2 Actors

The actors involved in the model are programmers and analysts. Detailed whose role is described below: a) the analyst interprets the legislation, writes the technical diagrams (via UML), publishes the requirements analysis, contributes to the writing of smart contracts. The function of evaluating the result of contract execution is shared between the analyst and the programmer. b) the programmer performs: the interpretation of requirements from the published requirements, evaluates the execution of smart contracts, performs the publication of contracts on the Bockchain and examines execution logs. These interactions between these actors and the model are represented in the use case diagram (Fig. 1). In the proposed model the analyst has the role of feeding, configuring and monitoring the processing throughout the model, while the programmer, his main assignment is in the evaluation of the feasibility and applicability of the generated code.

3.3 Legislation

Brazilian legislation is inserted into the model via text files (in TXT, HTML or XML formats). The legislation is inserted into the system by a pre-existing third-party system that contains the legislation and sends the text to a TCP port. Another form of input is via reading text files from a folder in the file system. The model provides for reading legislation files sequentially given the need to automate this step by reading multiple files deposited in the file system or receiving them via TCP port.

3.4 File Formats

Files are used as a means of input and output of the treated model in Fig. 2. One of the format will be text, text files must be plain text (or plain), containing text without an End of File (EoF) encoding them should be UNICODE UTF-8 standard, plain text (italic or bold). Other formats will be HTML in the WWW standard (starting with versions 3, to maintain compatibility with legacy systems) and XML (starting with version 1.0). The output file format will be TXT with extensions named according to the language used in the target Blockchain.

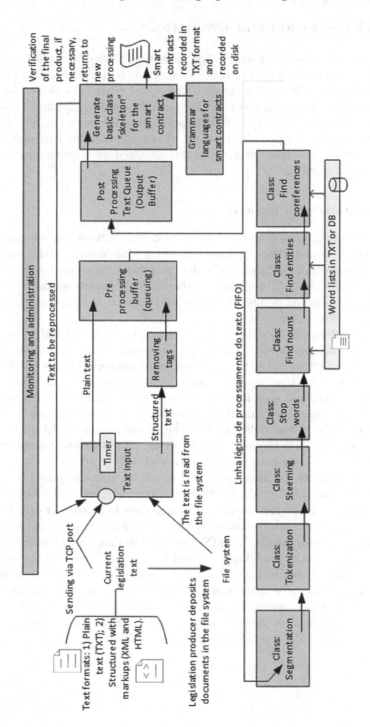

Fig. 2. Proposed model, highlighting data entry, intermediate stages of processing and data output.

4   Implementation and Evaluation

To evaluate the proposed model, a first prototype (called a "basic" prototype) was designed so that it is possible to evaluate basic text substitution operations. At the end of the tests begins the development of what is presented as a work in progress for the development of the second prototype (called "advanced" prototype), which will receive more operations to implement the full NLP. The basic prototype, performs only a subset of nlp steps, needed to perform text replacements, i.e. the basic prototype, was designed as a shorter path than is expected in a full NLP.

4.1   Prototype with Basic Features

The prototyping approach was adopted to test the previous concepts, that is, as parts of the software develop, they are evaluated. This resulted in the first version of the basic prototype developed. The first prototype aimed to test the following steps: 1) Receive text; 2) Separate sentences; 3) Separate words; 4) Removes stop words; 5) Loading nouns; 6) Find selected noun; 7) Generate classes from the selected word; 8) Save the class.

This prototype was developed in C Sharp, the idea was just to move and replace text, each method of the system performing a small operation. The system interface was designed so that the user could trigger each method via a button. The results of each method were placed on visual controls (such as textBox); This approach makes it easy for the operator to view each step. The basic prototype had only one screen (form). The use of several visual controls allowed the didactic execution of the walkthrough. The elements that make up the screen were numbered to guide the user during testing, all buttons had numbers in the textual properties along with the description of their purpose.

Figure 3, presents the main class, with methods, that were used in the basic prototype, described below: the method separateSentences used to separate sevens from paragraph blocks. 2. separatedWords used to separate words from sentences. 3. RemoveSBlank, removes blanks. 4. RemoveStopWords, remove the selected words as stop words. 5. findNouns, used to find words considered nouns you carry from a list. 6. loadStopWord and loadNouns, both load the words and stop words from a disk file. 7. generateClass, replaces the selected noun with the word "Template" by generating a new class. After the basic prototype has executed all the methods, one realizes the need to advance in the include of all the steps detailed in Fig. 3, so the basic prototype was finalized and the development of a new prototype was started as more functionality.

Fig. 3. Basic prototype class structure.

4.2 Prototype with Advanced Features

The second prototype is a work in progress. In study and development. In this prototype we suggest the identification and comparison of several API (Application Programming Interface), Frameworks and toolkits for natural language processing (among them: CoreNLP, NLP.js, Stanford NLP NET, OpenNLP, SharpNLP, MITIE, text2vec, Moses, NLTK). In this identification, one of the search criteria will be tools that can be adapted to languages other than English, in this case, Portuguese; for example the NILCS Corpora [21].

The NLP steps shown in Fig. 3. should be developed in a popular language so that code can be generated for the Ethereum, HyperLedger or Corda blockchain. New models should be trained, adapted or developed for the reality in which we use of brazilian portuguese [21]. Legal standards of varying sizes must be selected to be tested. The tests should focus on the alternative of generating code with the possibility of reuse and deployment in the blockchain environment, as this is where the smart contracts are. The purpose of this process is to minimize the programmer's work. Mainly performing the process of identifying entities (attributes and methods).

All steps in Fig. 3 can also be performed in parallel with a traditional system analysis process. It was the objective of the basic prototype and is still an objective of the ongoing project of the advanced prototype, to generate object-oriented code, because the languages supported by blockchains are OO (Object Oriented). In this sense, the advanced prototype can generate code with phrases "if-then" [22]. An alternative that requires user interaction has already been presented [23] and others generate UML models from [24,25] are related works that can contribute to the evolution of the advanced prototype. To test the feasibility of applying NLP to code generation services for a blockchain; a software can be built to enable the process; we suggest the following basic features:

1. Vision of the proposed solution:
   Following a software proposal for the advanced prototype, so that the ideas discussed in this article can be implemented, we guide an overview for a future implementation of prototypes. We describe its parts as well: a) An entry of legislation through the loadInitialText method; b) once loaded the text the main module (through a management class) will make the text go through the various steps of NLP processing. This system can implement one class for each step of the process; c) The system should have the ability to receive adaptations, or plugins, as an example we cite the specifications of languages, to support the generation of code for solidity, for example; d) during processing temporary data may be stored, for this XML files can be used; e) the system has at least 3 outputs (a miscellaneous report, generated code, and logs).

2. About classes:
   For each step of the NLP process, in a modular implementation, it is implemented in the form of a separate class. In addition to facilitating maintenance, it would allow collaborative development, and easy error detection. So for the 8 steps we will have 8 classes, each with at least one method to perform its main activities. These classes would be: segmentation, entities, steaming, stop words, coreferences, parsing, nouns and tokenization. See Fig. 4.

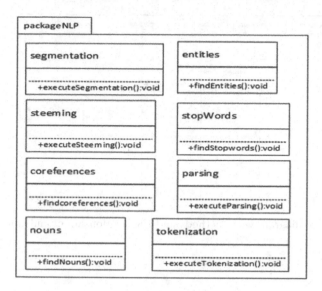

Fig. 4. Main classes of the advanced prototype.

A general control class or coordination of the software will be required; here called "manager". It is responsible for the initial receipt of the text to be processed and calls the other classes. Later these modules will generate the processing outputs. The most important method of the class is generatCode() code generation. This class can support multithreads to perform tasks in parallel when receiving data via data stream (similar to the RAPID [26, 27] approach). Other methods are: loadInitialText(), saveToFile(), loadPlugin(), dragndropFile().

3. Relationship between classes:
   Here we represent the sequencing of interactions between objects in a sequential order, equal to the steps of the NLP. One can observe the classes and methods, they are via message exchange connecting. It is important to note that the last class must return the value to the manager class, which can later direct the outputs. See Fig. 5.

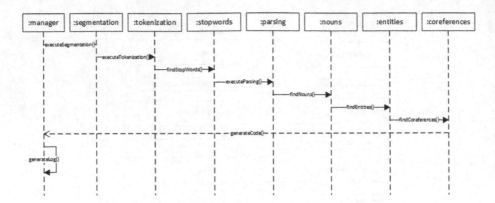

Fig. 5. Sequence diagram with methods and the relationship of calls for code generation.

4. Components:
   The system can be organized into components that help its structuring and modularity. Two types of groupings are interesting: a) one for language specifications (with modules for each target language), functioning as plugins; this modularization will allow the system to be adapted in the future to any blockchain language; (Initially it is suggested Solidity and Golang); b) a second groupings for the accessory files as reports, log and smartcontrat template.

5. Prototype interface:
   Implementation details may predict a way to view the initial text before processing. Each step can be represented by a screen or tabs on a form. The important thing in the user interface design is that each step has a representation. In this suggestion, we demonstrate the execution of each step in its own interface, so the user can understand the steps and their results.

5   Results Discussion

The basic prototype behaved as expected, performing basic localization, removal, and replacement of character strings for the inputs provided. It served as the basis for study and analysis of the steps of an NLP. We noticed that the literature on code generation versus diagrams is more abundant and contemporary to UML. We note that the growing legislation must be met by software that is more flexible and support NLP and its variations, because the technology is not so recent. In the advanced prototype, after processing the initial text, it is possible to have a list of words (verbs, nouns, etc.) that can be used by analysts and programmers for the development of the code of blockchain; The main expected post-processing result are classes (their attributes and methods) generated in the language of the installed plugin; To measure the effectiveness of the system we suggest the following measures: a) amounts of processed legislation; (b) the number of entities identified; (c) the number of nouns identified; d) number of generated code files; e) amount of plugins installed; f) space occupied on disk by the work files; g) size of the generated code; h) Processing time. The use of NLP and blockchain proved very effective to automate the legal code generation part. The set of classes, both of the basic and advanced prototypes, proved to be effective and pertinent in this sense. Also, we highlight the choice of NLP tools, as well as the definition of what should be implemented NLP for each of the prototypes. Finally, the authors also emphasize the graphical platform created, since it is through it that they are given the entries of configuration files and legislation data, and pseudo-codes of smart contracts are obtained to assist programmers.

6 Conclusion

At first, a research was developed in the literature on works involving NLP themes, blockchain, code generation and impacts on legislation. Subsequently, the need for assistance in the programming of smart contracts was identified and NLP enters as an alternative. Then a model to process legislation and generate smart contracts was elaborated and tested via prototypes. The applicability occurs in public or private institutions that need to generate computer programs (smart contracts) quickly to meet the increasing changes in legislation. The prototypes developed evaluated steps of an NLP. They were developed as applications that implement concepts treated in a simple and practical way. A basic prototype was the starting point for text manipulation, while an advanced prototype with greater implementation details was also described. The latter generates smart contracts and also serves as an implementation model for future developments with other smart contract languages. It is concluded that the results are encouraging, showing the relevance of blockchain and NLP to support the legal sector. Future works can be addressed from the prototypes, both in the sense of processing speed, as well as effectiveness in the treatment of texts, in the development of data patterns and word processing of other areas of knowledge. Still, another possibility is to explore cloud computing, exploring existing NLP and Blockchain libraries on platforms such as Amazon, Google, and Microsoft.

References

1. Sultan, K., Ruhi, U., Lakhani, R.: Conceptualizing blockchains: characteristics & applications (2018)
2. Maesa, D., Mori, P.: Blockchain 3.0 applications survey. J. Parallel Distrib. Comput. **138**, 99–114 (2020). https://doi.org/10.1016/j.jpdc.2019.12.019
3. Salah, K., Rehman, H.U., Nizamuddin, M., Al-Fuqaha, A.: Blockchain for AI: review and open research challenges. IEEE Access **7**, 10127–10149 (2018)
4. Butijn, B.-J., Tamburri, D., Heuvel, W.-J.: Blockchains: a systematic multivocal literature review. ACM Comput. Surv. **53**, 1–37 (2020). https://doi.org/10.1145/3369052

5. Porru, S., Pinna, A. Marchesi, M., Tonelli, R.: Blockchain-oriented software engineering: challenges and new directions. In: 2017 IEEE/ACM 39th International Conference on Software Engineering Companion (ICSE-C), Buenos Aires, pp. 169–171 (2017). https://doi.org/10.1109/ICSE-C.2017.142
6. do Amaral, G.L., et al.: Quantidade de normas editadas no Brasil: 28 anos da constituição federal de 1998. IBPT (2018). https://www.conjur.com.br/dl/estudo-ibpt-edicao-criacao-leis.pdf. Accessed 20 Mar 2020
7. Marques, C.: Ambiguidade no Direito: Algumas Considerações. Revista Diálogos. 74–82 (2011). https://doi.org/10.13115/2236-1499.2011v1n4p74
8. Gill, S.S., et al.: Transformative effects of IoT, blockchain and artificial intelligence on cloud computing: evolution, vision, trends and open challenges. Internet Things 8, 100118 (2019). https://doi.org/10.1016/j.iot.2019.100118
9. Almasoud, A.S., Eljazzar, M.M., Hussain, F.: Toward a self-learned smart contracts. In: 2018 IEEE 15th International Conference on e-Business Engineering (ICEBE), Xi'an, pp. 269–273 (2018). https://doi.org/10.1109/ICEBE.2018.00051
10. Khurana, D., Koli, A., Khatter, K., Singh, S.: Natural language processing: state of the art, current trends and challenges (2017). arxiv.org/abs/1708.05148
11. Nadkarni, P., Ohno-Machado, L., Chapman, W.: Natural language processing: an introduction. J. Am. Med. Inf. Assoc.: JAMIA. 18, 544–51 (2011). https://doi.org/10.1136/amiajnl-2011-000464
12. Meziane, F., Athanasakis, N., Ananiadou, S.: Generating natural language specifications from UML class diagrams. Requir. Eng. 13, 1–18 (2008). https://doi.org/10.1007/s00766-007-0054-0
13. Sureka, A., Mirajkar, P., Indukuri, K.: A rapid application development framework for rule-based named-entity extraction, p. 25 (2009). https://doi.org/10.1145/1517303.1517330
14. Deeptimahanti, D., Sanyal, R.: Semi-automatic generation of UML models from natural language requirements. In: Proceedings of the 4th India Software Engineering Conference 2011, ISEC 2011, pp. 165–174 (2011). https://doi.org/10.1145/1953355.1953378
15. Olajubu, O.: A textual domain specific language for requirement modelling. In: Proceedings of the 2015 10th Joint Meeting on Foundations of Software Engineering (ESEC/FSE 2015), pp. 1060–1062. Association for Computing Machinery, New York (2015). 2786805.2807562
16. Billings, J., McCaskey, A., Vallee, G., Watson, G.:Will humans even write code in 2040 and what would that mean for extreme heterogeneity in computing? (2017). arXiv:1712.00676
17. Lee, B.-S., Bryant, B.: Automated conversion from requirements documentation to an object-oriented formal specification language, p. 932 (2002). https://doi.org/10.1145/508969.508972
18. Jaramillo, C.M.Z.: Computational linguistics for helping requirements elicitation: a dream about automated software development. In: Proceedings of the NAACL HLT 2010 Young Investigators Workshop on Computational Approaches to Languages of the Americas (YIWCALA 2010), pp. 117–124. Association for Computational Linguistics, USA (2010)
19. Martinez, A.R.: Natural language processing. Wiley Interdisc. Rev.: Comput. Stat. 2, 352–357 (2010). https://doi.org/10.1002/wics.76
20. Sawai, S., et al.: Knowledge representation and machine translation. In: Proceedings of the 9th Conference on Computational Linguistics (COLING 1982), vol. 1, pp. 351–356. Academia Praha, CZE (1982). https://doi.org/10.3115/991813.991870

21. NILCS Corpora: Núcleo Interinstitucional de Linguistica Computacional (2000). http://www.nilc.icmc.usp.br/nilc/tools/corpora.htm. Accessed 06 Mar 2020
22. Quirk, C., et al.: Language to code: learning semantic parsers for if-this-then-that recipes. In: Proceedings of the 53rd Annual Meeting of the Association for Computational Linguistics (ACL-15), pp. 878–888, Beijing, China, July 2015
23. Osman, M.S., et al.: Generate use case from the requirements written in a natural language using machine learning. In: IEEE Jordan International Joint Conference on Electrical Engineering and Information Technology (JEEIT), Amman, Jordan, pp. 748–751 (2019). https://doi.org/10.1109/JEEIT.2019.8717428
24. Hamza, Z.A., Hammad, M.: Generating UML use case models from software requirements using natural language processing. In: 2019 8th International Conference on Modeling Simulation and Applied Optimization (ICMSAO), Manama, Bahrain, pp. 1–6 (2019). https://doi.org/10.1109/ICMSAO.2019.8880431
25. More, P.R., Phalnikar, R.: Generating UML diagrams from natural language specifications. Int. J. Appl. Inf. Syst. 1, 19–23 (2012)
26. Angstadt, K., Weimer, W., Skadron, K.: RAPID programming of pattern-recognition processors. SIGPLAN Not. 51(4), 593–605 (2016). https://doi.org/10.1145/2954679.2872393
27. Angstadt, K., Weimer, W., Skadron, K.: RAPID programming of pattern-recognition processors. SIGARCH Comput. Archit. News 44(2), 593–605 (2016). https://doi.org/10.1145/2980024.2872393

Algorithm and Related Applications

Fault Identification of Multi-level Gear Defects Using Adaptive Noise Control and a Genetic Algorithm

Cong Dai Nguyen, Alexander Prosvirin, and Jong-Myon Kim[✉]

School of Electrical Engineering, University of Ulsan, Ulsan 680-749, South Korea
daimtavn@gmail.com, a.prosvirin@hotmail.com,
jongmyon.kim@gmail.com

Abstract. This paper proposes a reliable fault identification model of multi-level gearbox defects by applying adaptive noise control and a genetic algorithm-based feature selection for extracting the most related fault components of the gear vibration characteristic. The adaptive noise control analyzes the gearbox vibration signals to remove multiple noise components with their frequency spectrums for selecting fault-informative components of the vibration signal on its output. The genetic algorithm-based feature selection obtains the most distinguishable fault features from the originally extracted feature pool. By applying the denoising during signal processing and feature extraction, the output components which mostly reflect the vibration characteristic of each multi-level gear tooth cut fault types allows for the efficient fault classification. Due to this, the simple k-nearest neighbor algorithm is applied for classifying those gear defect types based on the selected most distinguishable fault features. The experimental result indicates the effectiveness of the proposed approach in this study.

Keywords: Gearbox fault diagnosis · Adaptive noise control · Genetic algorithm · K-nearest neighbor · Multi-level gear tooth cut

1 Introduction

Gearbox fault diagnosis has been studied excessively and by analyzing its vibration characteristics it is possible to detect the hidden fault-related informative components [1, 2]. Vibration and acoustic signals are the two principal techniques for sensing the vibration characteristic of a gearbox. However, the vibration signal is most frequently used due to its easy data acquisition setup [3]. Nevertheless, there are several parasitic noise components observed in the vibration signal that appear from various sources. The most common causes of noise are the resonance processes ongoing in the shaft, gears, and other mechanical components as well as the bias hidden in data collection systems [4]. These factors cause difficulties when attempting to extract fault-related components. To overcome these issues, various techniques addressing the vibration signal processing and feature engineering areas have been proposed by researchers to improve the fault diagnosis capabilities of existing techniques.

© Springer Nature Switzerland AG 2021
M. Singh et al. (Eds.): IHCI 2020, LNCS 12615, pp. 325–335, 2021.
https://doi.org/10.1007/978-3-030-68449-5_32

Regarding signal processing, many studies focusing on approaches that analyze the signal in multi-domain have been introduced for gearbox fault diagnosis. For instance, those techniques include Hilbert transform with bandpass filters for envelope analysis [5], wavelet transform-based decomposition [6], Hilbert-Huang transforms using time-adaptive empirical mode decomposition (EMD) for decomposing vibration signals into intrinsic mode functions (IMF) [7], and a combined technique that utilizes wavelet transform and EMD [8]. Those methods could reduce the signal noise to some degree, but this noise reduction led to degraded magnitudes of the sideband and meshing frequency harmonics distorting the components of vibration signals related to gear faults. Due to this drawback, these methods are likely inapplicable for differentiating fault types of multilevel gear tooth cut (MGTC) defects. Therefore, in this paper, we apply the adaptive noise control (ANC) for denoising the vibration signals and preserving the original gearbox fault-related components in it [9].

Regarding the feature selection, this procedure from the feature engineering area is mainly used to select the most discriminative feature parameters from the feature set. In general, the feature set extracted from the vibration signals consists of both useful and low-quality features that can affect the classification accuracy of the classifier. To address this problem and remove the redundant features, a searching procedure for selecting the most distinguishable fault features (MDFF) is needed. In practice, several selection algorithms such as: independent component analysis (ICA) [10], principal component analysis (PCA) [11], a linear discriminant analysis (LDA) [12], and a genetic algorithm (GA) are used for selecting high-quality feature. The main performance of these approaches [10–12] is to find the important components (independent, principal or linear dependence) in the feature space by applying the statistical process or the statistical functions, it is useful when the fault-related informative components are in the relation with the space of important components, otherwise, the output can be lost fault useful components. GA was constructed based on the principles of natural generic systems and popularly used to output the result of high effectiveness [13], thence GA draws the attention of the researchers. In this paper, a GA implements a heuristic search algorithm on the original feature pool to select the most relevant and discriminative features related to MGTC defects, so the dimensions of feature vectors can be reduced which complements the accuracy of the fault classification process. In other words, GA accommodates a balance between the complexity of computation and the optimal selection. Finally, the MDFF subsets delivered by GA are inputted to the k-nearest neighbor (k-NN) classification algorithm to discriminate the health states of a gearbox system.

The remainder of this paper is organized as follows: Sect. 2 provides a problem statement of vibration characteristic of a healthy and a defect gearbox. The detail of the proposed methodology is presented in Sect. 3. Section 4 presents the experimental results and discussion, and the concluding remarks are provided in Sect. 5.

2 Problem Statements

In this paper, the operating faults of a one-stage transmission gearbox, which encompass a pinion wheel (on drive side) and a gear wheel (on non-drive side) are considered. For the healthy gear the operation of which generates a linear and periodical vibration signal [14], the vibration signal $x_P(t)$ under constant load speed is constructed by using the formula below [15], and the example of frequency spectrum illustrated in Fig. 1a:

$$x_P(t) = \sum_{i=0}^{M} X_i \cos(2\pi i f_m t + \xi_i),$$ (1)

where, ξ_i and X_i are the phase and amplitude of the i-th meshing frequency harmonics ($i = 1, \ldots, N$); f_m is the meshing frequency ($f_m = N_P \cdot f_p$, where f_p is a pinion rotational frequency, N_P is the number of teeth mounted on the pinion wheel; or $f_m = N_G \cdot f_g$ where f_g is a gear rotational frequency, N_G is the number of teeth mounted on the gear wheel); M is the total number of meshing frequency harmonics in the frequency range of a vibration signal.

In the case of faulty gear, the vibration signal, which is directly related to rotating acceleration, is non-linear and non-stationary (i.e. amplitude and phase-modulated signal) [14]. The Eq. (2) represents the vibration signal of the defected gear and its spectrum is depicted in Fig. 1b:

$$x_d(t) = \sum_{l=0}^{M} X_l(1 + s_l(t)) \cos(2\pi l f_m t + \xi_l + \psi_l(t)).$$ (2)

Here: $s_l(t) = \sum_{k=0}^{M} S_{lk} \cos(2\pi k f_g t + \Omega_{lk})$; $\psi_l(t) = \sum_{k=0}^{M} \Phi_{lk} \cos(2\pi k f_g t + \sigma_{lk})$, S_{lk} and Φ_{lk} are the amplitudes and Ω_{lk}, σ_{lk} are the phases of the k-th sideband frequency in the modulated signal around l meshing harmonic.

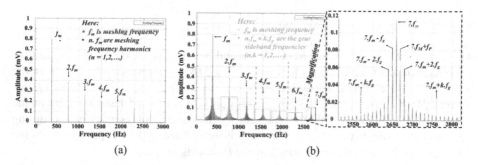

Fig. 1. The example of the frequency spectra corresponding to the vibration signal of (a) a healthy gearbox and (b) a defected gearbox.

3 Proposed Method

The proposed methodology is described in Fig. 2. It consists of five steps that are represented as follows: the vibration signal collection, adaptive noise control for signal preprocessing, feature extraction, feature selection based on GA, and the k-NN-based fault classification.

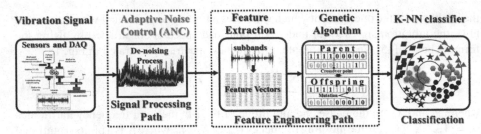

Fig. 2. Function block diagram of the proposed methodology.

3.1 Adaptive Noise Control (ANC)

This paper employed the adaptive noise reducer based Gaussian reference signal (ANR-GRS) which was proposed by Nguyen et al. in [9] as an adaptive noise control technique. This technique comprises four main processing modules: a digital filter, an adaptive algorithm, a reference signal generation, and a Gaussian parameter optimizer. The reference signal generation module creates the reference signal inputting to adaptive filter by combining the adjustable Gaussian (the mean value and standard deviation value are adaptably tuned) and white noise signals which are mostly homologous to the band noise and white noise in a vibration signal, respectively. The adjustable parameters of a Gaussian reference signal (GRS) are the functions of the defective wheel frequency (f_{DW}, which is proportional to a shaft rotational speed). The detail description of GRS is presented by the Gaussian window in the Eq. 3, and the flow chart diagram in the Fig. 3 [9]:

$$W_{Gau}(s) = \sum_{s=1}^{N_g} e^{-\frac{(s-F_C)^2}{2\sigma^2}},\tag{3}$$

where σ and F_C are the standard deviation and mean value of the Gaussian window (by linearization of the Gaussian function, $\sigma \cong 0.318 \cdot F_C$), respectively. These are the functions of the defective wheel frequency ($F_C = \xi \cdot f_{DW}$ and $\sigma \cong 0.318 \cdot \xi \cdot f_{DW}$). $N_g = \frac{2N_T}{f_s} \cdot f_{DW}$ is the total number of frequency bins in a sideband segment of the gearbox vibration signal with N_T representing the number of samples and sampling frequency f_s. To qualify the homologous condition defined in [9], the parameters of the Gaussian window are adaptively tuned in the following ranges:

$$0.25 \cdot f_{DW} \le F_C \le 0.75 \cdot f_{DW},\tag{4}$$

$$\sigma = \begin{cases} 0.318 * \xi * f_{DW} & when\ 0.25 \le \xi \le 0.5 \\ 0.318 * (1-\xi) * f_{DW} & when\ 0.5 < \xi \le 0.75 \end{cases}.\tag{5}$$

The digital filter module, which is used in the ANC, is an M-order FIR filter with the weight vector as $\mathbf{w}(n) \equiv [w_0, w_1, ..., w_{M-1}]^T$. The adaptive algorithm adjusts the weight vectors based on the convergence condition by least mean square (LMS) of error between the vibration signal and each filtered GRS (the reference signal generation module creates a set of GRS signals by adjusting σ and F_C) for fetching the optimal weight vector and local optimal subband signal in the output. The Gaussian parameter optimizer module collects all the local optimal subbands where each of them corresponds to a specific GRS signal to make the set of optimal subbands, then it selects the optimal subband that has a minimum mean square value as global optimal subband termed as an optimized subband, the output of ANC. The block diagram and signal processing flow are shown in Fig. 3.

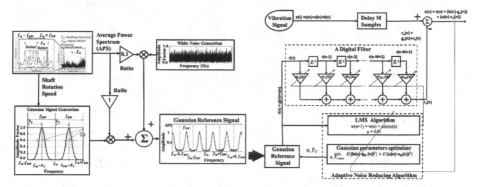

Fig. 3. A reference signal generation and function block diagram of the ANC module.

3.2   Feature Extraction

The optimized subband output from the ANC module represents a reduced-noise vibration signal which carries the intrinsic gear defective information of the MGTC faults. These outputs from ANC are then used to extract eighteen statistical feature parameters from the time and frequency domains to configure the feature pool [16]. The Table 1 illustrates the eighteen features with the specific name and calculated formulas [17]. From time-domain, the extracted features are as follows: peak (f1), root mean square (f2), kurtosis (f3), crest factor (f4), impulse factor (f5), shape factor (f6), entropy (f7), skewness (f8), square mean root (f9), energy of system (f10), mean (f11), shape factor of square mean root (f12), margin factor (f13), peak to peak (f14), kurtosis factor (f15). The features extracted from the frequency-domain are as follows: frequency center (f16), power spectral density or frequency spectrum energy (f17), and root variance frequency (f18).

3.3   GA-Based Feature Selection

By applying the procedures from evolution theory such as selection, crossover, mutation, and replacement, GA detects the MDFF based on the class-wise information embedded in the complete feature set. The degree of class discrimination (D_{dst}) is defined in Eq. (6) for creating separability among observed classes:

$$D_{dst} = \frac{I_{dst}}{W_{dst}} \tag{6}$$

Table 1. Definition of time and frequency domain statistical features.

Features	Equations	Features	Equations	Features	Equations
f1	$\text{Max}(|s|)$	f6	$\frac{s_{rms}}{\frac{1}{N}\sum_{n=1}^{N}|s_n|}$	f11	$\frac{1}{N}\sum_{n=1}^{N} s_n$
f2	$\sqrt{\frac{1}{N}\sum_{n=1}^{N} s_n^2}$	f7	$-\sum_{n=1}^{N} p_n \cdot log_2(p_n)$	f12	$\frac{s_{srm}}{\frac{1}{N}\sum_{n=1}^{N}|s_n|}$
f3	$\frac{1}{N}\sum_{n=1}^{N}\left(\frac{s_n - \bar{s}}{\sigma}\right)$	f8	$\frac{1}{N}\sum_{n=1}^{N}\left(\frac{s_n - \bar{s}}{\sigma}\right)^3$	f13	$\frac{max(s)}{s_{smr}}$
f4	$\frac{\text{Max}(|s|)}{s_{rms}}$	f9	$\left(\frac{1}{N}\sum_{n=1}^{N}\sqrt{|s_n|}\right)^2$	f14	$max(s) - min(s)$
f5	$\frac{\text{Max}(|s|)}{\frac{1}{N}\sum_{n=1}^{N}|s_n|}$	f10	$\sum_{n=1}^{N} s_n^2$	f15	$\frac{Kurtosis}{s_{rms}^4}$
f16	$\frac{1}{N_f}\sum_{f}^{N_f} S(f)$	f17	$\sum_{f}^{N_f} S(f)^2$	f18	$\sqrt{\frac{1}{N_f}\sum_{f}^{N_f}(S(f) - FC)^2}$

Here is an input signal (i.e., optimized subband), N is the total number of samples, S(f) is the magnitude response of the fast Fourier transform of the input signal s, N_f is total number of frequency bins, $\sigma = \sqrt{\frac{1}{N}\sum_{n=1}^{N}(s_n - \bar{s})^2}$, *and* $p_n = \frac{s_n^2}{\sum_{n=1}^{N} s_n^2}$.

where, I_{dst} is intercross classes discriminating parameter defining the distance between distinct classes, W_{dst} is the distance of the features inside the same class. As the distance between two vectors, the Euclidian distance $D_{Euclidian}(x, y) = \sqrt{\sum_{i=1}^{m}(x_i - y_i)}$ is used in this paper. The parameter D_{dst} tends to increase with the increase of the numerator value (I_{dst}) and the decrease of the denominator. I_{dst} is calculated based on the average distance of specific feature vector from different classes by the formula:

$$I_{dst} = \frac{1}{F.C.M}\sum_{i=1}^{F}\sum_{j=1}^{C}\sum_{k=1}^{m} D_{Euclidian}(i, j, k), \tag{7}$$

Equation (8) represents the computation of the average distance of a feature in the same class:

$$I_{dst} = \frac{1}{F.M}\sum_{i=1}^{F}\sum_{j=1}^{M} D_{Euclidian}(i, j). \tag{8}$$

Here, F is the total number of features ($F = 18$ in this study), C is the total number of categories or classes ($C = 6$), and M is the number of samples of each category.

The GA operates feature optimizing process following the flow chart shown in Fig. 4 to find out the features with maximum D_{dst} to select as MDFF.

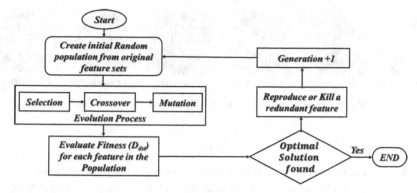

Fig. 4. The flowchart of the GA-based feature selection.

3.4 Gearbox Fault Classification Using k-Nearest Neighbor Algorithm (k-NN)

In practice, the k-NN algorithm faces problems such as the increase of computational complexity when the high dimensional feature vectors are used as input. Thus, the GA-based feature selection for reducing the dimensionality of the feature set is essential before the application of k-NN to perform fault diagnosis of the gearbox faults. In k-NN, data samples are classed by plurality votes of k-nearest neighbors, which are calculated by distance parameters [18]. Therefore, to complete the k-NN classification process, two important parameters such as the number of nearest neighbors (k) and the distance metric have to be selected. As the distance metric, the Euclidian distance is considered in this study. The appropriate value of k should be selected during the training process. The parameter k could be identified manually or through cross-validation progress. In this paper, the value of k is first arbitrarily assigned at beginning of the training process, and then its value has been changing during the cross-validation to achieve the optimal value of k.

4 Dataset, Experimental Results, and Discussion

4.1 Dataset Description

The experimental testbed for the gearbox fault diagnosis system is represented as a spur gearbox that consists of a pinion and gear wheels. The total number of teeth in the pinion wheel is 25 ($N_P = 25$) and 38 teeth ($N_G = 38$) in the gear wheel (the gearbox ratio is 1:1.52), respectively. The tooth failures of the gear wheel have been created by multiple levels of tooth cut seed fault (the total length of the tooth is 9 mm) and were termed as the fault states of the gearbox as follows: tooth cut 10% (D1), tooth cut 20% (D2), tooth cut 30% (D3), tooth cut 40% (D4), tooth cut 50% (D5), and a healthy gear (P). These types of faults are depicted in Fig. 5. The vibration sensor (accelerometer 622B01) was mounted at the end of the non-drive shaft for sensing the vibration signals of the gear wheel under constant shaft rotation speed at 900 RPM. The analog vibration signals were digitized by using a PCI-2 data acquisition device with a sampling frequency of 65536 Hz. For each fault state of the gearbox, 100 samples of 1-s length were acquired (further referred to as a 1-s sample).

No seeded fault   0.9 mm (10%)   1.8 mm (20%)   2.7 mm (30%)   3.6 mm (40%)   4.5 mm (50%)

Fig. 5. The description of the MGTC defects in a gear wheel.

4.2   Experimental Results and Discussion

As the ANC functions for noise reduction, white noise and band noise integrated into the 1-s vibration signal samples are used by accessing the spaces between two consecutive sideband frequencies. Then the ANC performs tuning of the model coefficients and the parameters of the Gaussian reference signal to simultaneously remove the noise from the vibration signal and preserve the sideband and meshing frequency harmonics at their original magnitudes. Figure 6a presents the effect of applying the ANC technique to one 1-s sample of fault state D3. It can be seen that the noise zones in the segment of the frequency spectrum (the green circles) are removed as the multitude in the same segment of the frequency spectrum of optimized subband output from ANC. On contrary, the magnitudes of informative components such as sideband frequencies and the meshing frequency (the black circles) are similar between the input vibration signal and output subband of the ANC.

The GA-based feature selection operates through the selection, crossover, and mutation operators with respect to the finest function of the maximum degree of distance evaluated for each feature in the extracted feature pool. In this experiment, GA is applied during 150 generations to achieve the optimal subset of two MDFF (f9, f17). f9 (square mean root) is a popular feature in the time domain, f17 is a power spectral density or frequency spectrum energy. The distribution zones of 6 MGTC fault states are sketched in Fig. 6b by a 2D plot with (f9, f17). The subsets of MDFF selected by GA are used as input data of the k-NN classifier to classify the data into their respective categories (i.e., P, D1, D2, D3, D4, D5). By plotting distribution areas of samples in Fig. 6b, we can see that samples of the same category are close to each other, whereas samples of different categories are separated in feature space.

The optimizing process for tuning k value of the k-NN classifier is implemented multiple times with cross-validation for obtaining the determination boundaries and evaluate the effectiveness of k-NN performance. After optimizing the procedure, the optimal value of k has been assigned as 8. The classification accuracy ($A_{accuracy}$) for assessing the fault diagnosis performance has been calculated as follows:

$$A_{accuracy} = \frac{Number\ of\ True\ Possitive\ +\ Number\ of\ True\ Negative}{Total\ number\ of\ Samples} \cdot 100\% \quad (9)$$

By using the optimal k value and GA-based selected MDFF subsets, the achieved an average classification accuracy of the proposed methodology was equal to 97.78%. The confusion matrix obtained by the proposed methodology is depicted in Fig. 6c. From this figure it can be seen that the classification result for each fault type of MGTC faults was achieved as follows: P (98%), D1 (100%), D2 (100%), D3 (94%), D4 (100%), D5 (94%).

From the experimental results, it can be concluded that the proposed model is capable of performing the diagnosis of MGTC gearbox faults at high classification accuracy.

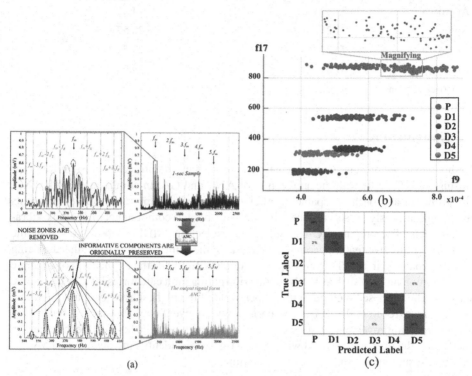

Fig. 6. The effectiveness of ANC, GA, and classification result: (a) the comparison of vibration signal frequency spectrum between input and output of ANC, (b) 2-D visualization of data samples of MGTC faults based on the MDFF, (c) confusion matrix of the k-NN classification result. (Color figure online)

To validate the effectiveness of the proposed method, the comparative experiments are conducted with other conventional feature extraction and classification approaches such as combined feature extraction [9] (further referred to as approach 1), ICA and k-NN (further referred to as approach 2), PCA and k-NN (further referred to as approach 3), and LDA (further referred to as approach 4). These comparisons are implemented by replacing the GA module of the fault identification system for multi-level gear defects in this paper with another one (ICA, PCA, LDA) for selecting the optimal feature space to attain comparing approaches (approach 2, 3, 4). The detail of the comparative results is demonstrated in Table 2. It is observed that the approach (GA+k-NN) achieves the highest accuracy (97.78%) outperforming the other state-of-the-art approaches. The improvements of the classification result of the proposed method are 14.26%, 26.37%, 23.51%, and 29.94% in comparison with four referenced approaches: 1, 2, 3, and 4, respectively. It is verified that the combined application of ANC and GA can construct an effective model for fault identification of the MGTC gearbox system.

334   C. D. Nguyen et al.

Table 2. The fault classification results of the proposed and referenced techniques.

The approaches	Average classification accuracy (%)	The improvement (%)
Approach 1	83.52	14.26
Approach 2	71.21	26.57
Approach 3	74.27	23.51
Approach 4	67.84	29.94
Proposed approach	97.78	-

5 Conclusions

In this paper, we present the new sensitive gearbox fault identification system for diagnosing MGTC faults with the application of ANC and GA. In combination with ANC, the proposed approach is capable of efficiently removing numerous noise components and simultaneously preserving the intrinsic fault-related components in gearbox vibration signals. The output of GA-based feature selection (MDFF subsets) contains the most discriminative feature parameters that make the samples of each health state being clearly separated in the 2D feature space graph and allow for the application of a simple classification method such as k-NN for discriminating fault categories into the respective classes. The proposed method in this paper yielded the highest average fault classification accuracy result of 97.78% in comparison with conventional approaches. It provides an accuracy improvement of at least 14.26% higher than the referenced techniques. In the future, the proposed method will be investigated for identifying defects of a MGTC gearbox system under varying speed conditions.

Acknowledgments. This research was financially supported by the Ministry of Trade, Industry & Energy (MOTIE) of the Republic of Korea and Korea Institute for Advancement of Technology (KIAT) through the Encouragement Program for The Industries of Economic Cooperation Region. (P0006123).

References

1. Chaari, F., Bartelmus, W., Zimroz, R., Fakhfakh, T., Haddar, M.: Gearbox vibration signal amplitude and frequency modulation. Shock Vib. **19**(4), 635–652 (2012)
2. Baxter, J.W., Bumby, J.R.: An Explanation for the Asymmetry of the Modulation Sidebands about the Tooth Meshing Frequency in Epicyclic Gear Vibration. Proc. Inst. Mech. Eng. Part C J. Mech. Eng. Sci. **199**, 65–70 (1995)
3. Ghodake, S.B., Mishra, P.A.K., Deokar, P.A.V.: A review on fault diagnosis of gear-box by using vibration analysis method. IPASJ Int. J. Mech. Eng. **4**(1), 31–35 (2016)
4. Shinde, K.U., Patil, C.R., Kulkarni, P.P., Sarode, N.N.: Gearbox noise & vibration prediction and control. Int. Res. J. Eng. Technol. **04**(2), 873–877 (2017)
5. Randall, R.B., Antoni, J., Chobsaard, S.: The relationship between spectral correlation and envelope analysis in the diagnostics of bearing faults and other cyclostationary machine signals. Mech. Syst. Signal Process. **15**(5), 945–962 (2001)

6. Aharamuthu, K., Ayyasamy, E.P.: Application of discrete wavelet transform and Zhao-Atlas-Marks transforms in non stationary gear fault diagnosis. J. Mech. Sci. Technol. **27**(3), 641–647 (2013)
7. Buzzoni, M., Mucchi, E., D'Elia, G., Dalpiaz, G.: Diagnosis of localized faults in multistage gearboxes: a vibrational approach by means of automatic EMD-based algorithm. Shock Vib. **2017**, 1–22 (2017)
8. Zamanian, A.H., Ohadi, A.: Gear fault diagnosis based on Gaussian correlation of vibrations signals and wavelet coefficients. Appl. Soft Comput. J. **11**(8), 4807–4819 (2011)
9. Nguyen, C.D., Prosvirin, A., Kim, J.M.: A reliable fault diagnosis method for a gearbox system with varying rotational speeds. Sensors (Switzerland) **20**(11), 3105 (2020)
10. Ariff, M.A.M., Pal, B.C.: Coherency identification in interconnected power system - an independent component analysis approach. IEEE Trans. Power Syst. **28**(2), 1747–1755 (2013)
11. Nguyen, V.H., Golinval, J.C.: Fault detection based on kernel principal component analysis. Eng. Struct. **32**(11), 3683–3691 (2010)
12. Ece, D.G., Baaran, M.: Condition monitoring of speed controlled induction motors using wavelet packets and discriminant analysis. Expert Syst. Appl. **38**(7), 8079–8086 (2011)
13. Kang, M., Kim, J., Kim, J.M., Tan, A.C.C., Kim, E.Y., Choi, B.K.: Reliable fault diagnosis for low-speed bearings using individually trained support vector machines with kernel discriminative feature analysis. IEEE Trans. Power Electron. **30**(5), 2786–2797 (2015)
14. Fakhfakh, T., Chaari, F., Haddar, M.: Numerical and experimental analysis of a gear system with teeth defects. Int. J. Adv. Manuf. Technol. **25**(5–6), 542–550 (2005)
15. Fan, X., Zuo, M.J.: Gearbox fault detection using Hilbert and wavelet packet transform. Mech. Syst. Signal Process. **20**(4), 966–982 (2006)
16. Caesarendra, W., Tjahjowidodo, T.: A review of feature extraction methods in vibration-based condition monitoring and its application for degradation trend estimation of low-speed slew bearing. Machines **5**(4), 21 (2017)
17. Kang, M., Islam, M.R., Kim, J., Kim, J.M., Pecht, M.: A Hybrid feature selection scheme for reducing diagnostic performance deterioration caused by outliers in data-driven diagnostics. IEEE Trans. Ind. Electron. **63**(5), 3299–3310 (2016)
18. Yigit, H.: A weighting approach for KNN classifier. In: 2013 International Conference on Electronics, Computer and Computation, ICECCO 2013, vol. 2, no. 2, pp. 228–231 (2013)

Applying Multiple Models to Improve the Accuracy of Prediction Results in Neural Networks

Hyun-il Lim[✉]

Department of Computer Engineering, Kyungnam University,
Changwon, Gyeongsangnam-do, South Korea
hilim@kyungnam.ac.kr

Abstract. The neural network is being widely applied to solve various types of real-life problems. Continuous researches are needed to improve the accuracy of the neural network for practical application. In this paper, the method of applying multiple neural network models as a way to improve the accuracy of the conventional neural network is proposed. Each of multiple neural networks is trained independently with each other, and each generated model is integrated to produce a single prediction result. The proposed prediction model with multiple neural network models is evaluated as compared to the conventional neural network model in terms of accuracy. The results of the experiment showed that the accuracy of the proposed methods was higher than that of the conventional neural network model. The use of neural networks is expected to be continuously increasing. The proposed prediction model with multiple neural network models is expected to be applied in prediction problems in real-life to improve the reliability of the prediction results.

Keywords: Neural network · Machine learning · Artificial intelligence · Multiple models

1 Introduction

The neural network is a popular approach in real-life problems that are difficult to solve in traditional algorithm-based approaches [1, 2]. A training process using a large number of training data can create a problem-solving model, even if the model does not provide specific algorithms for solving the problem. Thus, neural networks are actively applied to solve real-life problems in various fields, including language recognition, image analysis, voice recognition, finance, medical science, and robotics. The basic principle of neural networks is to create a model that can induce the prediction results by adjusting the connection strength of links between nodes that are connected between multistage layers through training. Despite the simple conceptual principle of neural networks, the accuracy of the prediction results may vary according to the various factors considered in neural network design.

© Springer Nature Switzerland AG 2021
M. Singh et al. (Eds.): IHCI 2020, LNCS 12615, pp. 336–341, 2021.
https://doi.org/10.1007/978-3-030-68449-5_33

Several areas of research are being carried out to reflect effective structure to improve the accuracy of the neural network. In this paper, an approach to applying multiple neural network models is presented to improve the accuracy of prediction results of trained models. Reflecting multiple models for a single prediction can be the basis for enhancing the reliability from the results agreed from multiple models. The proposed prediction model with multiple neural network models is evaluated as compared to the conventional neural network model in terms of the accuracy of results. The proposed model with multiple neural network models is expected to be applied in prediction problems in real-life to improve the reliability of the prediction results.

This paper is organized as follows. Section 2 describes the motivation of this study in applying neural networks in prediction problems. Section 3 presents an approach to applying multiple models to improve the accuracy of results in neural networks. Section 4 shows the experimental results of the proposed approach. Finally, Sect. 5 concludes this paper.

2 Applying Neural Network in Prediction Problems

The neural network is a way to solve problems by learning from training data [3]. The structure of a neural network is originated from the learning process of the human brain that consists of connected neurons. The neural network is organized with many connected nodes operating independently, and information is distributed on many connected links between nodes. So, the failure of some nodes does not significantly affect the entire system. The neural network models are used to solve problems in the area of artificial intelligence, such as text recognition, image recognition, natural language processing, and several types of prediction problems. The typical neural network model is a multilayer computation system for modeling complex nonlinear and multidimensional relationships between input and output data. The model is statistical models that can be applied to solve real-life problems by modeling the human brain which can learn prediction rules through experience.

The neural network is a mathematical model of several layers of nodes, and there are various types of models according to the theory of machine learning. Figure 1 shows an example of the structure of a neural network model. The structure generally consists of an input layer, several hidden layers, and an output layer. The nodes in each layer are connected to every node in their adjacent layer, and the weight values of the connection links are adjusted in the process of learning from training data to predict the correct result for input data. So, applying the neural network model is used to solve a prediction problem by adjusting the weight values of connected links between layers through training with data so that the trained model can predict the results from other input data.

3 Designing Multiple Neural Network Models

The neural network is being used as a model for predicting solutions for various real-life problems, and the study is in progress to improve the prediction results. The accuracy of the neural network requires the management of well-organized training data and the design of the effective structure of a neural network model. The conventional neural

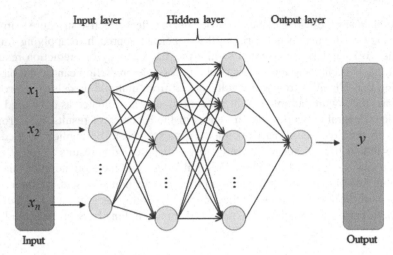

Fig. 1. An example of the structure of a deep neural network model.

network tries to train only a single model for predicting the correct results from previously generated training data by adjusting weights on the connected links between layers of the model. However, there are some limits on increasing the accuracy of the trained model due to the effect of noise information that may be contained in training data or possible constraints on the structural characteristic of a neural network model. To mitigate this problem, multiple independent neural network models for training data can be considered in designing a prediction model to improve the reliability in results by reflecting the results of the multiple models together.

Figure 2 shows the basic design of a prediction model with multiple neural network models for input and output data. To generate independent models for predicting results, several features that are selected randomly from the original input features are added to the original data in duplicate. So, the randomly added features of input data can diversify the neural network models for applying multiple models in prediction problems. Each embedded neural network model is trained independently with each other, and the prediction results of the embedded models may be different from each other. The results from the independent models are brought together and combined into one result. Because the embedded models are trained independently, the prediction results may differ from each other according to the process of training the models. In such a case, the majority of the results are reflected in the final results. Training errors that may appear in a single model can be reduced by reflecting the majority of results that are predicted from the independent models. So, it is important to train models independently to complement each other with errors in the model's prediction results.

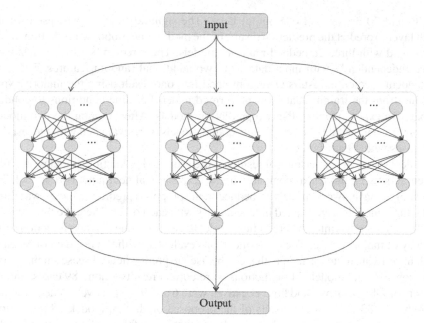

Fig. 2. The design of a prediction model with multiple neural network models.

4 Experimental Results

In this section, the accuracy of the proposed prediction model with multiple neural network models is evaluated as compared to the conventional approach of the neural network [4–6]. This experiment was performed on a heart disease prediction problem with the Cleveland heart disease data set [7–9]. Table 1 shows the specification of the data set. The data has 303 instances with information on heart disease status and 13 attributes for predicting heart disease, including age, sex, chest pain type, and blood pressure. Among 303 data, 214 cases were used as training data, and the remaining 89 cases were used to evaluate the accuracy of heart disease prediction results. The proposed prediction model and the conventional model were implemented in Python [10] and Keras [11] with TensorFlow backend.

Table 1. The specification of the Cleveland heart disease data set used to evaluate the proposed prediction model with multiple neural networks.

Data set name	# of instances	# of input attributes	# of output attributes
Cleveland heart disease data set	303 (214 + 89)	13 (measured attributes)	1 (diagnosis attribute)

The implemented two prediction models were made using neural network models of identical layer structures, those were organized with one input layer, two hidden layers with 30 and 20 nodes, and one output layer. The sigmoid function was used for the output layer to predict the prediction results of the model. The proposed prediction model was applied with three embedded multiple models. The three embedded models were trained independently with input data with two additional random features. The three independent prediction results were combined into one result using the majority vote. Learning from the training data was performed under the same conditions to conduct comparative experiments on the two prediction models. After training the two models with training data with 214 cases, the generated models were evaluated with the test data of 89 cases with 13 attributes.

Table 2 shows the experimental results of the proposed prediction model with multiple neural network models as compared to the conventional neural network model. The number of neural network models embedded in the proposed prediction model was three. So, the three models were trained independently with each other. Then, the three trained models are integrated into one prediction model to calculate one agreed prediction result by a way of majority vote. The experiment was evaluated with the test data of 89 cases which have 13 attributes as factors that can be used to predict heart disease. In the results of the conventional model, 74 predictions were correct results among 89 cases, and the numbers of false positives and false negatives were 6 and 9, respectively. The prediction accuracy was 83.15%. In the results of the proposed prediction model, 78 predictions were correct results, and the numbers of false positives and false negatives were 4 and 7, respectively. The accuracy of the proposed method was 87.64%, and it was about 4.4% points higher than that of the conventional approach.

Table 2. The experimental results of the proposed prediction model with multiple neural network models as compared to the conventional neural network model.

	The conventional neural network model	The proposed prediction model
# of neural network models	1	3
# of test data	89	89
# of correct predictions	74	78
# of false positives	6	4
# of false negatives	9	7
Overall accuracy	83.15%	87.64%

To improve the prediction accuracy of a neural network model, the structure of the neural network model is needed to be studied. Just as designing the structure of layers and nodes in a single neural network model, the method of applying multiple neural network models is expected to improve the reliability from the multiple analyses of models. The experimental results of the proposed method showed that the prediction model with multiple neural networks can improve accuracy as compared to a single neural network. In future studies, further research and analysis will be done on how to effectively integrate multiple neural networks to improve the accuracy and efficiency of the prediction model. The research for improving the accuracy of the neural network could contribute to making life more convenient for the future.

5   Conclusion

The neural network is being widely applied to solve various types of real-life problems. Extensive researches are in progress to improve the accuracy of prediction results of the neural network. In this paper, the method of applying multiple neural network models as a way to improve the accuracy of the conventional neural network is proposed. Each of multiple neural networks is trained independently with each other, and each generated model is integrated to produce a single prediction result. The comparative experiments with the existing neural network were conducted to evaluate the accuracy of the proposed method. The results of the experiment showed that the accuracy of the proposed methods was higher than that of the conventional neural network model. The use of neural networks is expected to be continuously increasing, and various approaches need to be studied to improve the accuracy of prediction results of neural networks.

Acknowledgments. This work was supported by the National Research Foundation of Korea (NRF) grant funded by the Korea government (Ministry of Education) (No. NRF-2017R1D1A1B03034769).

References

1. Shalev-Shwartz, S., Ben-David, S.: Understanding Machine Learning: From Theory to Algorithms. Cambridge University Press, Cambridge (2014)
2. Domingos, P.: A few useful things to know about machine learning. Commun. ACM **55**(10), 78–87 (2012)
3. Murphy, K.P.: Machine Learning: A Probabilistic Perspective. The MIT Press, Cambridge (2012)
4. Chollet, F.: Deep Learning with Python. Manning Publications, New York (2017)
5. Gulli, A., Pal, S.: Deep Learning with Keras: Implementing Deep Learning Models and Neural Networks with the Power of Python. Packt Publishing, Birmingham (2017)
6. Hope, T., Resheff, Y.S., Lieder, I.: Learning TensorFlow: A Guide to Building Deep Learning Systems. O'Reilly, Newton (2017)
7. UCI Machine Learning Repository, Heart Disease Data Set. https://archive.ics.uci.edu/ml/datasets/heart+disease/
8. Almustafa, K.M.: Prediction of heart disease and classifiers' sensitivity analysis. BMC Bioinform. **21**, 278 (2020)
9. Gárate-Escamila, A.K., El Hassani, A.H., Andrès, E.: Classification models for heart disease prediction using feature selection and PCA. Inform. Med. Unlocked **19**, 100330 (2020)
10. Python Programming Language. https://www.python.org/
11. Keras: The Python Deep Learning API. https://keras.io/

OST-HMD for Safety Training

Christopher Koenig, Muhannad Ismael$^{(\boxtimes)}$, and Roderick McCall

Luxembourg Institute of Science and Technology (LIST),
4362 Esch-sur-Alzette, Luxembourg
{christopher.koenig,muhannad.ismael,roderick.mccall}@list.lu

Abstract. Despite improvements in technologies and procedures, lab safety training remains problematic and often too theoretical. Challenges include providing training as close to the real-situation as possible and which can simulate hazards such as fires and explosions. In this paper we present SmartLab which uses augmented reality using an OST-HMD which allows for training as close to reality as possible without the need to train without using real hazards. In this prototype we simulate and implement two critical cases through OST-HMD for a chemical laboratory safety: 1) acid manipulation and 2) mix water and acid. The system should improve the ability for trainees to memorise the situation. Novel user interface techniques such as the attention funnel and a vocal assistant are used to assist trainees. Two use cases are presented which illustrate the relevancy of the approach.

Keywords: HMD · OST-HMD · Safety training · Mixed reality

1 Introduction

According to the U.S. Chemical Safety and Hazard Investigation Board, 490 chemical accidents have been recorded in US public and industrial laboratories in the period from 01.2001 to 07.2018. These incidents resulted in 262 injuries and 10 fatalities [1]. Therefore, improving the awareness of the risks posed by chemicals in laboratories using efficient methods is highly relevant. Safety training is currently performed using 1) in-person learning, including lectures and lab tours, 2) E-learning, and 3) serious games [2]. Lab users may also be asked to fill in a questionnaire evaluating their theoretical knowledge on lab safety. Many of the current training methods are either too theoretical and/or overwhelming trainees with too much information, or information which is not relevant for their future needs. In addition, the lack of critical cases simulations, experience with the safety tools and the working environment reduces the ability of trainees to respond avoid accidents and respond appropriately when an incident occurs. Our proposition is to build a new training tool, simulating realistic and specific environments and incidents using an **A**ugmented **R**eality (AR) application.

Augmented reality provides one key advantage over many techniques including virtual reality. In the case of the latter, the trainees can remain aware of the

© Springer Nature Switzerland AG 2021
M. Singh et al. (Eds.): IHCI 2020, LNCS 12615, pp. 342–352, 2021.
https://doi.org/10.1007/978-3-030-68449-5_34

wider real world environment while interacting with virtual objects etc. which represent the risks. This means they are still able to touch, see and interact with the rest of the world but face no risks with respect to the hazardous objects they are learning about. It is also potentially cheaper to develop as there is no need for developers to create a complex 3D model of for an example an entire laboratory. That said, care still needs to be taken to design a training scenario which blends the real and virtual aspects effectively and in a way which is safe. In comparison to more traditional techniques it is more interactive than books and questionnaires and provide an intermediate step before people are trained with or use real chemicals. Furthermore, as noted later in the paper AR allows novel interface techniques can be used to enhance learning and attract attention. Such approaches are not available with more traditional methods.

The paper contains a review of literature, including the nature of the augmented reality technology used and existing examples of where augmented reality has been used for related forms of training. It then details the scenarios and technologies used to build the system ending with a discussion and conclusions.

2   Background

Azuma et al. [3] identify the AR applications with three mains characteristics: a) combine virtual and real image such that both can be seen at the same time b)interactive in real time c) registered in 3d which refers that virtual objects appear fixed to a given location. In the recent years, significant research has been undertaken on the development of sensors such as (IMU, depth, camera .. etc.) integrated within mobile of **H**ead **M**ounted **D**isplay (HMD). We can classify AR applications into three main groups: Mobile based, Projector-based and HMD applications. This latter is also be classified into two types:

- **V**ideo **S**ee **T**hrough HMD, VST-HMD where the real environment is used as input to generate an AR scene. Virtual objects are merged with the information captured from one or two cameras integrated and fixed on the visor frame to bring the outside world in within HMD. Vuzix WRAP 920AR [14], Visette45 SXGA [15] and TriVisio SXGA61 3D [16] are some of commercial products using VST-HMD technique.
- **O**ptical **S**ee **T**hrough HMD, OST-HMD is based on half-silvered mirrors technique to merge the view of virtual and real objects. The advantage of this technique is the ability to directly view to the real world as is and not via a computer rendering as is the case with VST. Epson Moverio [17], Lumus optical [18], Microsoft HoloLens [20], Optinvent ORA-1 [19] are examples of commercial optical see-through head-mounted displays.

This paper focuses on using OST-HMDs to train individuals within a chemical laboratory setting.

2.1 Training Using OST-HMD

Evaluating OST-HMD in the field of training (Education, Industry,... etc.) is still in the early stages. Werrlich et al. [4] proposed an engine assembly training scenario using paper-based and OST-HMD-based training. The author concludes that participants performed significantly faster and significantly worse using paper-based instructions. González et al. [5] evaluated participants performance and the system usability between OST-HMD and a desktop interface for Scenario-Based training (SBT). They found no significant difference between the two interfaces in time taken to accomplish tasks. However, the desktop interface remained the favorite for the participants. This study is interesting because it demonstrates that some tasks are not suitable for OST-HMD interfaces. Holobrain project[1] is AR interactive teaching tool for medical students. It is a collaboration between Microsoft and British Columbia University and uses an AR HMD and visualisation to help students understand complicated spatial relationship between different parts of human brain. It makes use of AR-HMD 3D volumetric reconstructions from MRI scans. Another example uses Microsoft Hololens 1 to train operatives for radiological incidents and visualises invisible risks [11]. The referenced work provides an indication of the relevance of using OST-HMD based AR for safety training and that some issues and problems remain.

2.2 Training to Improve Safety

In this paper, we outline the use of an OST-HMD for training within a chemical laboratory. This poses challenges in combining the interaction between the virtual and real environments. Zhu et al. [6] proposed a method to create an AR application using an OST-HMD to teach biochemistry laboratory safety instead of traditional lecture-based training methods. The application has two modes, the first is called "Trainer" where the virtual objects like fire extinguisher, telephone, etc. are added to the real scene (i.e. the physical laboratory). The second is "Trainee" which is used by the students who should find the locations of items (virtual objects) in the laboratory, move to them and listened to the holograph narrator. Another proposition by Huwer et al. [7] called "HR Reveal" using handheld AR technology to familiarize students with laboratory rules specially the safety symbols and their meaning on a first visit to the laboratory. The user scans a symbol and augmented information is displayed on the mobile device. In 2019, Mascarenas et al. [8] used the AR in nuclear infrastructure to improve the safety, security, and productivity of employees. They proposed using OST-HMD to show information associated with canisters and to display a video of the process of loading the canister. The previous examples focus on providing information and making people aware of guidelines. In contrast, our approach focuses on enhancing training through improved interaction and the use of an OST-HMD.

---

[1] https://hive.med.ubc.ca/projects/holobrain/.

3   Proposition

3.1   Idea

In order to improve the effectiveness of safety training session for chemical's laboratory using OST-HMD, we propose the concept of SmartLab which consists of using mixed reality techniques in training session. We had chosen two cases of study as a proof of concept to be simulated through Microsoft HoloLens 1 as OST-HMD.

- **Concentrated acids neutralization:** Concentrated acids are very dangerous and corrosive chemicals that are commonly encountered in the laboratory. Concentrated acids may be fatal if inhaled; and can cause severe eye and skin burns, severe respiratory and digestive tract burns. Contact with other materials may cause a fire. Therefore, it is important to train on how to naturalize an concentrated acids.
- **Mix acid and water safely:** In chemical laboratory, adding water to acid could generate an extremely concentrated solution of acid initially. Indeed, this latter could cause a more or less violent heating which projects droplets of acid. For this reason, we should add the acid to the water and not the other way around when mixing an acid with water

3.2   Technical Implementation

The system was implemented using a Microsoft Hololens 1 and Unity3D as the development environment. Due to limitations, standard fluid dynamics libraries such as Obi Fluid [13] could not be used. Instead, we used a particle systems implemented in Unity3D to simulate liquid, fire and powder. However, Unity3D doesn't allow the direct interaction between multiple particles systems. Therefore, we developed a script to permit direct interaction between different particle systems. This operates to a satisfactory level on a Hololens 1 and a moderately equipped PC.

3.3   Scenarios

1st Case Study. In this first case, a trainee learns to neutralize an acid using a specific powder called "Trivorex"[2]. Figure 1 illustrates our mixed reality scene captured from Microsoft HoloLens 1 and implemented using Unity 3D. The scene is composed of virtual table, upon which there are different virtual bottles of powder including Trivorex and two virtual beakers. The first is filled by concentrated acid whereas the second is empty. The scenarios starts by asking a trainee to pour an acid into an empty beaker. Once a beaker is completely filled, an accident occurs and this is shown by a broken beaker which spreads out concentrated

---

[2] https://environnement.prevor.com/fr/produits/trivorex-absorbant-neutralisant-pour-produits-chimiques/.

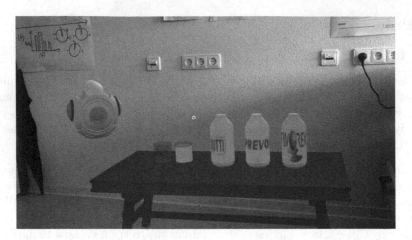

Fig. 1. Mixed reality scene captured from Microsoft HoloLens 1 with 2 beakers, three powder bottle and the vocal assistant. The last bottle on the right is Trivorex powder bottle

acid on a table. From this moment, the trainee needs to take a Trivorex bottle and apply a powder on the concentrated acid spilled on table. Once a powder touches a concentrate acid, its color is modified and trainee is informed that the acid is neutralized and that the training is over. In our safety training application, we modeled our environment including the table, two beakers and bottles of powder. Liquid inside containers like beakers or bottles is computed using a shader developed which updates depending on the amount of liquid inside the container by adjusting height size of shader's texture. The liquid outside containers is modeled using a particle system implemented in Unity3D. This particle system allows to apply an effect close to the real dynamic of fluids. The powder is also simulated thanks to a particle system in Unity3D but with different parameters of those used within liquid outside containers. The chemical reaction between acid and powder is realized by a particle interaction script which computes the position between two different particles.

2nd Case Study. The order in which acid and water is mixed can result in a dangerous chemical reaction which could result in a fire and droplets of acid. When a lab user adds water to a beaker filled by a concentrated acid, no reaction occurs and the manipulation is safe. In contrast, if the lab user adds acid to water, an exothermic reaction occurs which results in flames. In this case, the lab user needs to take a fire extinguisher to apply it on the fire. Figure 2 illustrates our mixed reality scene captured in a Microsoft HoloLens 1. The scene consists of a virtual table, an empty beaker and two other beakers which are filled by acid and water, respectively. The trainee starts the experience by pouring out the acid on the empty beaker and adding water to concentrated acid. As a result, exothermic reaction will be created and a fire on the top of the beaker will appear. The trainee needs to take the right fire extinguisher to put out the fire. Afterwards, the trainee repeats the same experience by adding firstly water and secondly acid in order to understand the safe way of mixing these two materials. As in the first case study, we use a particle system in Unity to simulate the fire as well as powder extinguishers.

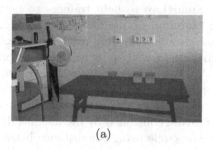 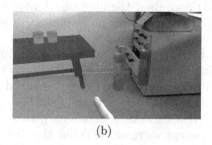

(a)   (b)

Fig. 2. Mixed reality scene captured in a Microsoft HoloLens 1. (a) and (b) represent the same scene for second case study (Mix acid and water safely). The scene is composed of Vocal assistant, virtual table, three virtual beakers as well as two fire extinguisher

4   Comprehension

One of the goal of the SmartLab is to guide the user among the experience by explaining different steps and procedures. Having a clear view of the objectives and a visual support is helpful to improve the understanding of session and increase the trainee's enthusiasm. Zhu et al. [6] had developed an open-source AR biochemistry laboratory safety application. They had tested it on a group of 29 undergraduate students to demonstrate the advantages of AR compared to traditional lectures. First group had a basic traditional lecture whereas the second had visual indicator on Microsoft HoloLens 1 with holographic narrator who is guided them explaining all the features. The experiment concludes that students had enjoyed more the Microsoft HoloLens course and recalled the location of more items compared to those with the lecture-based format. Within SmartLab, vocal assistant had implemented for the purpose of explaining all required tasks for trainees. In order to increase the performance of the training session on the one hand, and draw attention of the trainee in an effective manner on the other hand, we had also chosen to add an attention funnel with the vocal assistant.

4.1   Attention Funnel

Biocca et al. [9] had presented a study about the attention funnel which is an AR interface technique that interactively guides the attention of a user to any object, person, or place in space. To understand the benefits of attention funnel, fourteen participants were asked to do some scenarios where the main goal was to find objects in the environment. They finally found that the attention funnel increased the consistency of the user's search by 65%, decreased the visual search time by 22% and decreased the mental workload by 18%. In 2018, Werrlich et al. [4] had proposed an engine assembly training by OST-HMD using an augmented tunnel which guides the user to right box. Based on these two references [9] and [4], we used the attention funnel in SmartLab to help trainees to better understand the order of the tasks during training. As illustrated in the Fig. 3, the basic components of the attention funnel are: a moving arrow to indicate the object, a dynamic set of attention funnel circles and an invisible curved path linking the eye of the user to the object. We defined the path using a Hermite curve which is a cubic curve segment defined by a start location, end location, and derivative vectors at each end. The curve describes a smoothly changing path from the user to target the object with curvature controlled by the magnitude of the tangent vectors. We define the size of the circle using the distance between the head of user and the target object. Each circle is distanced from the others by a pre-defined value of about 0.2 m. To have an unobstructed view of all circles, they are constantly aligned with the user's gaze. The start point for the Hermite curve is located at a defined distance in front of the user's head. Whereas, the end point is the targeted object.

(a) 1st case study

(b) 2end case study

Fig. 3. Attention funnel in SmartLab, the green circles make the path between the head of the user and the targeted object. The green arrow also points the targeted object. (Color figure online)

4.2   Vocal Assistant

To improve the global understanding of the two cases studies, a vocal assistant was added to the mixed reality scene to guide the trainee during the manipulations. The vocal assistant is firstly a text to speech speaker which explains each task one by one. It also interacts with the trainee thanks to the vocal commands. Kim et al. [10] investigate how visual embodiment and social behaviors influence the perception of the Intelligent Virtual Agents (IVA) to determine whether or not a digital assistant requires a body. Results indicate that imbuing an agent with a visual body in AR and natural social behaviors could increase the user's confidence in the agent's ability to influence the real world. However, creating a complete virtual assistant is a complicated task due to all modeling and animation process behind. Indeed, having a humanoid virtual assistant with bad animation or low texture could lead to the uncanny valley effect. Frank E. Pollick [12] describes the uncanny valley as something equivocal as it depends on people's perceptions. However, getting close to this valley by low textured 3D models, bad or incomplete animations can cause discomfort and reducing then the overall enthusiasm of the training. In this version of the SmartLab application, we have choices to simulate our vocal assistant as a simple flying robot avoiding the uncanny valley. At the beginning of the scenario, flying robot called "Zyra" is instantiated on the left of table in the direction of user as illustrated in the Fig. 1. To make the visualisation more realistic, flying animation on the robot is created.

5 Experimental Test and Discussion

5.1 Performances

We had evaluated the performance of Microsoft HoloLens 1 with our application using Frame Per Second (FPS) to test smooth manipulations during two cases studies. The most computational costly components of our final application are particle systems. According to mobile and gaming console performance, an application with good performance should run at least at a constant 30 FPS. For the two cases studies, we had gathered FPS data in the menu, during the explanation phase managed by vocal assistant, during the beginning of manipulation and when particles are instantiated. We found that the performance goal for our application was successfully reached as illustrated in Fig. 4

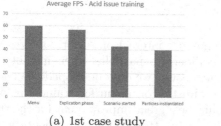

(a) 1st case study   (b) 2end case study

Fig. 4. Evaluation of Microsoft HoloLens 1 with our application using average FPS through different phases

5.2 Difficulties

The main difficulty encountered when we developed the application is due to the limitation of Microsoft HoloLens 1 with its specifications. Fluid implementations in Unity3D such as Obi Fluid [13] were not compatible with Microsoft HoloLens 1 because it is not supported Universal Windows Platform (UWP) which is computing platform created by Microsoft to help develop universal apps that run on HoloLens and other Microsoft's products. We then had to compute fluids using a method allowing us to have better performances on Microsoft HoloLens 1. Moreover Microsoft HoloLens 1 was also limited in term of interaction compare to Microsoft HoloLens 2 which had hand tracking. We thus had chosen to implement interactions with voice to overcome complex manipulation issues. From other side due to the limitation of Unity3D and its particle system that doesn't allow direct particle to particle interaction, we had created a script to allow interactions between particles from two different particle systems. However, as this script can easily run on Microsoft HoloLens 1 as well as PC with good performances for our two cases of study thanks to a small amount of instantiated particles, the script has a quadratic time complexity ($O(n^2)$) for the "Mix acid and water safely" case of study and a cubic time complexity ($O(n^3)$) for the "Concentrated acid neutralization" case of study. Having this kind of time complexity can quickly overload any type of devices as long as we compute new particles. This time complexity is explained by the fact that the script navigate through multiple list of particles to calculate their positions between each others. Moreover, we find on the first case of study that a powder particle can react with multiple acid particles, destroying then all the liquid particles. In normal situations, a powder particle must only react with one acid particle before being inactive. One of our final goal for developing this proof of concept was to test it with participants to compare different safety training methods using OST-HMD. However, due to covid-19 pandemic, we had chosen to avoid the step of user study validation for sanitary reasons.

5.3 Improvements

We propose to work on the particle to particle interaction script. One research axis would be to directly implement colliders to particles when instantiating them to detect more precise collisions. Moreover, we only developed the trainee part of the application. Creating an administration menu for trainers to place virtual objects and control the level would be a great improvement for the training method. Another suggestion would be to attempt the implementation on Microsoft HoloLens 2 or a newer version to see the performance improvement but also to work on new interaction like detecting the entire hand and fingers which would allow to improve manipulation like rotating objects. We propose to add augmented information to virtual object like the name, the composition or the hazard symbols. Finally, we suggest to use Azure Remote Rendering which offers highest quality 3D content and interactive experiences to mixed reality devices, such as Microsoft HoloLens 2. This service uses the computing power of

Azure to render even the most complex models in the cloud and streams them in real time to user's device.

6   Conclusion

This paper has presented an innovative approach using OST-HMD augmented reality to train people in lab safety. The system was designed to illustrate a prototype with a limit feature set in response to the desire to bring training closer to the area in which it is needed but without the inherent risks. Two case studies are presented which provided a motivation of the design involved and in response to this additional user interface techniques were added to improve the experience for trainees, these include the attention funnel and the vocal assistant. From a technical perspective 30 FPS was achieved, although there remains limitations with the HoloLens 1. The plan is to move to testing the system on HoloLens 2 and use MS Azure for in-cloud processing. These two approaches should help to overcome many of the current technical limitations. In summary, this system illustrates the relevancy of using OST-HMD based augmented reality to assist in lab training.

References

1. Laboratory Chemical Safety Incidents. https://synergist.aiha.org/201811-lab-chemical-safety-incidents. Accessed Jan 2019
2. Di Raddo, P.: Teaching chemistry lab safety through comics. J. Chem. Educ. **83**(4), 571 (2006)
3. Azuma, R.T.: A survey of augmented reality. Presence: Teleoper. Virtual Environ. **6**, 355–385 (1997)
4. Werrlich, S., Daniel, A., Ginger, A., Nguyen, P., Notni, G.: Comparing HMD based and paper-based training. In: 2018 IEEE International Symposium on Mixed and Augmented Reality (ISMAR), pp. 134–142 (October 2018)
5. Gonzalez, A.N.V., Kapalo, K., Koh, S., Sottilare, R., Garrity, P., Laviola, J.J.: A comparison of desktop and augmented reality scenario based training authoring tools. In: 2019 IEEE Conference on Virtual Reality and 3D User Interfaces (VR), pp. 1199–1200 (March 2019)
6. Zhu, B., Feng, M., Low, H., Kesselman, J., Harrison, L., Dempski, R.E.: Increasing enthusiasm and enhancing learning for biochemistry laboratory safety with an augmented-reality program. J. Chem. Educ. **95**(10), 1747–1754 (2018)
7. Huwer, J., Seibert, J.: A new way to discover the chemistry laboratory: the augmented reality laboratory-license. World J. Chem. Educ. **9**, 124–128 (2018)
8. Mascarenas, D., et al.: A new way to discover the chemistry laboratory: augmented reality for enabling smart nuclear infrastructure. Front. Built Environ. **5**, 82 (2019)
9. Biocca, F., Tang, A., Owen, C., Xiao, F.: Attention funnel: omnidirectional 3D cursor for mobile augmented reality platforms, pp. 1115–1122 (January 2006)
10. Kim, K., Boelling, L., Haesler, S., Bailenson, J., Bruder, G., Welch, G.F.: Does a digital assistant need a body? The influence of visual embodiment and social behavior on the perception of intelligent virtual agents in AR. In: 2018 IEEE International Symposium on Mixed and Augmented Reality (ISMAR), pp. 105–114 (2018)

11. Huynen, J., McCall, R., Griffin, M.: Towards design recommendations for training of security critical agents in mixed reality environments. In: Proceedings of the 32nd International BCS Human Computer Interaction Conference (HCI) (2018)
12. Pollick, F.: In search of the uncanny valley, vol. 40, pp. 69–78 (December 2009)
13. Obi Fluid Asset. https://assetstore.unity.com/packages/tools/physics/obi-fluid-63067. Accessed 21 Sep 2020
14. Vuzix WRAP 920AR. https://www.vuzix.com/
15. SXGA HMD. https://cinoptics.com/
16. TriVisio SXGA61 3D. https://www.trivisio.com/
17. Epson Moverio. https://tech.moverio.epson.com/en/bt-2000/
18. Lumus Optical. https://lumusvision.com/
19. Microsoft Hololens. https://lumusvision.com/
20. Optinvent ORA-17. http://www.optinvent.com/

One-Dimensional Mathematical Model and a Numerical Solution Accounting Sedimentation of Clay Particles in Process of Oil Filtering in Porous Medium

Elmira Nazirova[1] , Abdug'ani Nematov[2] , Rustamjon Sadikov[1(✉)] ,
and Inomjon Nabiyev[1]

[1] Department of Multimedia Technologies, Tashkent University of Information Technologies
named after Muhammad al-Khwarizmi, Tashkent, Uzbekistan
magistr_uz@bk.ru
[2] Multimedia Technologies, Tashkent University of Information Technologies named
after Muhammad al-Khwarizmi, Tashkent, Uzbekistan

Abstract. The article deals with the unsteady process of filtration of highly contaminated oil in a heterogeneous porous medium in a two-dimensional setting. A detailed analysis of scientific works related to the problem of mathematical modeling of the filtration process of heavily contaminated oils in porous media is given. The arbitrariness of the configuration of the field and the scatter of reservoir parameters over the reservoir area, the unevenness of the location of oil wells in the oil-bearing region and their variability in production rates, which are not limiting factors for the use of numerical modeling in calculations of the development of oil fields. A mathematical model has been developed that takes into account such factors as the sedimentation rate of fine particles, the change in porosity and filtration coefficients over time. The mathematical model is reduced to the joint solution of a system of differential equations of parabolic type, describing the filtration processes in reservoirs separated by a low-permeability bulkhead, with appropriate initial and boundary conditions. Computational experiments are given for various reservoir parameters and the cost of two production wells. The results of the developed software are presented, where the results of calculating the main indicators of field development are presented in tabular and graphical form. All the results of computational experiments are presented in a visual form.

Keywords: Mathematical model · Numerical method · Filtration ·
Non-stationary · Oil-water system

1 Introduction

In recent years, much attention has been paid to mathematical modeling, improvement and development of fluid filtration processes in porous media around the world using numerical methods and communication technologies for solving linear and nonlinear problems of non-stationary filtrations. One of the main tasks in this area is the creation of automated systems for identifying and predicting key performance indicators for oil and gas fields, studying the processes of unsteady filtrations based on modern information technologies. In many countries of the world, including the USA, France, China, United Arab Emirates, Iran, Russia, Ukraine, Kazakhstan, Azerbaijan and other countries, much attention is paid to the development of mathematical models, computational algorithms and software for non-stationary oil and gas filtration processes.

These authors discuss the most significant of all reservoir models known as the black-oil model or beta model. More sophisticated models can be understood using the techniques used to create solid oil models. In practice, the following equivalent terms are used: mathematical models, numerical models, grid models, finite difference models and reservoir models. In reality, three types of models are used in the development of a reservoir simulation program [1].

The work [2] describes a technique for the numerical solution of the equation of single-phase unsteady gas filtration in a porous medium. The linearization of the classical Leibenson equation is carried out. To solve the obtained linear equation, an efficient numerical algorithm is constructed without saturation in spatial variables and time.

Mathematical simulation of stress-strain state of loaded rods with account of transverse bending is considered in the paper. The urgency, the correctness of mathematical statement of the problem, the mathematical model, the computational algorithm and the computational experiments of the problem set are given in the paper. Results are presented in a graphical form. The oscillations of spatially loaded rods are shown. The analysis is made in linear and non-linear statements [3].

The problem [4] of transfer of a three-phase mixture "water-gas-oil" in a porous medium in the case when the water contains a finely dispersed gas phase in the form of micro- or nanosized bubbles is considered. It is assumed that the transfer of bubbles is mainly determined by the flow of the dispersed phase (water). In this case, large accumulations of the gas phase in the pore space, as well as water and oil, are transferred in accordance with the modified Darcy's law for multiphase mixtures. A mathematical model of the mixture movement has been built, when the main phases (water, gas, oil) obey the equations of filtration, and the finely dispersed gas phase is described by a kinetic equation of the Boltzmann type. Some one-dimensional numerical solutions of transport problems are obtained and analyzed.

The paper [5] presents a numerical model, computational algorithm and software for studying the process of gas filtration in porous media in order to determine the main indicators of gas fields.

Real-time field experiments on the filtration of liquids and gases in a porous medium are very laborious and costly. Sometimes full-scale experiments are impossible. In this case, to carry out computational experiments on a computer, it is necessary to apply numerical simulation using effective numerical methods and modern computer technology to implement the developed computational algorithms for solving problems [6–20].

Designing the operation of oil and gas fields in the republic, mathematical modeling of heterogeneous fluid filtration processes, developing algorithms for studying non-linear gas-hydrodynamic processes, calculating and evaluating key performance indicators for oil and gas fields using modern computer technologies, as well as automated software.

2 Mathematical Model

The mathematical model of non-stationary oil filtration, taking into account the fine dispersion particles that enter the pore medium during oil filtration, is written as follows:

$$\begin{cases} \beta h(x)\frac{\partial P}{\partial t} = \frac{\partial}{\partial x}\left(\frac{k(x)h(x)}{\mu}\frac{\partial P}{\partial x}\right) - Q, & 0 < x < l \\ \frac{d\eta}{dt} = \lambda(\theta_0 - \gamma\eta), & 0 < t < T \end{cases} \tag{1}$$

The initial and pick conditions can be obtained as follows:

$$\begin{cases} P(x) = P_H, \quad \eta(t) = \eta_0, \quad t = 0; \\ \frac{\partial P}{\partial x} = 0, \quad x = 0; \\ \frac{\partial P}{\partial x} = 0, \quad x = L; \\ \oint\limits_{S_{i_q}} \frac{k(x)h(x)}{\mu}\frac{\partial P}{\partial x}dx = -q_{i_q}(t), \quad i_q = 1, N_q, \quad Q = \sum_{i_q=1}^{N_q} \delta_{i_q}q_{i_q} \\ m = m_1 + \eta(m_0 - m_1), \quad k = k_0\left(1 - \sqrt{\eta}\right)^3, \quad \beta = m\beta_H + \beta_c. \end{cases} \tag{2}$$

Here, P – reservoir pressure; P_H – initial pressure; μ – oil viscosity; k – layer permeability coefficient; k_0 – initial permeability coefficient of the layer; h – reservoir thickness; β – coefficient of elasticity; β_H – oil compression coefficient; β_c – reservoir compression coefficient; m – porosity coefficient; m_0 – primary porosity; m_1 – porosity of the settling mass; q_{i_q} – i_q – well flow rate; s_{i_q} – i_q – well contour; n – Γ boundary normal condition; N_q – number of wells; η – particle deposition rate in porous media; γ – filter parameter; λ – kinematic mass deposition coefficient.

To solve (1)–(2) problem by dividing, we include the following dimensionless variables [2]:

$$P^* = P/P_0, \quad x^* = x/L, \quad y^* = y/L, \quad k^* = k/k_0, \quad h^* = h/h_0,$$

$$\eta^* = \eta/\eta_0, \quad \tau = \frac{k_0 t}{\beta \mu L^2}, \quad q^* = \frac{q\mu}{\pi k_0 P_0 h_0}, \quad \lambda^* = \frac{k_0\lambda}{\beta\mu\eta_0 L^2}, \quad \alpha^* = \frac{\mu L}{k_0 h_0}.$$

Then, for convenience, drop the asterisk and write down problem (1)–(2) for dimensionless variables as follows:

$$\begin{cases} h(x)\frac{\partial P}{\partial \tau} = \frac{\partial}{\partial x}\left(\frac{k(x)h(x)}{\mu}\frac{\partial P}{\partial x}\right) - Q, \\ \frac{d\eta}{d\tau} = \lambda(\theta_0 - \gamma\eta), \end{cases} \tag{3}$$

$$\begin{cases} P(x) = P_H, \quad \eta(t) = \eta_0, \quad t = 0; \\ \frac{\partial P}{\partial x} = 0, \quad x = 0; \\ \frac{\partial P}{\partial x} = 0, \quad x = 1; \\ \oint\limits_{S_{i_q}} k(x)h(x)\frac{\partial P}{\partial x}dx = -q_{i_q}(t), \quad i_q = 1, 2, ..., N_q, \\ m = m_1 + \eta(m_0 - m_1), \quad k = k_0\left(1 - \sqrt{\eta}\right)^3, \quad \beta = m\beta_H + \beta_c. \end{cases} \tag{4}$$

3   Numerical Modeling

To solve the dimensionless problem (3)–(4), we replace the rectangular filtering field Γ with the outer boundary G by the same step.

$$\overline{\omega}_{h\tau} = \left\{ x_i = ih, \ \ i = 0, 1, \ldots, N, \ \ h = \frac{l}{N}, \ \ t_j = j\tau, \ \ j = 0, 1, \ldots, N_0, \ \ \tau = \frac{T_0}{N_0} \right\}$$

In this case, we will solve the problem of this discrete field by approving the equation and its boundary conditions as follows:

$$-3P_2 + 4P_1 - P_0 = 0;$$
$$a_i P_{i-1} - b_i P_i + c_i P_{i+1} = -d_i;$$
$$3P_n - 4P_{n-1} + P_{n-2} = 0;$$
$$i = 1, 2, \ldots, n - 1$$

Here

$$a_i = T_{i-0.5}; \quad c_i = T_{i+0.5}; \quad b_i = T_{i-0.5} + T_{i+0.5} + \frac{h_x^2}{\tau};$$

$$d_i = T_{i-0.5}\hat{P}_{i-1} - (T_{i-0.5} + T_{i+0.5})\hat{P}_i + T_{i+0.5}\hat{P}_{i+1} + \frac{h_x^2}{\tau}\hat{P}_i;$$

$$T(x) = \frac{k(x)h(x)}{\mu}; \quad T_{i-0.5} = \frac{T_{i-1} + T_i}{2}; \quad T_{i+0.5} = \frac{T_i + T_{i+1}}{2}.$$

$\hat{P}_i$ - The value of the pressure function in the k-th time layer;
P_i - The value of the pressure function $k + 1$ in the time layer;

Using the sweep method to solve this finite-difference problem, we find the value of the pressure function P_i in each time layer with sufficient accuracy.

To determine the particle deposition rate η in porous media, we obtain the following formula using the second equation τ of the system of Eqs. (1).

$$\eta_i = \frac{\lambda \Delta \tau \theta_0 + \hat{\eta}_i}{1 + \gamma \lambda \Delta \tau},$$

Here, $\hat{\eta}_i$ − is the value of η_i in one direct time layer, the initial value (3) of which is obtained from the initial condition.

The value of η_i is used to determine the porosity coefficients m and the permeability coefficient k in each time layer.

4   Algorithm for Solving the Problem

The algorithm for solving the problem by the numerical method is as follows:

- enter the values of the coefficients involved in the differential equation;
- calculate the coefficients of the partial differential equation, a_i, b_i, c_i, d_i $(i = 1, 2.. N)$.
- determining the initial values of the run α_0 and β_0 from the pick condition;
- calculate the values of the coefficients α_i and β_i $(i = 1, 2.. N_j - 1)$;
- calculate the value of the pressure function P_N from the right pick condition;
- calculation of pressure function values P_i $(i = N - 1, N - 2..0)$;
- calculate the rate of deposition of particles η_i in a porous medium;
- calculate the values of permeability m and permeability coefficients k;

This algorithm is repeated at each time interval. In this case, the obtained solution is the initial value of the time interval.

5 Computational Experiments on a PC

The computational algorithm and the program were developed on the basis of the developed numerical model, with the help of which computational experiments were carried out on a computer to study the filtration process in porous media and the results were obtained in the Delphi program. The program interface is shown in Fig. 1. It consists of the following blocks: input data for the program, numerical results of the program, coordinates and flow rates of oil wells, graphical visualization of the results and calculation of the main indicators of oil fields.

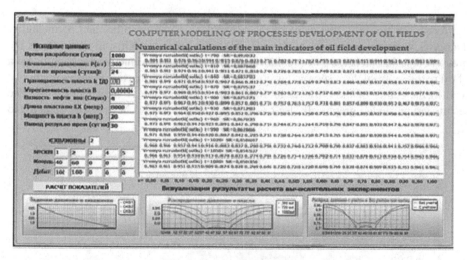

Fig. 1. Application interface.

The computer performs computational experiments on the main parameters of oil fields to study fine clay particles in porous media. The following basics are used for computational experiments in software: $m_0 = 0.4$; $k_0 = 0.05$; $\eta_0 = 0$; $\gamma = 0.0008$; $\lambda = 0.00045$; $P_0 = 300$; $\mu = 4$; $L = 8000$ m.

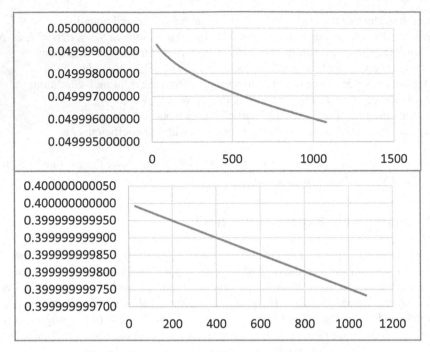

Fig. 2. Change in permeability and porosity with time.

As shown in Fig. 2, the value of the coefficient of permeability and permeability decreases with increasing rate of deposition of clay particles. This, in turn, affects the pressure distribution in the pore medium and leads to a decrease and further complication of oil recovery from the reservoir system (Fig. 3 and 4). The pressure distribution in one well in the center of the oil layer at 1080 days of operation is shown in Fig. 3. The results are presented graphically for cases where clay particles have been counted and

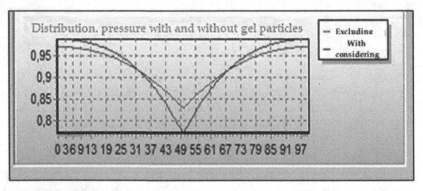

Fig. 3. Pressure distribution of the oil layer. 1-clay particles are excluded. 2-clay particles were counted.

excluded. As shown in the graphs, the pressure distribution is slowed down due to the deposition of clay particles in the oil layer.

The pressure distribution for 1080 days of operation of two symmetric wells in the center of the oil layer is shown in Fig. 4. At the same time, the graphs show that when clay particles were deposited in the oil layer, the pressure distribution slowed down and a pressure drop was observed in the wells.

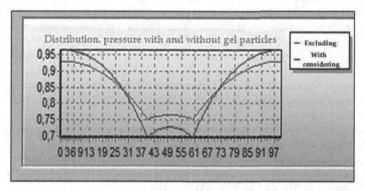

Fig. 4. Pressure distribution of the oil layer. 1-clay particles are excluded. 2-clay particles were counted.

6   Conclusion

As a rule, clay particles must be considered during oil filtration to ensure that the results of calculations in the pore medium of the oil are close to the actual result of the calculation of the main parameters of the field. This is due to the fact that the coefficients of porosity and permeability in the oil layer decrease over time (Fig. 2). This, in turn, requires the calculation of porosity and permeability coefficients over time. As a result, the pressure distribution in the oil pore medium slows down over time.

The developed mathematical model and software can be used to calculate the main parameters of the field, process analysis, predicting and engineering.

References

1. Aziz, H., Sattari, E.: Mathematical Modeling of Reservoir Systems. Moscow-Izhevsk, Moscow (2004)
2. Algazin, S.D.: On the calculation of the eigenvalues of the transport equations. Appl. Mech. Tech. Phys. **45**(4), 113 (2004)
3. Anarova, S., Ismoilov, S.: Mathematical simulation of stress-strain state of loaded rods with account of transverse bending. In: Journal of Physics: Conference Series, vol. 1260, no. 10 (2019). https://doi.org/10.1088/1742-6596/1260/10/102002. статья № 102002
4. Demyanov, A.Yu., Dinariev, O.Yu., Ivanov, E.N.: Modeling of water transfer with a finely dispersed gas phase in porous media. Eng. Phys. J. - Minsk **85**(6), 1145–1154 (2012)

5. Nematov, A., Nazirova, E.Sh.: Numerical modeling of gas filtration in a porous medium. Int. Acad. Bull. **1**(13), 52–56 (2016)
6. Singh, D., Singh, M., Hakimjon, Z.: One-dimensional polynomial splines for cubic splines. In: Signal Processing Applications Using Multidimensional Polynomial Splines. SpringerBriefs in Applied Sciences and Technology, pp. 21–26. Springer, Singapore (2019). https://doi.org/10.1007/978-981-13-2239-6_3
7. Singh, D., Singh, M., Hakimjon, Z.: Requirements of MATLAB/Simulink for signals. In: SpringerBriefs in Applied Sciences and Technology, pp. 47–54. Springer, Heidelberg (2019). https://doi.org/10.1007/978-981-13-2239-6_6
8. Zaynidinov, H., Zaynutdinova, M., Nazirova, E.: Digital processing of two-dimensional signals in the basis of Haar wavelets. In: ACM International Conference Proceeding Series, pp. 130–133. Association for Computing Machinery (2018). https://doi.org/10.1145/3274005.3274023
9. Ravshanov, N., Nazirova, E.: Numerical simulation of filtration processes of strongly polluted oil in a porous medium. Ponte, vol. 74, №. 11/1, pp. 107–116 (2018). (№ 1, Web of science, Impact Factor 0.814)
10. Sadullaeva, S.A.: Numerical investigation of solutions to a reaction-diffusion system with variable density. J. Sib. Fed. Univ. - Math. Phys. **9**(1), 90–101 (2016)
11. Sadullaeva, Sh.A., Khojimurodova, M.B.: Properties of solutions of the cauchy problem for degenerate nonlinear cross systems with convective transfer and absorption. Algebra Complex Anal. Pluripotential Theory **264**, 183–190 (2018)
12. Zayinidinov, K.N., Turapov, U.U.: Mathematical model of a non-invasive glucometer sensor. J.: High. Sch. **21**(1), 44–53 (2016)
13. Ravshanov, N., Nazirova, E.Sh., Pitolin, V.M.: Numerical modelling of the liquid filtering process in a porous environment including the mobile boundary of the "oil-water" section. In: Journal of Physics: Conference Series, vol. 1399, p. 022021. IOP Publishing (2019).https://doi.org/10.1088/1742-6596/1399/2/022021
14. Nurgatin, R.I., Lysov, B.A.: The use of 3D modeling in the oil and gas industry. Bull. Sib. Branch Sect. Earth Sci. Russ. Acad. Nat. Sci. **1**(44), 66–73 (2014)
15. Vasil'ev, V.I., Vasil'eva, M.V., Laevsky, Yu.M., Timofeeva, T.S.: Numerical simulation of the two-phase fluid filtration in heterogeneous media. J. Appl. Ind. Math. **11**(2), 289–295 (2017). https://doi.org/10.1134/S1990478917020156
16. Aripov, M.: Asymptotic of the solution of the non-newton polytropical filtration equation. ZAMM **80**(suppl. 3), 767–768 (2000)
17. Sadullaeva, Sh.A.,Beknazarova, S.S., Abdurakhmanov, K.: Properties of solutions to the cauchy problem for degenerate nonlinear cross-systems with absorption "Researcher", vol. 12, №. 6 (2020). ISSN 1553-9865 (print); ISSN 2163-8950 (online). https://www.sciencepub.net/researcher. https://doi.org/10.7537/marsrsj120620.02
18. Nazirova, E.Sh.: Mathematical modeling of filtration problems three phase fluid in porous medium. Inf. Technol. Model. Control Sci. Tech. J. **1**(109), 31–40 (2018)
19. Ravshanov, N., Nazirova, E.Sh.: Mathematical model and algorithm for solving the problem of oil filtration in two-layer porous media. Probl. Comput. Appl. Math. **4**(16), 33–46 (2018)
20. Nazirova, E.Sh.: Numerical modeling of oil filtration processes in multi-layer porous media with dynamic connection between layers. Descend. Muhammad Al-Khorezmi **4**(6), 10–14 (2018)

A Novel Diminish Smooth L1 Loss Model with Generative Adversarial Network

Arief Rachman Sutanto[1] and Dae-Ki Kang[2]([⊠]) [iD]

[1] Department of Computer Engineering, Dongseo University, 47 Jurye-ro, Sasang-gu, Busan 47011, Republic of Korea
arief_r_s@yahoo.com
[2] Division of Information and Communication Engineering, Dongseo University, 47 Jurye-ro, Sasang-gu, Busan 47011, Republic of Korea
dkkang@dongseo.ac.kr

Abstract. The training process of GAN can be regarded as a process in which the generation network and the identification network play against each other and finally reach a state where it cannot be further improved if the opponent does not change. At the same time, the start of the gradient descent method will choose a direction to reduce the defined loss. The loss function plays a key role in the performance of the model. Choosing the right loss function can help your model learn how to focus on the correct set of features in the data to achieve optimal and faster convergence. In this work, we propose a novel loss function scheme, namely, Diminish Smooth L1 loss. We improve a robust L1 loss called Smooth L1 loss by lowering the threshold so that the network can converge to a lower minimum. From our experimental results on several benchmark data, we found that our algorithm often outperforms the previous approaches.

Keywords: Pix2Pix · WGAN-GP · Smooth L1 loss · Image translation

1 Introduction

One of the areas of research in the field of deep learning is an image to image translation. It has been proposed in [1] that CNN successfully merges the style from a source image with the content of another image. This process is difficult for humans but very easy for computers. But there are also a few drawbacks to this method. Firstly, it takes a while to train the network. Secondly, the network can only learn one style. These are due to the need to iteratively calculate the difference of each pixel from the source and content image with the combined image.

Generative Adversarial Networks (GAN) [13] is composed of two networks, called generator and discriminator. The generator generates an image to fool the discriminator, while the discriminator distinguishes the generated image from the ground truth image. GAN learn the distribution of data that is similar to original data. But it is difficult to train multiple classes because it will mix the data distribution. The solution is by adding labels [9] in the training process of GAN so that the distribution of data is not mixed one after another.

M. Singh et al. (Eds.): IHCI 2020, LNCS 12615, pp. 361–368, 2021.
https://doi.org/10.1007/978-3-030-68449-5_36

Regularization [16] is any supplementary technique that aims at making the model generalize better, i.e. produce better results on the test set. This can include various properties of the loss function, the optimization algorithm, or the other techniques. Note that this definition is more in line with machine learning literature than with inverse problem literature, the latter using a more restrictive definition.

Previous researches have discovered that adding standard loss, like L1 [10] or L2 [18], to the GAN objective is advantageous. By adding these penalties will force the generated image to be closer to the ground truth. L1 loss is more preferred than L2 loss because the results are less blurry but the L1 loss is not differentiable at zero. These are due to the sparsity properties of L1 loss. To alleviate this problem, we explore a more robust L1 loss called Smooth L1 loss which combines the advantages of both L1 and L2 loss. This loss is robust to outliers and smoothes near zero. Furthermore, Smooth L1 loss characteristics match the experiment from [2], where even though a network reached convergence by L1 loss function and this network is trained again with L2 loss function, the network can achieve lower loss value. These are related to the behavior of L1 and L2 loss, where L2 loss can easily get stuck in local optimum while L1 loss can achieve a lower minimum faster.

Smooth L1 loss has a threshold that separates between L1 and L2 loss, this threshold is usually fixed at one. While the optimal value of the threshold can be searched manually, but others [4, 15] found that changing the threshold value during training can improve the performance. Different value of fixed threshold corresponds to different gradient and loss, and the lower the threshold is, the faster the convergence rate will be. The previous method [15] changes the threshold by utilizing mean and variance with momentum. The middle point of the difference between mean and variance is used as a focal point. To mitigate the imbalanced during training, clipping is implemented.

Inspired by the problem of searching the optimal value of the threshold, therefore, in this paper, we propose a modification to the threshold of Smooth L1 loss called Diminish Smooth L1 loss. We lower the threshold by a step after each epoch to improve the quality of the generated image. The penalty that is used will be different every time depending on the value of the threshold and the absolute difference between the generated and original image. We also show the effectiveness of this approach both quantitatively and qualitatively using styled text and facades dataset.

2 Background and Related Work

2.1 Pix2Pix

Image translation in deep learning becoming easier with the invention of GAN. The Pix2pix module is one of the examples that utilize conditional GAN [9] as a framework. Figure 1 showing an example of Pix2Pix, using a label as input to assist the generator to generate an image that cannot differentiate from ground truth image. The generator architecture is based on encoder-decoder networks with skip connections [8]. In this model, the information flow downsampled in the encoder and upsampled in the decoder and the skip connections act as a bridge to pass important information in each corresponding layer.

The discriminator tries to determine whether an input image is real or fake in patch scale not by pixels. The generator is combined with the L1 loss in this architecture. The objective function for conditional GAN is in Eq. (1), while the full objective function is shown in Eq. (2).

$$\mathcal{L}_{cGAN}(G, D) = \mathbb{E}_{x,y}\big[\log D(x, y)\big] + \mathbb{E}_{x,z}\big[\log(1 - D(x, G(x, z)))\big] \quad (1)$$

$$G^* = \arg \min_G \max_D \mathcal{L}_{cGAN}(G, D) + \lambda \mathcal{L}_{L1}(G) \quad (2)$$

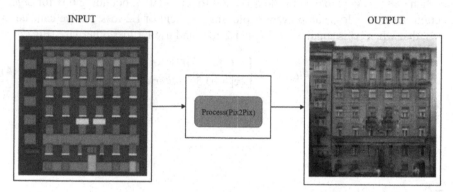

Fig. 1. Example of Pix2Pix

3   WGAN-GP

In [17], Wasserstein GAN minimizes a distance called Earth-mover, which is assumed as a cost to move from one distribution to another distribution. The discriminator in WGAN is called critic because it does not distinguish between real and fake but minimize the distance. Weight clipping is implemented so that this critic can calculate the distance.

This clipping looks simple but leads to other problems such as the generated images are low quality, it is slow to converge, and suffer unstable training. The gradient penalty [12] is proposed to change the clipping. In the differentiable Lipchitz function, the highest gradient is one so the penalty occurs when the gradient is greater than one. The objective function is shown in Eq. (3) as follow:

$$\mathcal{L}_{wgan-gp} = \underbrace{\underset{\tilde{x} \sim \mathbb{P}_g}{\mathbb{E}} [D(\tilde{x})] - \underset{x \sim \mathbb{P}_r}{\mathbb{E}} [D(x)]}_{\textit{Original critic loss}} + \lambda \underbrace{\underset{\hat{x} \sim \mathbb{P}_{\hat{x}}}{\mathbb{E}} \Big[\big(\|\nabla_{\hat{x}} D(\hat{x})\|_2 - 1 \big)^2 \Big]}_{\textit{gradient penalty}} \quad (3)$$

4 Smooth L1 Loss

There are two most popular loss functions in this field which are L1 and L2 loss and each loss has its advantage and disadvantage. The disadvantage of L1 loss is that it has break-points and is not smooth, leading to instability. The derivative (gradient) of L2 loss contains the difference between the predicted value and the target value. When the predicted value is very different from the target value, the gradient of L2 will explode. However, L2 loss is more sensitive to outliers than L1 loss.

However, there is a loss that combined both advantages of L1 and L2 loss, that loss is called Smooth L1 loss. This loss was introduced by [14] and it is derived from Huber loss when delta is equal to one, this loss is used for calculating bounding box for object detection. Also, this loss can prevent exploding gradient of L2 loss. As we can see in Eq. (4), that when $|x|$ is small, we can use L2 loss and use L1 loss otherwise (Fig. 2).

$$smooth_{L_1}(x) = \begin{cases} 0.5x^2 & if\,|x| < 1 \\ |x| - 0.5 & otherwise \end{cases} \tag{4}$$

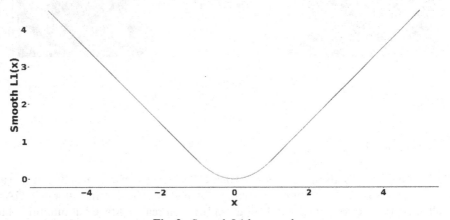

Fig. 2. Smooth L1 loss graph

5 Evaluation Metric

Many evaluation metrics are used to evaluate the result of GAN. In this paper, we used three from the frequently used. They are Peak Signal to Noise Ratio (PSNR), Structural Similarity (SSIM), and Fréchet Inception Distance (FID) [3, 5, 11].

PSNR is used to compute quality between ground truth and generated image. The higher this value corresponds to the better quality of the generated image. This metric calculates the log of the maximum possible pixel value of image divided by Mean squared error (MSE) loss between ground truth and generated image.

SSIM is also used for measuring the image quality between ground truth and generated image. The higher this value, it will mean that the generated image is more similar to the ground truth image. Because this metric compares the similarity of structure and perceptual difference between two images.

FID is used to calculate the distance of distribution between ground truth and generated image. The lower this value the closer the distance between these distributions. Because GAN tries to match distribution with original image distribution so this means that the lower the FID value the more similar the distribution which related to better performance of that GAN.

6 Proposed Method (Diminish Smooth L1)

Both [4, 15] mention that setting the threshold value into less than one can improve the performance. However, altering the threshold during the training process can raise the performance even more. The previous method [15] calculates the mean and variance of absolute loss in a minibatch, and a middle value between these two is used as a focal point. Because the mean value is not stable, clipping is implemented to limit the threshold value.

To find optimal threshold value, we propose an improved Smooth L1 loss (called Diminish Smooth L1 loss) where we decrease the threshold so that the loss function keeps changing between L1 and L2 loss even at later epoch. As we know, the L1 loss will set many features to zero while L2 did not. So, this will make the network learn more optimal representation of the data. We assume that decreasing the threshold can guide the reduction of loss during training. For example, if $|x|$ is bigger than beta, it will use the L1 loss to significantly decrease the loss, and if $|x|$ is smaller, it will use the L2 loss to smoothly decrease the loss. The method that we used is like learning rate decay, where we subtract the value of threshold after each epoch with a fixed predetermined value shown in Eq. (6) and we get this value is by dividing one by the total number of epochs shown in Eq. (5). Our updated loss function as shown in Eq. (7).

$$\text{step} = \frac{1}{num\,of\,epoch} \tag{5}$$

$$\beta_{new} = \beta_{old} - step \tag{6}$$

$$diminishsmooth_{L_1}(x) = \begin{cases} 0.5x^2/\beta_{new} & if\,|x| < \beta_{new} \\ |x| - 0.5 * \beta_{new} & otherwise \end{cases} \tag{7}$$

In our implementation, this proposed method will act as a regularization combined with the GAN loss function and Pix2pix architecture. The GAN function that we used is WGAN-GP, this GAN function is commonly used due to its stable performance. Equation (8) shows the generator's objective function.

$$G = \lambda_1 \mathcal{L}_{WGAN-GP} + \lambda_2 \mathcal{L}_{diminish\,smooth\,L1} \tag{8}$$

7   Experiment and Result

To explore the effectiveness of our proposed method, we experiment with a styled text dataset [6] and facades dataset [7]. The styled text dataset consists of 60 variety of text effects in 52 English letters with 19 fonts type which bring the total images to around 59k images, where 51k images are used for training and the rest 8k for testing. The facades dataset consists of 400 training images and 106 testing images. It consists of a picture of architecture and a corresponding label. In our implementation, we match the network architecture and implementation details from the original setting. In this section, we evaluate our proposed method with L1 and Smooth L1 loss. We also add quantitative results to evaluate which generated images are better. These evaluation metrics are often used for image quality, the peak signal-to-noise ratio (PSNR), and structural similarity index (SSIM). We also add the FID evaluation metric to calculate the performance of GAN.

Table 1. Quantitative comparison between L1, Smooth L1 loss and Diminish Smooth L1 Loss in Text Effect Dataset

Loss Function	PSNR	SSIM	FID
L1 loss	32.038251	0.970475	3.33972
Smooth L1 loss	32.380051	0.970487	4.58997
Diminish Smooth L1 Loss (Ours)	32.624572	0.973798	3.33652

Table 2. Quantitative comparison between L1, Smooth L1 loss and Diminish Smooth L1 Loss in Facades Dataset

Loss Function	PSNR	SSIM	FID
L1 loss	12.723859	0.245888	156.22170
Smooth L1 loss	12.540056	0.236877	149.96268
Diminish Smooth L1 Loss (Ours)	12.642780	0.247033	149.78192

As shown in Tables 1 and 2, our proposed method achieves almost all the best scores in both datasets. These correlate with the quality of the image which is showing in Figs. 3 and 4. Therefore, our proposed model will also perform better results in other datasets and architectures.

In Fig. 3, even though all the images generated using L1, Smooth L1 loss, and our method contain artifact. Our method generates a clearer image compared to other methods. In Fig. 4, the L1 loss failed to generate a whole image while our method and Smooth L1 can generate a whole image. Our method also able to generate a pillar-like structure like in the label.

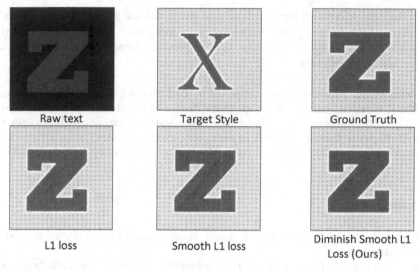

Fig. 3. Ground truth image and generated images using text effect dataset.

Fig. 4. Ground truth image and generated images using facades dataset.

8 Conclusion

In this paper, we presented a way to modify the threshold for the Smooth L1 loss which helps the training process. Specifically, we propose Diminish Smooth L1 loss which lowers the threshold by a step after each epoch. Through this process, we show that by alternating L1 and L2 loss during the generator training process, our method can learn a better representation of data and can generate images with better image quality with fewer artifacts. We provide supplementary evidence both quantitatively and qualitatively on two datasets. Moreover, the distribution of the generated images by our method is closer to the distribution of ground truth images. In future work, we are going to improve our method by including other aspects, such as previous losses in the calculation of a step to make it more flexible.

Acknowledgment. This research was supported by Basic Science Research Program through the National Research Foundation of Korea (NRF) funded by the Ministry of Education (NRF-2018R1D1A1A02050166) and Institute for Information and Communications Technology Promotion (IITP), South Korea grant funded by the Korea Government (MSIT) (No. 2018–0-00245, Development of prevention technology against AI dysfunction induced by deception attack).

References

1. Gatys, L.A., Ecker, A.S., Bethge, M.: Image Style Transfer Using Convolutional Neural Networks (2016)
2. Zhao, H., Gallo, O., Frosio, I., Kautz, J.: Loss Functions for Image Restoration with Neural Networks, 28 November 2015
3. Zhang, L., Zhang, L., Mou, X., Zhang, D.: A Comprehensive Evaluation of Full Reference Image Quality Assessment Algorithms (2012)
4. Zhang, H., Chang, H., Ma, B., Wang, N., Chen, X.: Dynamic R-CNN: towards high quality object detection via dynamic training. In: Vedaldi, A., Bischof, H., Brox, T., Frahm, J.-M. (eds.) ECCV 2020. LNCS, vol. 12360, pp. 260–275. Springer, Cham (2020). https://doi.org/10.1007/978-3-030-58555-6_16
5. Wang, Z., Bovik, A.C., Sheikh, H.R., Simoncelli, E.P.: Image quality assessment: from error visibility to structural similarity. IEEE Trans. Image Process. **13**, 600–612 (2004)
6. Wang, W., Liu, J., Yang, S., Guo, Z.: Typography With Decor: Intelligent Text Style Transfer (2019)
7. Tyleček, R., Šára, R.: Spatial pattern templates for recognition of objects with regular structure, pp. 364–374 (2013)
8. Ronneberger, O., Fischer, P., Brox, T.: U-Net: Convolutional Networks for Biomedical Image Segmentation, 18 May 2015
9. Mirza, M., Osindero, S.: Conditional Generative Adversarial Nets, 6 November 2014
10. Isola, P., Zhu, J.-Y., Zhou, T., Efros, A.A.: Image-to-Image Translation with Conditional Adversarial Networks, 21 November 2016
11. Heusel, M., Ramsauer, H., Unterthiner, T., Nessler, B., Hochreiter, S.: GANs trained by a two time-scale update rule converge to a local nash equilibrium. In: Advances in Neural Information Processing Systems (NIPS 2017), vol. 30, 26 June 2017
12. Gulrajani, I., Ahmed, F., Arjovsky, M., Dumoulin, V., Courville, A.: Improved Training of Wasserstein GANs, 31 March 2017
13. Goodfellow, I.J., et al.: Generative Adversarial Networks, 10 June 2014
14. Girshick, R.: Fast R-CNN, 30 April 2015
15. Fu, C.-Y., Shvets, M., Berg, A.C.: RetinaMask: Learning to Predict Masks Improves State-of-the-Art Single-Shot Detection for Free, 10 January 2019
16. Kukacka, J., Golkov, V., Cremers, D.: Regularization for Deep Learning: A Taxonomy, CoRR, vol. abs/1710.10686 (2017)
17. Arjovsky, M., Chintala, S., Bottou, L.: Wasserstein GAN, CoRR, vol. abs/1701.07875 (2017)
18. Pathak, D., Krähenbühl, P., Donahue, J., Darrell, T., Efros, A.A.: Context Encoders: Feature Learning by Inpainting. CoRR, vol. abs/1604.07379 (2016)

Interactive Machine Learning Approach for Staff Selection Using Genetic Algorithm

Preethi Ananthachari[1]([⊠]) [iD] and Nodirbek Makhtumov[2]

[1] Endicott College of International Studies, Woosong University, Daejeon, South Korea
preethi.ga@wsu.ac.kr
[2] Technology Studies, Woosong University, Daejeon, South Korea
nodirbekmakhtumov@gmail.com

Abstract. Machine learning is meant to extract knowledge from vast data. It is believed to retrieve useful data from enormous raw data. It is applied in several fields in which if there is a big dataset which consists of information about a person's name, age, address, height, weight, medical test details etc.,. Statistics department may be interested to get the attributes such as name, age and address. Health department may be interested in medical details. They may need to get the ratio of old aged people and wanted to give vaccination to prevent Covid-19. From these raw data, they predict the percentage of Old aged people and the exact number of people who needs more attention. Mere dependence on machine models for benevolence sometimes leads to huge malfunction or loss. So it is advisable to supervise the machine models by having check points. Interaction with the machine is cumbersome and it is fully dependent on the context of the model. In this paper, a different approach is presented for employee work planning schedule. In a moderate firm, employee plan schedule on everyday basis requires a machine model. Genetic Algorithm is used for staff planning and interactive machine learning approach is used for supervision. Human supervision reduces the cost and the algorithm converges rapidly.

Keywords: Genetic algorithm · Optimization · Fitness function · Interactive machine learning · Crossover · Mutation

1 Introduction

A group of researchers, doing their experiment explores a new intelligence, then that knowledge could be used for beneficial purpose. Even though automatic machines are fast paced and gives accurate results, it is the human intelligence which is on its technology back. Fully automated Machine Learning (ML) activities have some specific negative impacts also. Machine intelligence has some disadvantages though; one best example is US governments, Affordable Care Act (ACA), which is also known as Obamacare, which included many health care choices with lower charges. As a result of huge number of enrollment in the health care webpage, software failure caused the website inactive which resulted in a big repercussion for the administration in the White House for around

© Springer Nature Switzerland AG 2021
M. Singh et al. (Eds.): IHCI 2020, LNCS 12615, pp. 369–379, 2021.
https://doi.org/10.1007/978-3-030-68449-5_37

two hours [4]. So when the sole machine intelligence is inefficient to conquer the target, the human intelligence could help in that context.

The incorporation of machine intelligence and human intelligence believed to effect in a better way. Not to worry about the training data, as the human interference will modify the bias value or the training input to receive the expected output. The usage of Machine Learning concepts for data extraction, pre-processing the data is done non-parallel with the human experts who experiment the algorithm. Cognitive intelligence is expected to bridge the gap between the machine intelligence and human intelligence, So that a fine line could be created to integrate and give a whole system of interactive machine learning. One best example is AlphaGo versus Lee Sedol game, which also known as the Google DeepMind Challenge Match. It was a five-game Go match between 18-time world champion Lee Sedol and AlphaGo, a computer Go program developed by Google DeepMind, played in Seoul, South Korea between the 9th and 15th of March 2016. AlphaGo won all but the fourth game [1]; it is proven that the human intelligence is no par with the machine intelligence. The fourth game won by Lee Sedol is a good example. Cognitive ability and computational thinking could be connected via machine learning or machine intelligence. Therefore ML promises a feasible solution for the huge data processing problems, multiple criteria decision making problems and so on. However ML approach will not give us a required outcome when the proper information is not provided while learning/training. So in this loop hole human intelligence could be utilized and positive target would be acquired. The rest of the paper is structured as in Sect. 2, related work; Sect. 3 consists of Genetic algorithm explanation and its implementation in Sect. 3.1. In Sect. 3.2, Interactive Machine Learning Algorithm implementation is presented. Section 4 consists of Results and Discussion and Sect. 5 concludes the work.

2 Related Work

In [7], Authors showed an experimental evidence for interactive machine learning by implementing it in the optimal search technique – Ant colony optimization. By introducing human in the travelling snakesman problem, the algorithm improved dramatically and it showed better convergence compared to traditional approach. In [5], crayons tools have been used by the authors for image processing. Authors created a perpetual interface and used the classifier for skin detection, paper card tracking, robot car tracking and laser car tracking. Interactive approach showed less time consumption and fast convergence in iterations. In [9], author presented an extensive view on interactive ML which enables an "expert user" to create a precise model than the automatic machine algorithm. Authors used 5 different datasets for evaluation. Comparison is made between c4.5, LTree, OCI and other variants. Kiwifruit has been used for classification. In [3], authors explain about the significance of human loop in self-adaptive systems. As the machine outcome is not corresponding to the human intentions and requirement. MUSIC_Middleware based approach is suggested and tested with different kinds of users with MEET-U application. In [10], authors signify that trust in machines bears from various partialities than believing in human. Explanatory interactive learning enforces the interaction to improve belief and direct approach. Here the authors introduce a method CAIPI which incorporates active learning and co-active learning in a better way.

The search for Nano-particle alloys for material discovery is carried out extensively with machine learning approach and compared with traditional genetic algorithm in [8]. The former approach proved to reduce the search space from manifolds. Here authors proposed an ML accelerated GA which provides good results for materials discovery. In [12] a combined presentation of neural networks with genetic algorithm has been shown. Neurons are selected as a group of population and their fitness is based on the cost. Author presented a genetic natural selection way for understanding neural networks. Genetic Algorithm approach towards ML is presented in [2]. Chess game has been selected for implementation and fitness value is calculated based on Re-enforcement learning algorithm.

3   Genetic Algorithm

Genetic Algorithm (GA) is an optimization technique widely used to enhance parameters for a given problem. It is one of the well-known bench-mark algorithms. It is based on the idea of natural selection which suggests random choices which follows the biological nature. It follows the rule of survival of the fittest which is always the caption for genetic algorithm. Given a number of solutions for a particular problem, the best fitting solution will be considered for selection. So in every generation, new group of values (genes) are created for manipulation [6].

3.1   Algorithm (GA)

1. *Genetic Algorithm starts with a group of values (populations)*
2. *For each value, corresponding weights will be added (weights are like bias values)*
3. *Based on the logic of higher the better, best fitting values will be selected (fitness function selection depends on the context of the problem)*
4. *Group of best outcome will be selected for further mating (making generation)*
5. *The core of the Genetic Algorithm which is crossover will be carried out by exchanging some bits of two best fit functions.*
6. *Mutation is carried out which is a modification in the obtained offspring from the crossover and this process will get repeated till the algorithm converges to an optimal solution.*

In this work, the staff planning example is taken for implementing the Genetic Algorithm, in which it consists of staff id, start time of work and duration of work. The working nature (hours of work) could be calculated from the previous day workload. Employees will work for 5week days and 2 days weekend they get off. Based on this, Data representation is shown as, the dataset courtesy and Genetic algorithm approach inspired by Joos korstanje [11].

3.2 Implementation

Staff= [

[[1,0,7], [2,0,7], [3,0,7], [4,0,7], [5,0,7], [6,0,7],

[[1,0,7], [2,0,7], [3,0,7], [4,0,7], [5,0,7], [6,0,7],

[[1,0,7], [2,0,7], [3,0,7], [4,0,7], [5,0,7], [6,0,7],

[[1,0,7], [2,0,7], [3,0,7], [4,0,7], [5,0,7], [6,0,7],

[[1,0,7], [2,0,7], [3,0,7], [4,0,7], [5,0,7], [6,0,7]

]

Now, the hourly basis staff requirement is given below,

Hourly= [

[0, 0, 0, 0, 0, 0, 0, 2, 2, 3, 3, 4, 4, 5, 5, 5, 4, 4, 4, 3, 3, 2, 2, 2],

[0, 0, 0, 0, 0, 0, 0, 2, 2, 3, 3, 4, 4, 5, 5, 5, 4, 4, 4, 3, 3, 2, 2, 2],

[0, 0, 0, 0, 0, 0, 0, 2, 2, 3, 3, 4, 4, 5, 5, 5, 4, 4, 4, 3, 3, 2, 2, 2],

[0, 0, 0, 0, 0, 0, 0, 2, 2, 3, 3, 4, 4, 5, 5, 5, 4, 4, 4, 3, 3, 2, 2, 2],

[0, 0, 0, 0, 0, 0, 0, 2, 2, 3, 3, 4, 4, 5, 5, 5, 4, 4, 4, 3, 3, 2, 2, 2]

]

GA checks for the staff presence and absence for duty. Based on this data, employee will be selected for shifts. Hours have been given on 24 h basis counting. Nevertheless the business hour doesn't include late-night. The working hour starts from 7.A.M as per the above shown data. The start time of each employee is a random choice and each employee will get instructions through mobile every day. Duration of workload will also be sent through message for each employee.

As here in GA, the start time of the employee is randomly chosen, there are chances of early morning allocation of start time which leads to a confusion. Here the human interference is needed which paves way for the interactive machine learning approach. There are also chances that same employee may get same set of working hours repeatedly. Cost function is defined for allocating more staff or allocating less staff whenever needed. This is considered to be the core concept as the profit of a firm relies partly on the perks offered for their employees.

The number of generations is which results in choosing different set of employees. Parents are created randomly using system implementation and their cost function is shown below,

```
Best cost value is: 214,Worst cost value is: 317
Best cost value is: 217,Worst cost value is: 280
Best cost value is: 206,Worst cost value is: 274
Best cost value is: 201,Worst cost value is: 257
Best cost value is: 198,Worst cost value is: 246
Best cost value is: 181,Worst cost value is: 248
Best cost value is: 189,Worst cost value is: 238
Best cost value is: 180,Worst cost value is: 231
Best cost value is: 178,Worst cost value is: 227
Best cost value is: 172,Worst cost value is: 230
Best cost value is: 172,Worst cost value is: 220
Best cost value is: 173,Worst cost value is: 224
Best cost value is: 169,Worst cost value is: 221
Best cost value is: 166,Worst cost value is: 222
Best cost value is: 169,Worst cost value is: 214
Best cost value is: 164,Worst cost value is: 207
Best cost value is: 161,Worst cost value is: 205
Best cost value is: 158,Worst cost value is: 208
Best cost value is: 157,Worst cost value is: 202
Best cost value is: 157,Worst cost value is: 207
Best cost value is: 158,Worst cost value is: 202
Best cost value is: 155,Worst cost value is: 199
Best cost value is: 152,Worst cost value is: 201
Best cost value is: 154,Worst cost value is: 198
Best cost value is: 153,Worst cost value is: 200
Best cost value is: 151,Worst cost value is: 211
Best cost value is: 149,Worst cost value is: 197
Best cost value is: 150,Worst cost value is: 199
Best cost value is: 147,Worst cost value is: 193
Best cost value is: 146,Worst cost value is: 192
Best cost value is: 144,Worst cost value is: 187
Best cost value is: 145,Worst cost value is: 195
Best cost value is: 144,Worst cost value is: 188
Best cost value is: 143,Worst cost value is: 183
Best cost value is: 144,Worst cost value is: 183
```

```
Best cost value is: 141,Worst cost value is: 187
Best cost value is: 143,Worst cost value is: 181
Best cost value is: 141,Worst cost value is: 186
Best cost value is: 141,Worst cost value is: 184
Best cost value is: 136,Worst cost value is: 181
Best cost value is: 138,Worst cost value is: 175
Best cost value is: 133,Worst cost value is: 169
Best cost value is: 133,Worst cost value is: 169
Best cost value is: 133,Worst cost value is: 166
Best cost value is: 133,Worst cost value is: 169
Best cost value is: 131,Worst cost value is: 161
Best cost value is: 131,Worst cost value is: 161
```

From the above data (output), it is clear that when the algorithm executes initially, it takes the value randomly and at the end, it ought to give the same cost function repeatedly. Obviously, human intervention is needed in such scenarios. By introducing human, to analyze the parent creation, crossover and mutation, it significantly improved the results.

3.3 Interactive Machine Learning Approach

Algorithm : Interactive GA

P<- Population
W <- Weights
Initialize P and W (Randomly)
While !Convergence:

> *Pick group-population* $\sum_{i=1}^{n} P_i$
>
> *If* $P_i \in P$ *and* $w_j \in W$: $\sum_{i,j=1}^{n} P_i * w_j$
>
> *Select best parents for mating*
>
> $p_1 X \ p_2 = P_{12} \mid p_3 \ X \ p_4 = P_{34}$
>
> *Mutate* P_{12} *and* P_{34}
>
> *Mutated Offsprings -* P_{21}^m, P_{43}^m
> *Loop-over (human supervision) through feedback*

End

In the above Algorithm, Interactive ML approach is explained, where random initialization of population and weights are done. Best parents are selected for mating. p_{12} is the crossover result of $p_1 X p_2$ and p_{34} is the crossover of $p_3 X p_4$. Mutate the crossover results, which gives p_{21}^m and p_{43}^m. If the outcome is not up to the requirement, then human supervision is done by having the feedback and the population is analyzed and modified. It is based on MAPE-K feedback loop [3], which follows Monitoring, Analyzing, Planning and Executing. Different parents are selected and crossover is executed. Finally the offspring is mutated to get the better results. This sequence of process is carried over till the algorithm reaches the optimal level.

From the below Fig. 1, the cost for 50 days has been calculated on every day basis. The cost gradually decreases after each day. This is the cost function which is calculated as best based on the genetic algorithm random selection with human interaction. In below Fig. 2, Worst cost function has been shown. Compared to best cost function, Worst cost function shows more cost and that should not be considered as good option for staff selection.

Fig. 1. Best cost function

Since the selection of staff is randomly chosen, there are chances that same staff could be selected often within short period of time. So in this case, human feedback is needed to shuffle and select other staff. This is termed as interactive staff selection, which results in more feasible way. Without human intervention, the cost seems to increase gradually which is not a suggested idea for optimal staff selection. This could be considered as Human Computer Interaction (HCI). From the below Fig. 3, Random staff selection with cost function is shown.

When there are more number of population, there are chances that same set of generation is selected as best option after training the data which is shown above in Fig. 4. When the number of population is increased, the cost function gets reduced considerably and stays constant.

Fig. 2. Worst cost Function

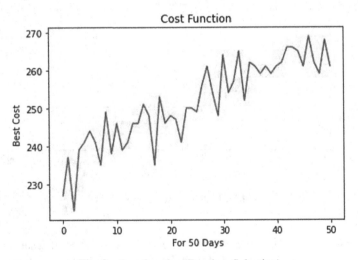

Fig. 3. Cost function (Random Selection)

4 Results and Discussion

GA optimization is a bench mark algorithm. But in some cases it works as a brute force approach and needs some attributes to be examined. As initially when the number of generation was 200 and in the final stage the cost function resulted in the similar set of values. Through scrutiny, the parent selection is modified and mating (crossover) carried out. After mutation it showed enhanced results. As the interactive model suggests the human intervention, it is time consuming compared to automatic model. Since the

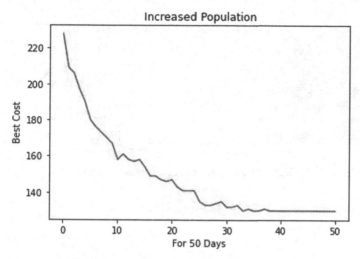

Fig. 4. Increased population best cost function

automatic model may end in huge loss in terms of selecting same employee all the time and also assigning the same schedule may end in chaos. When the population is increased, there are more chances of different genes being selected. Here it results in different people selected randomly.

To inspect the metamorphosis between the (human base) iML and the GA, frequent measure analysis of variance (ANOVA) [7] for the different populations are conducted. The ANOVA generated a substantial core effect of the influence level [I (3,170) = 71456.172, $p < .001$]. This effect seems trifling, though, since the populations require random selection of staff to reduce the cost as well. Even more interesting is the best population selection, where a significant main effect implied [I (1,100) = 13.689, $p <$ .001]. At first execution, the genetic algorithm yielded 4298902.18 (SD = 1066928.451), the population mean of human intervention was 4175070.665 (SD = 92340.842). At second level, the mean of genetic algorithm was 35218248.76 (SD = 844200.243), the population mean of human based implementation was 34381925.72 (SD = 710450.534). Transversely in both executions, human supervision resulted with lesser cost. On the basis of the entire executions (implementations), a relative comparison is not possible since the population selection varies.

From the above Fig. 5, different types of plots are shown based on the obtained results. In bar chart, the difference is not obvious whereas in scatter and base plot, it is clearly seen that interactive human base supervision of the implemented algorithm yields lesser cost.

From the Fig. 6, the mean population of the 2nd execution is shown. Here the term population simply refers to the offspring. Interactive model converges to lesser cost function compared to traditional search algorithm (GA).

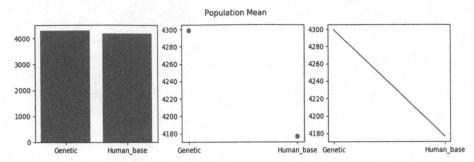

Fig. 5. Mean value of the generated population

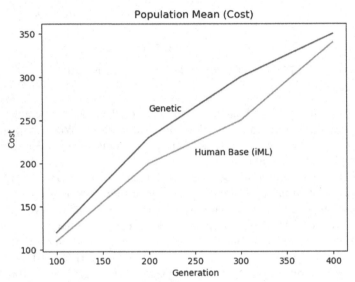

Fig. 6. 2nd generation mean cost function.

5   Conclusion

Staff selection on daily basis is a cumbersome task. It requires an automatic system model. GA is a fundamental algorithm which has proven enhanced optimization. Though, it depends on the nature of problem. Population selection is the key concepts of GA. Automatic systems have benefits as well as problems. When machine model suffers from inefficient data, human interaction is needed. As human cannot interact with all sorts of system models, it is purely based on the nature of problem. Human interaction is easy and the problem could be corrected in this particular case. Interactive Machine Learning approach shows enhanced results compared to GA approach. ANOVA measurement results proved to be better for evaluating both the implementation. MAPE-K feedback loop in iML yielded enhanced results.

Reference:s

1. BBC News Online: Artificial intelligence: Go master Lee Se-dol wins against AlphaGo program. Accessed 13 Mar 2016
2. Bhasin, H., Bhatia, S.: Application of genetic algorithms in machine learning. Int. J. Comput. Sci. Inf. Technol. (IJCSIT) 2(5), 2412–2415 (2011)
3. Evers, C., et al.: The user in the loop: enabling user participation for self-adaptive applications. Fourth Gener. Comput. Syst. 34, 110–123 (2014). https://doi.org/10.1016/j.future.2013.12.010
4. Digital Initiative Homepage. https://digital.hbs.edu/platform-rctom/submission/the-failed-launch-of-www-healthcare-gov/. Accessed 18 Nov 2016
5. Jerry, F., Dan, O.: Interactive machine learning. In: Proceedings of The 8th International Conference On Intelligent User Interfaces, pp. 39–45, January 2003. https://doi.org/10.1145/604045.604056
6. Holland, J.H.: Adaptation in Natural and Artificial Systems (The University of Michigan Press, Ann Arbor, M), p. 211 (1975)
7. Holzinger, A., et al.: Interactive machine learning: experimental evidence for the human in the algorithmic loop. Applied Intelligence 49(7), 2401–2414 (2018). https://doi.org/10.1007/s10489-018-1361-5
8. Jennings, P.C., Lysgaard, S., Hummelshøj, J.S., et al.: Genetic algorithms for computational materials accelerated by machine learning. NPJ Comput. Mater. 5, 46 (2019). https://doi.org/10.1038/s41524-019-0181-4
9. Ware, M., Frank, E., Holmes, G., Hal, M., Witten, I.H.: Interactive machine learning: letting users built classifiers. Int. J. Hum. Comput. Stud. 55, 281–292 (2001). https://doi.org/10.1006/ijhc.2001.0499
10. Teso, S., Kersting, K.: Explanatory interactive machine learning. In: AIES19 Proceedings of the AAAI/ACM Conference on AI, Ethics, and Society. AAAI (2019)
11. Towards Data Science Homepage. https://towardsdatascience.com/a-simple-genetic-algorithm-from-scratch-in-python-4e8c66ac3121. Accessed 1 June 2020
12. Towards Data Science Homepage. https://towardsdatascience.com/gas-and-nns-6a41f1e8146d. Accessed 26 Mar 2018

Software of Linear and Geometrically Non-linear Problems Solution Under Spatial Loading of Rods of Complex Configuration

Sh. A. Anarova[1]([⊠]) [iD], SH. M. Ismoilov[2]([⊠]) [iD], and O. Sh. Abdirozikov[3]([⊠])

[1] Tashkent University of Information Technologies, 100200 Tashkent, Republic of Uzbekistan
shahzodaanarova@gmail.com
[2] Namangan Engineering-Construction Institute, 160103 Namangan, Republic of Uzbekistan
shohsoft@gmail.com
[3] Information and Communication Technology and Communication MOD,
111811 Tashkent, Uzbekistan
abotabek0679@mail.ru

Abstract. A software has been developed for calculating linear and geometrically nonlinear problems of rods under spatial loading. The software system structure for calculating linear problems of spatial loading of rods of complex configuration was improved by the R-function method (RFM) and successive approximations . Based on the developed software, the oscillation processes of spatially loaded rods in linear statements, the statics and dynamics of spatially loaded rods in linear and geometrically nonlinear statements were investigated; the problems of constrained torsion of a prismatic body of arbitrary section with different cavities were solved.

Keywords: R-functions method (RFM) · Constrained torsion · Stress-strain state (SSS) · Structure of a software package · Geometrically nonlinear · Bubnov-galerkin method

1 Introduction

In this paper is based on the generalized variational Hamilton-Ostrogradskiy's principle, the theory of elastic strains and the refined Vlasov-Dzhanelidze-Kabulov theory, the generalized mathematical models for the statics and dynamics, for linear and geometrically nonlinear problems of rods under spatial loading were developed. These models serve for a detailed description of the processes of geometrically linear and nonlinear deformation of rods, taking into account the combined action of longitudinal, transverse and torsional forces. Mathematical models of spatially loaded rods were developed with account for constrained torsion with the corresponding natural initial and boundary conditions [1]. These models were used to describe the processes of the stress-strain state of a prismatic body in problems of constrained torsion [1, 2, 4, 9, 11–15].

Mathematical models of linear and geometrically nonlinear problems of rods taking into account transverse bending under spatial loading were developed in [12, 14].

These models serve for a detailed description of the processes of linear and geometrically nonlinear deformation of rods under flexural-torsional, longitudinal-flexural and longitudinal-torsional oscillations. Based on the method of central finite differences, a computational algorithm was developed to calculate the statics and dynamics of linear and geometrically nonlinear problems of rods under spatial loading. The test examples were solved based on this algorithm, the results obtained were evaluated according to the criteria of reliability and accuracy [1–3, 7, 10–12].

2 Integration of Algorithm

Algorithms for integrating systems of resolving equations based on the R-function method (RFM) and successive approximations for constrained torsion were developed. With the developed algorithms, the test problems were solved, the results obtained can be evaluated according to the criteria of reliability and accuracy [1, 2, 9]. Using the computational algorithm [14, 15] developed by the mathematical model [8, 12, 14], a software package (SP) was created for solving boundary value problems of the systems of linear and nonlinear partial differential equations, designed to calculate the stress-strain state of elastic prismatic bodies of arbitrary section with a cavity. Taking into account the theoretical foundations, the SP was created using the methods of finite differences, Bubnov-Galerkin, R-functions method (RFM) and successive approximations. The method of successive approximations was used to solve linear and nonlinear problems at each iteration step. The structure of the program complex consists of the following blocks [4–6, 8]:

1. ROP - a library of modules for R-operations and card operations.
2. COMM - a library of types and constants.
3. INTEG - a library for defining integrands.
4. STERJOBL - a library for domain geometry functions.
5. BASPOL – a library for structural formulas.
6. ITERKRUCH - a library for iteration processes.
7. RAZR - a library for generating elements of the resolving equation.
8. RESHRU - a library of modules for solving resolving equations.
9. OFORM - a library of modules for registration of calculation results.
10. Block "Control program".

   The structure of the software (SW) is shown in Fig. 1. It was built based on a software package [4–6, 8], designed to solve the boundary value problems of the mechanics of deformable rigid body (MDRB). An extension allows using in calculations prismatic bodies of arbitrary section with a cavity.
   Each block of the program complex consists of several modules, designed in the form of procedures and functions. Routine libraries are created from these modules. This SP is implemented in the Java language in the MS WINDOWS environment. The developed SP makes it possible to automate the solution of the boundary value problems of the stress-strain state of elastic prismatic bodies of arbitrary section with a complex shape in plan for the systems of partial differential equations, which many problems of continuum mechanics are reduced to.

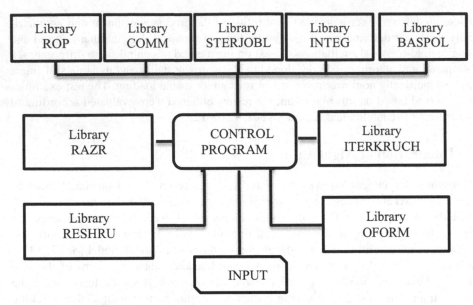

Fig. 1. Scheme of the structure of the program complex.

Consider in detail the description of the main modules of SP, intended to calculate the stress-strain state of elastic prismatic bodies of an arbitrary section of a complex shape. The theory of R-functions is used to calculate the stress-strain state of elastic prismatic bodies of arbitrary cross-section of complex configurations. When applying this theory, it is necessary to use R-operations (R-disjunctions, R-conjunctions and R-negation). The implementation of this task on a computer is performed by the library for R-operations and card operations.

The STERJOBL library for domain geometry equations contains procedures-functions and procedures that describe the equations of the domain geometry and their derivatives of the required order. This module has been updated with related functions and procedures for new complex domains.

To calculate the values of structural formulas and their derivatives of the required order, the software package includes the procedures that implement structural formulas in accordance with the boundary conditions.

As noted above, the Gauss point method was used for approximate calculation of the integrals values. The corresponding procedures were included in SP to generate Gaussian nodes, used to calculate the values of integrals for complex domains that provide a high degree of accuracy. The library for generating points and weights of integrals contains these procedures. The library for structural formulas contains procedures for calculating basic polynomials and their derivatives of the required order, described in [4–6, 8].

The procedures for generating matrices of the resolving equation are contained in the library for generating the elements of the resolving equation. The library of modules for the solution of resolving equations includes modules intended for solving resolving equations. Depending on the problem statement, they can be a system of algebraic, differential or integro-differential equations. In the case of algebraic equations, the methods of

Gauss and other methods can be used. To obtain the results of numerical calculations in a form convenient and required for the user, appropriate procedures and functions were created, contained in the library of modules for the registration of calculation results. The developed software package was tested in the calculation of a new class of problems of the stress-strain state of elastic prismatic bodies of arbitrary cross-section of various shapes.

The following is a brief description of each software library, including the subroutines (procedures and functions) developed. Some imported libraries are generic and include: ROP (a library that implements R-operations and structural formulas), COMM (a library of types and constants), INTEG (a library for generating Gaussian nodes and weights), BAZPOL (a library for calculating polynomial values and their derivatives of the n-th order), RESHRU (a library for solving resolving equations, in this case, the systems of algebraic equations), OFORM (a library for registering calculation results). Libraries RAZR, STERJOBL are supplemented with appropriate modules, taking into account the specifics of the problem: ITERKRUCH is designed to implement the algorithm described in [4–6, 8]. Further, to calculate the integrands and form the elements of the resolving equations, the functions FBB and FFX were developed, where FBB is for the left-hand part, and FFX is for the right-hand part of the equation. They enter the RAZR library. The created procedures and functions (D1f, NEYMAN, FBB, FFX), as well as the ITERKRUCH library, allow automating the problem solution. Description of the software package for calculating linear and geometrically nonlinear problems of spatial loading of rods of complex configuration. Consider the procedures and functions used to solve the tasks.

- Procedure header D1f looks like a procedure: D1f (int np; int so; mat5 F, mat5 w; mat5 c); where np is the maximum order of the differential present in the differential equation of the boundary value problem; so is the order of the differential operator present in the structural solution; F(тип:mat5) is an array containing the values of operator D1 functions and the values of partial derivatives of the required order, specified in the form

$$F; \frac{\partial F}{\partial z}, \frac{\partial F}{\partial y}; \frac{\partial^2 F}{\partial z^2}, \frac{\partial^2 F}{\partial z \partial y}, \frac{\partial^2 F}{\partial y^2}; ...; \frac{\partial^n F}{\partial y^n}; \tag{1}$$

w is an array that contains the values of the equations of the boundary and geometry of the domain and their derivatives; C - an array where the results of the calculated values D1f and partial derivatives of the required order are written. The results are presented in the form (1). The type Mat5 is defined as float mat5 [dl], where dl = (np + so + 1) × (np + so + 1) div 2. The NEYMAN procedure header is void Neyman (int np, int so; int I; int j;float z; float y; tpex pc2; imjpolpro imjapolpro; mati5 wi; mat5 un), Here i, j are the powers of the polynomial in variables z and y, respectively; pc2 - an information array of type (float tpex [6]) containing data on the left lower and right upper boundaries along the oz and oy axes: pc2 [1]-Zmin; pc2 [2]-Zmax; pc2 [3]-Ymin; pc2 [4]-Ymax; Imjapolpro takes the value stepm2 or cheb2, or trigon2 depending on whether a power polynomial or a Chebyshev polynomial or a trigonometric one is used in the structure. Un is an array containing the values of coordinate systems and partial derivatives up to the required order at the point (z, y).

3 Mathematical Modeling

- The equation for the domain geometry is defined as the following function:
  bool <name> (float z; float y), which defines the belonging of the specified function inside the domain or on the boundary, and outside the domain, where <name> is user-defined; (z, y) are the coordinates of the given point. If the point is in the domain $D \vee \Gamma$ (D is the domain, Γ is its boundary), then <name> takes the value true, otherwise - false.
  The function header for calculating the values of the equation of the domain geometry and their partial derivatives in formula (1) at a given point has the form void <имя> (int np; int so;float z; float y; tpex pc2; mati5 WI); where WI an array of type mati5 contains the values of the equations of the sub-domain boundaries and their derivatives of the required order at the point (z, y). This procedure enters the STERJOBL library.
- The mati5 type is defined as (this type enters the COMM library) mat5 Mati5 [kolobl]; where kolobl is the number of mat5 subdomains, and the purpose of the np, so parameters is described above.
- Arbitrary constants in (2) are:

$$\Phi = \sum_{i=0}^{n} \sum_{j=0}^{i} C_{ij} \, Z_i(z) \, Y_j(y), \tag{2}$$

where C_{ij} are the unknown coefficients to be determined; $Z_i(z) \, Y_j(y)$ - a complete system of basic polynomials (power, trigonometric, Chebyshev ones, etc.), determined using the CIPR procedure; its header has the form.

- void CIPR (int r; int l; inttip Inti; typci CI);
  where l is the length of the rod; Inti - an array of type inttip (float inttip [5]; containing the values of integrals $I_p, I_d, I_k, I_{\phi\phi}$ I2H; CI (long long typCI [5]) is the array, where the values of coefficient C_i (i = 1, 5) are formed. The VICHINT procedure is designed to calculate the values of double integrals; its header of is void VICHINT (rez nch;int clutoch; DIFOB DifObl; DifObl TIP; korftip korfun; mat4 ZG; inttip intMas); where clutoch is the number of Gaussian nodes; DifOb, korfun are described above; ZG - an array containing the solution of a system of algebraic equations; INTmas is an array containing integral values. The STR procedure is designed to calculate the values of structural formulas and their derivatives of the required order at a given point. The header of the procedure that solves the system of algebraic equations by the Gauss method has the form void STR (float z; float y; char nch; MAT4 XG; DIFOBTIP DIFOBL; KORFTIP KORFUN; mat5 rez); where z, y are the coordinates of the calculated point; ch takes the values of "c", "z", "y", "v" depending on the complete symmetry or symmetry along the oz and oy axes and the asymmetry of the domain, respectively: depending on the symmetry of the problem or the symmetry of the oz or oy axis, or the asymmetry of ZG array (int mat4 = array [1..skm]), skm is the number of coordinate functions that contain the solution of the system of algebraic equations:
- void difobtip (int np; int so; float z; float y; MATI5 ww);
- void KOrFtip (int np; int SO;int i; int j; tpex pc2;

- float z;float y; imjpolpro imjapolpro; mati5 wi; mat5 Un);
- void imjpolpro (int so;int i; int j; tpex pc2/* tpex s*/; float z; float y; mat5 Un);

DIFOBL is the name of the procedures (entered the STERJOBL library) designed to calculate the equation values of the domain geometry and their n-order derivatives. Its description is given above, KORFUN is the name of the structural formula. In this case, it is Neymafist.

The function ψ_1, ψ_2 is intended to form the values of $\overline{\psi}_1$ (array omegchar) present in the formula (3).

$$r^2 = \frac{I_{\kappa p}}{I_{\varphi\varphi}}, \quad \beta_{1H} = \frac{M_{\kappa p}}{I_{\varphi\varphi}}, \quad \overline{\psi}_1 = \frac{\psi_2}{\psi_3}. \tag{3}$$

where

$$I_{\varphi\varphi} = \int_F E\varphi^2 dF, \quad I_{\kappa p} = \int_F G\left[\left(z + \frac{\partial\varphi}{\partial y}\right)^2 + \left(-y + \frac{\partial\varphi}{\partial z}\right)^2\right]dF,$$

$$\psi_3 = -\int_0^\ell \left(\frac{\partial\theta}{\partial x}\right)^2 dx, \quad \psi_2 = \int_0^\ell G\left(\frac{\partial^2\theta}{\partial x^2}\right)dx, \quad \psi_5 = -\psi_3,$$

$$M_{\kappa p} = \int_F (z\varphi_2 - y\varphi_3)dF.$$

- The function header looks like void ψ_1 ψ_2 (float R; TYPSI CI); where the values of the variable r are calculated by formula (4). Array Ci contains the values of the coefficients from formula (2).

$$r^2 = \frac{I_p + 2I_d + I_k}{I_{\varphi\varphi}}, \tag{4}$$

- Formation of the value of r2 according to the formula (4) has the form: Flost rkv (typci CI). The purpose of CI is described above. Checking the convergence of the computational process was carried out using the PROVERKA procedure; its header has the form
- Void Proverka (real Eps; typfile oldfile; typfile newfile; bool Log); where Eps is the specified accuracy; oldfile and newfile are the files where the values of the previous and current iterations are set, respectively; typfile = file of real; Log can be true or bool depending on the numerical convergence.
- The ITERKRUCH procedure is a control program for solving a problem.
- The initial value is determined by the variable OMEGCHER: = 0, then, the values of the arrays are formed.

  - pc2 [1] = -ag /*Zmin*/; pc2 [2] = ag /*Zmax*/;
  - pc2 [3] = -bg /*Ymin*/; pc2 [4] = bg /*Ymax*/;

- An obtip OB variable (defined in the COMM library) is assigned a domain name (user-defined), and a DIFOB is assigned the name of a procedure (entered the STERJOBL library) that computes the equation values of the domain geometry and their n-th order derivatives.

Further, when calling the OBLGAU procedure (included in the INTEG library), the Gaussian nodes and weights are generated. The number of nodes is formed in clutoch, and the value is determined in COMM. Then, the values of the variables NP, SO, NK and the call to the procedure are formed in this library (included in the RAZR library), where the elements of the matrix of resolving equations are formed. The resulting system of algebraic equations is solved by the Gauss method and the values of unknown coefficients in the structures are determined. Substituting these values into the structures, we calculate the values of the sought functions (φ) and their derivatives at a given point. Then the values of the integrals I_p, I_d, I_k, $I_{\phi\phi}$ I2H are calculated. This is done using the VICHINT procedure (included in the ITERKRUCH library). These values of the integrals are contained in the IntMas array (its type is float IntMas [5]). Further, forming the value of the variable r, we turn to the Qf procedure, where the values of the array C_i ($i = 1$, 2, 3, 4) are formed and θ, X_z, X_y, X_x and other parameters are calculated. Convergence is checked using the PROVERKA procedure. If the appropriate accuracy is reached, the process continues. Otherwise, the value of the OMEGCHER variable is re-generated, and the above processes must be started anew.

4   Results

Instructions for using the software package for calculating linear and nonlinear problems of rods under spatial loading. In this paragraph, we provide instructions for using the software package described above. Instructions for using the software package for calculating linear and nonlinear problems of rods under spatial loading. In this paragraph, we provide instructions for using the software package described above.

1. The case is considered when two ends of the rod are rigidly fixed. Then the conditions are:

$$\vec{U}\Big|_{\bar{x}=0} = 0, \quad \frac{\partial \vec{U}}{\partial x}\Big|_{\bar{x}=0} = 0, \quad \vec{U}\Big|_{\bar{x}=l} = 0, \quad \frac{\partial \vec{U}}{\partial x}\Big|_{\bar{x}=l} = 0.$$

2. The case is considered when one end is rigidly fixed and the other one is free. Then the conditions are:

$$\vec{U}\Big|_{\bar{x}=0} = \begin{pmatrix} 0 \\ 0 \\ 0 \\ 0 \\ 0 \\ 0 \end{pmatrix}, \quad \frac{\partial \vec{U}}{\partial \bar{x}}\Big|_{\bar{x}=0} = \frac{\partial}{\partial \bar{x}} \begin{pmatrix} 0 \\ 0 \\ 0 \\ 0 \\ 0 \\ 0 \end{pmatrix}, \quad \vec{U}\Big|_{\bar{x}=l} = \begin{pmatrix} u \\ \alpha_1 \\ \alpha_2 \\ v \\ w \\ \theta \end{pmatrix}, \quad \frac{\partial \vec{U}}{\partial \bar{x}}\Big|_{\bar{x}=l} = \frac{\partial}{\partial \bar{x}} \begin{pmatrix} u \\ \alpha_1 \\ \alpha_2 \\ v \\ w \\ \theta \end{pmatrix},$$

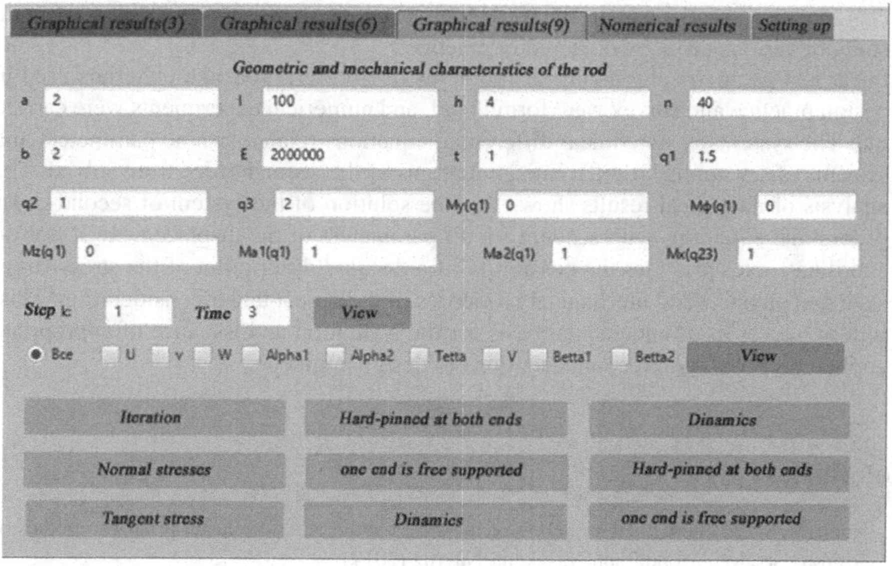

Fig. 2. User interface "Calculation of linear and geometrically nonlinear problems under spatial loading of rods with nine equations"

After starting the program, the user interface "Calculation of linear and geometrically nonlinear problems under spatial loading of rods" of the types, shown in Fig. 2, appears.

Next, we can select one of the commands "Calculation of the statics and dynamics of a rod rigidly fixed at two ends", "Calculation of the statics and dynamics of the rod, one end of which is fixed and the other is free", "Calculation of the dynamics" to view the results of one parameter or all parameters of the equation in graphical and numerical forms. The geometrical and mechanical characteristics of the rod can be changed.

5 Conclusions

The main results are as follows:- the software for calculating linear and geometrically nonlinear problems of rods under spatial loading was developed. This software was used to study the stress-strain state of the processes of geometrically linear and nonlinear deformation of rods, taking into account the combined action of longitudinal, transverse and torsional forces.

- based on the developed software, the processes of oscillation of spatially loaded rods in linear statement, the statics and dynamics of spatially loaded rods in geometrically linear and nonlinear statements were investigated, the problems of constrained torsion of a prismatic body of arbitrary section with different cavities were solved. These results serve for an accurate study of various tasks encountered in practice.
- the systems of differential equations of the fourth order for the problems of constrained torsion with different cavities of an arbitrary section of a prismatic body were solved. Analytical and numerical results obtained with the R-function methods (RFM) for

classical cross-sections in profile were compared. These results serve for an accurate study of various tasks encountered in practice.

– linear and geometrically nonlinear problems of structural rod-like materials used in design practice and survey were formulated, and numerical experiments were carried out. The systems of nine linear differential equations related to nine parameters and systems of six nonlinear differential equations of the second order were solved. The analysis of numerical results shows that the solution of the system of second-order differential equations with account for all parameters of the displacement vector on coordinate axes provides the possibility of a detailed description of the stress-strain state and physical and mechanical properties of the object under consideration. This, in turn, serves as a fundamental basis for the formation and issuance of appropriate applied proposals and recommendations to design engineers.

References

1. Anarova, S.: Mathematical modeling of the stress-strain state of elastic prismatic bodies of arbitrary section. Monograph, Tashkent, Navruz (2017)
2. Kabulov, V.K., Anarova, Sh.A.: Comparison of the cavity effect on the stress state of prismatic bodies of arbitrary cross-section in the problems of constrained torsion. J. AS RUz **6**, 7–10 (2003)
3. Kurmanbaev, B.: Unified technology for constructing discrete models of three-dimensional static and dynamic problems of elastic and inelastic bodies, algorithms and their solutions, software and numerical analysis of results. Dis.Doc. Phys.-Math. Sci. - Tashkent (1992)
4. Nazirov, Sh.A., Piskorsky, L.F.: Complex of software tools for calculating and optimizing plate structures of a complex shape. In: Algorithms: Collection of scientific works. - Tashkent, IK AN RUz, vol. 81, pp. 41–54
5. Nazirov, Sh.A., Sadikov, Kh.S.: Complex of software tools for solving boundary value problems by variational methods. In: Algorithms: Collection of scientific works. - Tashkent, IK AN RUz, no. 65, pp. 38–48 (1988)
6. Nazirov, Sh.A., Nuraliev, F.M, Amanov, O.T.: The structure of the software complex for solving boundary value problems for systems of partial differential equations. Uzbek Mag. "Probl. Inform. Energy" – Tashkent **6**, 40–43 (1998)
7. Nazirov, Sh.A., Anarova, Sh.A.: Modeling the stress-strain state of prismatic bodies of arbitrary cross-section in problems of constrained torsion. Uzbek Mag. "Probl. Inform. Energy" –Tashkent **3**, 22–25 (2003)
8. Nazirov, Sh.A., Abduazizov, A.A.: Description of a set of programs for calculating radio-electronic devices. In: Issues of Calculated and Applied Mathematics: Collection of Scientific Works. - Tashkent, RPP and AIC Center, no. 128, pp. 64–84 (2012)
9. Nazirov, Sh.A., Nuraliyev, F.M.: Anarova, Sh.A.: Study of numeric convergence of the method of R - functions in problems of constraint torsion. Am. J. Comput. Appl. Math. **2**(4), 89–196 (2012). https://doi.org/10.5923/j.ajcam.20120204.07
10. Anarova, Sh.A., Nuraliev, F.M.: Study of stressed state of elastic prismatic bodies of arbitrary section with a cavity in problems of constraint torsion. Int. J. Sci. Innov. Math. Res. (IJSIMR) **3**(2), 1–15 (2015). ISSN: 347–307X (Print) & ISSN: 2347–3142. www.arcjournals.org
11. Anarova, Sh.A., Nuraliyev, F.M.: Usmonov, B.Sh., Chulliyev, Sh.I.: Numerical solution of the problem of spatially loaded rods in linear and geometrically nonlinear statements. Int. J. Eng. Technol. **7**(4), 4563–4569 (2018). www.sciencepubco.com/index.php/doi:10.14419/ijet.v7i4.22669

12. Anarova, Sh., Ismoilov, Sh.: Mathematical simulation of stress-strain state of loaded rods with account of transverse bending. J. Phys.: Conf. Ser. **1260**, 102002 (2019). https://doi.org/10.1088/1742-6596/1260/10/102002
13. Anarova, S., Isomidinov, A., Ismoilov, S.: Numerical analysis of vibration processes of spatially loaded rods. Probl. Comput. Appl. Math. Tashkent **5**, 29–45 (2019)
14. Anarova, Sh.A., Abdirozikov, O.Sh.: Mathematical and numerical model of thermo elastic plates with complex form. Mukhammad al-Khorazmiy avlodlari ilmiy-amaliy va axborot tahliliy jurnal. Tashkent **5**(10), 121–124 (2019)
15. Nuraliev, F., Anarova, Sh., Amanov, O., Jumaev S. & Abdirozikov O.: Mathematical model and computational experiments for the calculation of three-layer plates of complex configuration. J. Phys.: Conf. Ser. **1546**, 012094 (2020). https://doi.org/10.1088/17426596/1546/1/01204

Mathematical Modeling of Pascal Triangular Fractal Patterns and Its Practical Application

Sh. A. Anarova[1]($\boxtimes$) ⓘ, Z. E. Ibrohimova[2]($\boxtimes$) ⓘ, O. M. Narzulloyev[3]($\boxtimes$) ⓘ, and G. A. Qayumova[3]($\boxtimes$) ⓘ

[1] Tashkent University of Information Technologies, 100200 Tashkent City, Republic of Uzbekistan
shahzodaanarova@gmail.com
[2] Samarkand Branch of Tashkent University of Information Technologies Named After Muhammad al-Khwarizmi, Tashkent, Uzbekistan
zulayhoibrohimova90@gmail.com
[3] Tashkent University of Information Technologies Named After Muhammad al-Khwarizmi, Tashkent, Uzbekistan
oybek.88.13@gmail.com, gulshanqayumova@gmail.com

Abstract. This article focuses on the mathematical modeling of triangular fractal patterns and carpet design in textiles, demonstrating the classic and new arithmetic, combinatorial features of the generalized arithmetic triangles of Pascal's triangle. Binomial, three-term, and other combinatorial numbers constructed based on recurrence relations have been investigated. The software "Pascal's Triangle" was developed in the C# programming language. A convenient user interface of the program has been developed. Based on the software implementation, arithmetic triangles are constructed from residues in the form of Pascal's triangle and combinatorial numbers. The developed software can be used in industry, when drawing patterns, for their subsequent stamping on carpets, fabrics, ceramic tiles, etc.

Keywords: Fractal · Pascal's triangle · Arithmetic triangles · Binomial coefficients · L-system · RFM · IFS · Software · Interface · Binomial coefficients · Combinatorial numbers

1 Introduction

Fractal geometry originated in the 19th century. Cantor by using a simple repeating procedure, turned the line into a set of unconnected points, thus obtained the so-called Cantor dust [1].

The word "fractal" was introduced by Benoit B. Mandelbrot from the Latin word "fractus", which means broken, i.e. divided into parts [1]. One of the definitions of a fractal is as follows: A fractal is a geometric shape made up of parts that can be divided into parts, each of which will represent a reduced copy of the whole. Namely, a fractal is an object for which it does not matter with what magnification it is viewed through a magnifying glass, but with all its magnifications, the structure remains the same. Large-scale structures completely repeat smaller-scale structures.

Self-similarity is one of the main properties of fractals. The dimension of an object shows by what law its internal area grows. Similarly, the "volume" of a fractal increases with an increase in its size, but its dimension is not an integer value, but a fractional one. Therefore, the border of a fractal figure is not a line: at high magnification, it becomes clear that it is blurred and consists of spirals and curls, repeating the figure itself on a small scale.

Today fractals are widely used in science, technology and practice, for instance, in light industry, textiles, architecture, computer graphics, computer systems, telecommunications, radio engineering, cinema, television, animation, fluid mechanics, surface physics, astronomy, medicine, biology, music, art, etc. [1, 7, 9, 13, 15–17].

Textile design emerged as a new and growing technology area, became known for its creativity, science and technology. Many textile design concepts might be analyzed along with mathematical concepts such as geometric transformations and formulas and setting the level of sharpness and their reflection in the fabric. These concepts represent a typical process in the textile design industry. However, the methods used to date are not effective enough as an improved method for creating and coloring multifaceted designs with innovative patterns. An attempt was made to study mathematical functions and create software using fractal models to create innovative designs to solve this problem. Fractals are a new branch of mathematics with precise fractal dimensions, one of nature's greatest design secrets, and the use of this technique in the textile industry has ushered in a new era in design. The article is devoted to an investigation of the many concepts and patterns that are opening up new directions in the field of textile design.

The textile industry is very competitive and manufacturers are constantly looking for additional advantages to achieve better results and stay one step ahead of other manufacturers. It is important for designers to know the market that is focused on their goals, the level of demand or the age of people. Textile and pattern design is a key research factor in both the textile industry and computer graphics. Pattern design can use any object and landscape, including man-made abstract objects. Previously, the techniques used to create traditional designs have created only by specialists and the complexity level of patterns was limited. Today, the design of any product amazes people, since there is an appeal in product design. Designs and patterns represent history and experiences passed down from culture to culture, from artist to artist, and from generation to generation.

Over time, every day needs change, such as the patterns and colors of clothes and the decor of home. Textile design is a growing area among all design fields, It covers the fields of fashion design, carpet production and fabrics. Since ancient times, textile has played an important role in various geographic regions of the world and in all climatic conditions. Traditionally, weavers have used special forms of graphic paper to create patterns on fabrics, however there are several design software's available today to create a design or pattern. Fractal geometry research has attracted a lot of attention over the past 20 years due to its theoretical significance and practical application and it was the next step towards to new design ideas. In the mid-1970s, Benoit Mandelbrot first used the word fractal to describe the chaotic shapes of objects in nature.

Fractals are beautiful designs and can delight with their colors and images. This is a great scientific discovery that changed the way people think about life and helped them understand nature and the universe. Many textile patterns have fractal elements due to their self-similarity. These designs are a bit complex and lead to the development of decorative design.

In the design industry, fractals are used to compress images by reducing data redundancy and creating an ideal platform for textile design. Fractal creation software's create images by repeating the following three steps:

- setting the parameters of the corresponding fractal programs;
- perform the long possible calculations
- product evaluation.

Textile designers need to understand the link between creativity, science and technology in order to offer new and innovative solutions that meet global, multicultural needs.

Design of patterns based on fractal geometry is an important area of textile engineering.

Traditional fractal design has some disadvantages associated with the inability to effectively reflect the characteristics of real landscapes and textures. In this article, we aim to offer a great textile design solution by combining pattern design techniques with fractal geometry using a Pascal Triangles software.

One of the main ideas of fractal geometry is the idea of non-integer values for the number of existing dimensions. B. Mandelbrot [1] called the non-integer dimension 2.76 the fractal dimension. Simple Euclidean geometry states that the universe is flat and smooth. The property of such a being is represented by points, lines, angles, triangles, cubes, spheres, tetrahedrons, and so on.

The fractal equation is constructed using different methods. These are as follows, IFS (Itered function systems - Iterative function systems) method [1, 7, 9, 14], L - systems method [1, 7, 13], method of the theory of binomial polynomials based on arithmetic properties [2–6, 9, 10], R-function (RFM) method [14–17] set theory method [7, 9] and then others [18–20].

This article discusses the construction of Pascal's triangles using the method of binomial polynomial theory based on arithmetic properties and s-generalized Pascal triangles.

Interest in the study of the Pascal triangle is still widespread today. This is due to the division and distribution of p-module elements, the discovery of new and often unexpected features associated with the tasks of constructing and studying graphs, as well as their application in various practical problems.

Arithmetic structures are arithmetic tables consisting of combinatorial numbers, their remainders in prime and mixed modules, or the level of primes divisible by other combinatorial numbers. This paper investigates triangles constructed based on binomial, trinomial, and other combinatorial numbers constructed based on recursive relationships.

One of the most famous tables in the history of mathematics is the arithmetic triangle, named Pascal's triangle after the great French mathematician and philosopher of the seventeenth century. The research findings of Blaise Pascal (1623–1662) were presented in the booklet Traite du Triangle Arithmetique, published after the author's death. In this pamphlet, he summarized the known features of the Pascal triangle and introduced many new features [5, 11].

Pascal wrote the results of various properties of the numbers that make up the arithmetic triangle as a whole, without algebraic notation. Some of Pascal's fundamentally important discoveries are directly related to Pascal's arithmetic and probabilistic investigations, for instance, complete mathematical induction method, application of the arithmetic triangle to problems of probability theory, etc.

The law of addition of an arithmetic triangle and its members was known in India even before our era.

2 Problem Statement

Interest in the arithmetic triangle does not subside today. This module deals with the division and distribution of p elements, the discovery of new and often unexpected functions related to the problems of constructing and studying graphs, as well as their application in various practical problems.

An arithmetic triangle is often written as an equilateral triangle, it has "ones" on the sides, each remaining number is the sum of the two numbers to the left and right of the previous line.

Properties of arithmetic triangles and binomial coefficients, as well as some questions about division, can be found in V.A. Uspensky [3], in the literature on combinatorial analysis and number theory, as well as in mathematical reference books. Arifmetik uchburchakning ko'plab elementar xususiyatlarining eng batafsil tavsifi T. M. Green, K.L. Hamberg [5].

It is known that the elements of Pascal's triangle are binomial coefficients, known even before the appearance of Pascal's triangle. However, he was the first to use them and gave Pascal's description [6, 10]. Historical information about binomial coefficients and binomial theorems can be found in [2–6].

Main part. Arithmetic triangles and binomial coefficients. The arithmetic triangle appeared in Europe long before the publication of Pascal's treatise. This is due to the division and distribution of the elements of the module p, the discovery of new and often unexpected possibilities associated with the problems of constructing and studying graphs, as well as their application to various practical problems.

Let's focus on new arithmetic triangles and rectangles, pyramids and other arithmetic tables:

An arithmetic triangle is often described as an equilateral triangle (Fig. 1), it has ones on the sides, and each of the remaining numbers is equal to the sum of the two numbers on the left and right in the previous row. The line number consists of the binomial expansion coefficients $(1 + x)n$, which are shown differently in the literature. Here they are indicated by symbols introduced by Euler instead of the symbol that appeared in the 19th century.

Properties of Pascal's triangle and binomial coefficients, as well as some questions about division, can be found in the book by V. A. Uspensky [3], in the literature on combinatorial analysis and number theory, as well as in mathematical reference books.

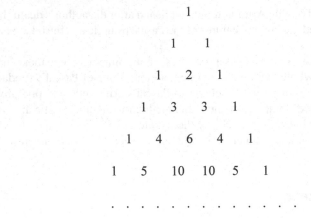

Fig. 1. Arithmetic triangle

The most detailed description of the many elementary properties of the Pascal triangle is given in the book "Pascal's Triangle" by T.M. Green, C.L. Hamberg [5].

It is known that the elements of Pascal's triangle are binomial coefficients, known even before the appearance of Pascal's triangle. However, he was the first to apply them and defined the Pascal triangle [3–5].

Binomial coefficients are the simplest combinatorial objects that determine the number of different combinations of n elements greater than m. The binomial coefficients can be obtained by extending the generation function to the binomial level:

$$(1+x)^n \sum_{m=0}^{n} \binom{n}{m} x^m,$$

where

$$\binom{n}{m} = \frac{n!}{m!(n-m)!}, \quad \ldots n = 0, 1, 2, \ldots, \quad m \le n, \quad n! = 1, 2, 3 \ldots n.$$

The binomial coefficients satisfy the following recurrence relations:

$$\binom{n+1}{m} = \binom{n}{m-1} + \binom{n}{m}, \quad \binom{0}{0} = 1 \tag{1}$$

as well as the simplest equations:

$$\binom{n}{0} = \binom{n}{n} = 1, \quad \binom{n}{m} = \binom{n}{n-1}, \quad \sum_{m=0}^{n} (-1)\binom{n}{m} = 0, \quad \sum_{m=0}^{n} \binom{0}{0} = 2^n. \tag{2}$$

There are different characteristics and relationships between binomial coefficients. Some new features of binomial coefficients are described in [10].

$$
\begin{array}{c}
1 \\
1\ \ 1\ \ 1 \\
1\ \ 2\ \ 3\ \ 2\ \ 1 \\
1\ \ 3\ \ 6\ \ 7\ \ 6\ \ 3\ \ 1 \\
1\ \ 4\ \ 10\ \ 16\ \ 19\ \ 16\ \ 10\ \ 4\ \ 1 \\
1\ \ 5\ \ 15\ \ 31\ \ 45\ \ 51\ \ 45\ \ 31\ \ 15\ \ 5\ \ 1 \\
1\ \ 6\ \ 21\ \ 51\ \ 91\ \ 127\ \ 141\ \ 127\ \ 91\ \ 51\ \ 21\ \ 6\ \ 1 \\
\cdots \cdots \cdots \cdots
\end{array}
$$

$$
\begin{array}{c}
1 \\
1\ \ 1\ \ \ \ 1\ \ 1 \\
1\ \ 2\ \ 3\ \ 4\ \ 3\ \ 2\ \ 1 \\
1\ \ 3\ \ 6\ \ 10\ \ 12\ \ \ \ 12\ \ 10\ \ 6\ \ 3\ \ 1 \\
1\ \ 4\ \ 10\ \ 20\ \ 31\ \ 40\ \ 44\ \ 40\ \ 31\ \ 20\ \ 10\ \ 4\ \ 1 \\
\cdots \cdots \cdots \cdots
\end{array}
$$

Fig. 2. Generalized (arithmetic) Pascal triangles.

Their specificity and binomial coefficients of various ratios play an important role in solving many problems in mathematics and physics. This is the basis for various generalizations of binomial coefficients.

Generalized Pascal Triangles and Generalized Binomial Coefficients. An expression consisting of coefficients distributed over levels x looks as follows:

$$
\left(1 + x + x^2 + \dots + x^{s-1}\right)^n = \sum_{m=0}^{(s-1)n} \binom{n}{m}_s x^m, \quad s \geq 2. \tag{3}
$$

This is called generalized s-order Pascal triangles or triangular tables.

The $\begin{pmatrix} n \\ m \end{pmatrix}_s$ coefficients are called s-orderly generalized binomial coefficients. For s

$= 2$, they become constant binomial coefficients, i.e. $\begin{pmatrix} n \\ m \end{pmatrix}_2 = \begin{pmatrix} n \\ m \end{pmatrix}$ and triangle table

corresponding to Pascal's triangle. In some literatures, generalized Pascal triangles are called s-arithmetic triangles. Generalized s-order Pascal triangles, like Pascal's triangles, can be written as right-angled or equilateral triangles. Arithmetic triangles of the sixth and fourth orders are represented as equilateral triangles (Fig. 2):

In the first triangle ($s = 3$), each element is the sum of the three numbers on the previous line: the number above it and the two numbers to the left of it.

The elements of the second triangle ($s = 4$) are calculated in a similar way, each element of which is equal to the sum of four numbers of the previous line. The lines of the generalized Pascal triangle of any order are filled in the same way.

The generalized binomial coefficient $\begin{pmatrix} n \\ m \end{pmatrix}_s$ m should be the number of different

ways of placing the same elements in n cells, and each cell should have s-1 elements.

Primarily, we write the recursive relation for the generalized binomial coefficients:

$$\begin{pmatrix} n+1 \\ m \end{pmatrix}_s = \sum_{k=0}^{s-1} \begin{pmatrix} n \\ m-k \end{pmatrix}, \quad \begin{pmatrix} n \\ 0 \end{pmatrix}_s = 1. \tag{4}$$

It corresponds to a recursive relation for binomial coefficients (1) at $s = 2$. Generalized binomial coefficients satisfy many equations, identities, and other relations, as well as similar relations for binomial coefficients. For instance,

$$\left. \begin{array}{l} \begin{pmatrix} n \\ 0 \end{pmatrix}_s = \begin{pmatrix} n \\ n \end{pmatrix}_s = 1, \quad \begin{pmatrix} n \\ m \end{pmatrix}_s = \begin{pmatrix} n \\ (s-1)\,n-m \end{pmatrix}_s, \\[4mm] \displaystyle\sum_{m=0}^{(s-1)n} \begin{pmatrix} n \\ m \end{pmatrix}_s = s^n, \quad \sum_{m=0}^{(s-1)n} (-1)^m \begin{pmatrix} n \\ m \end{pmatrix}_s = \begin{cases} 0, & s = 2t \\ 1, & s = 2t+1. \end{cases} \end{array} \right\} \tag{5}$$

The recursive connection between the generalized binomial coefficients in terms of s is as follows:

$$\begin{pmatrix} n \\ m \end{pmatrix}_{s+1} = \sum_{k=0}^{n} \begin{pmatrix} n \\ k \end{pmatrix} \begin{pmatrix} k \\ m-k \end{pmatrix}_s,$$

where, $s \geq 2$, $k < \dfrac{m}{s}$, $\begin{pmatrix} k \\ m-k \end{pmatrix}_s = 0.$

Generalized binomial s-order coefficients can be expressed by binomial coefficients as follows:

$$\begin{pmatrix} n \\ m \end{pmatrix}_s = \sum_{k=0}^{[m/s]} (-1)^k \begin{pmatrix} n \\ k \end{pmatrix} \begin{pmatrix} n+m-sk-1 \\ n-1 \end{pmatrix}. \tag{6}$$

Pascal's Triangles Software. A software tool was developed based on the ideas outlined above. Pascal's Triangles software tool which is based on six modules is developed in C # [8, 11] and has a very user-friendly interface.

The following describes a software tool implemented in the C # programming language and consisting of six main modules. Its block diagram is shown in Fig. 3.

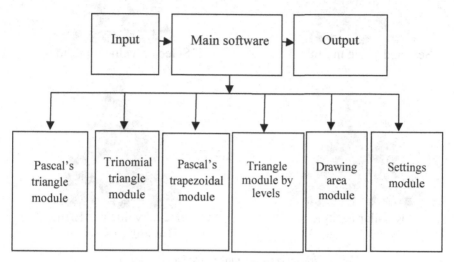

Fig. 3. Structure of the software tool

The software includes 6 basic modules:

- Pascal's triangle module includes 9 functional procedures;
- Trinomial triangle module contains of three functions;
- Pascal's trapezoidal module includes 8 functional procedures;
- Triangle module by levels (bin) includes 7 functional procedures;
- Drawing area module contains of three functions;
- Settings module - includes two functional procedures used to adjust color.

In the Triangle Module field, it is possible to enter values from 2 to 9. For example, Entering a value of 4 in the module field gives the triangle depicted in Fig. 4. Besides, In the "Components" section, it is possible selecting one of the options "Constant Hexagon", "Narrow Hexagon", "Squares". Also, by choosing "Regular Hexagonal Lines" as components, might be specified the boundaries of the elements. The element border area allows to more clearly define the boundaries of the selected regular hexagonal, tapered hexagonal lines and squares of elements. Here the modulus of the triangle is 4 and "Simple hexagons" are selected as components of an arithmetic triangle (Fig. 4).

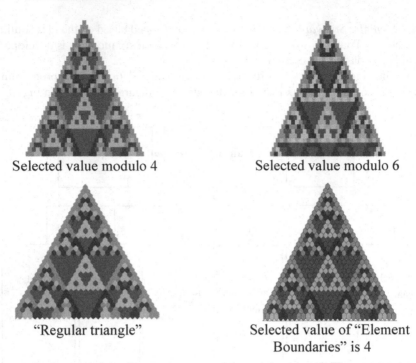

Selected value modulo 4   Selected value modulo 6

"Regular triangle"   Selected value of "Element
                                    Boundaries" is 4

Fig. 4. Fractal constructions of the Pascal's Triangle

3   Conclusion

The main results are as follows:

- Arithmetic triangles and generalized arithmetic triangles of order s have been investigated;
- It is shown that the generalized elements of the arithmetic triangle are generalized binomial coefficients of order s;
- It was found that generalized s-order arithmetic triangles can be written as right-angled or equilateral triangles, such as arithmetic triangles;
- It has been found that binomial coefficients, their identities and various relationships play an important role in solving many mathematical problems;
- The structure of the software "Pascal Triangles" has been developed.

From the above analysis, fractal-based apps prove that they have ushered in a new era by combining mathematical formulas with design. Although this is a purely mathematical software, no math knowledge is required to use it. Interested people can create their own projects using predefined functions, templates and customizing various program settings.

References

1. Mandelbrot, B.B.: Fraktalnaya geometriya prirodi. Per. s angl. – M.: Institut kompyuternix issledovaniy, p. 656 (2002)

2. Nayanzin, N.G., Adilov, A.A.: Obobshenniy m – arifmeticheskiy graf . Avtomatizirovannie sistemi upravleniya, Vip. 90. TashPI, Tashkent, pp. 45–48 (1973)
3. Uspenskiy, V.A.: Treugolnik Paskalya. Izd. 2–e. M.: Nauka, p. 48 (1979)
4. Gould, H.W., Greig, W.E.A.: Lucas triangle primality criterion dual to that of Mann-Shanks. Fibonacci Quart 23(1), 66–69 (1985)
5. Green, T.M., Hamberg, C.L.: Pascal's triangle. Dale Seymour, Palo Alto, p. 280 (1986)
6. Bondarenko, B.A.: Obobshennie treugolniki Paskalya, ix fraktali, grafi i prilojeniya. -Tashkent: Fan, p. 192 (1990)
7. Feder, E.: Fraktali. Per. s angl. M.: Mir, p. 254 (1991)
8. Abramyan, M.E.:Visual C# na primerax. SPb.: BXV-Peterburg, p. 496 (2008)
9. Edgar, G. (ed.): Measure, Topology, and Fractal Geometry. Springer, New York (2008). https://doi.org/10.1007/978-0-387-74749-1
10. Bondarenko, B.A.: Generalized Pascal Triangles and Pyramids, their Fractals, Graphs, and Applications, 3rd ed. Fibonacci Associations, Santa Clara, p. 296 (2010)
11. Flenov, M.E.: Bibliya C#. – 2-e izd. Pererab. i dop. SPb.: BXV- Peterburg, p. 560 (2011)
12. Maksimenko-Sheyko, K.V., Tolok, A.V., Sheyko, T.I.: R-funksii v fraktalnoy geometrii // Informasionnie texnologii. M.: Izdatelstvo "Novie texnologii", №. 7, pp. 24–27 (2011)
13. Kenneth, F.: Fractal Geometry: Mathematical Foundations and Applications, p. 400. Wiley, Hoboken (2014)
14. Qi, X.: Fixed Points Fractals, Iterated function systems and Generalized Support Vector Machines, p. 26 (2016)
15. Anarova, Sh.A., Nuraliev, F.M., Narzulloev, O.M.: Construction of the equation of fractals structure based on the rvachev r-functions theories. In Journal of Physics: Conference Series, vol. 1260. Institute of Physics Publishing. https://doi.org/10.1088/1742-6596/1260/7/072001
16. Murodillayevich, N.F., Amanbayevna, Sh. A., Mirzayevich, N.O.: Mathematical and software of fractal structures from combinatorial numbers. In: International Conference on Information Science and Communications Technologies: Applications, Trends and Opportunities, ICISCT 2019. Institute of Electrical and Electronics Engineers Inc. (2019). https://doi.org/10.1109/ICISCT47635.2019.9012051
17. Anarova, Sh.A., Narzulloev, O.M., Ibragimova, Z.E.: Development of fractal equations of national design patterns based on the method of R-Function. Int. J. Innov. Technol. Explor. Eng. 9(4), 134–141 (2020). https://doi.org/10.35940/ijitee.D1169.029420
18. Singh, D., Singh, M., Hakimjon, Z.: Geophysical application for splines. In: Singh, D., Singh, M., Hakimjon, Z. (eds.) Signal Processing Applications Using Multidimensional Polynomial Splines, pp. 55–63. Springer, Heidelberg (2019). https://doi.org/10.1007/978-981-13-2239-6_7
19. Zaynidinov, H., Zaynutdinova, M., Nazirova, E.: Digital processing of two-dimensional signals in the basis of Haar wavelets. In: ACM International Conference Proceeding Series, pp. 130–133. Association for Computing Machinery (2018). https://doi.org/10.1145/3274005.3274023
20. Singh, D., Singh, M., Hakimjon, Z.: One-dimensional polynomial splines for cubic splines. In: Singh, D., Singh, M., Hakimjon, Z. (eds.) Springer Briefs in Applied Sciences and Technology, pp. 21–26. Springer, Heidelberg (2019). https://doi.org/10.1007/978-981-13-2239-6_3

Crowd Sourcing and Information Analysis

An NLP and LSTM Based Stock Prediction and Recommender System for KOSDAQ and KOSPI

Indranath Chatterjee[1] , Jeon Gwan[2], Yong Jin Kim[2], Min Seok Lee[2], and Migyung Cho[2(✉)]

[1] Department of Computer Engineering, Tongmyong University, Busan 48520, South Korea
[2] Department of Game Engineering, Tongmyong University, Busan 48520, South Korea
mgcho@tu.ac.kr

Abstract. Stock market prediction is one of the complex analysis of all time. Different expert analysts, as well as computer scientists, are working for the development of a stable and robust platform for the prediction of future stock value. The primary challenge is the nature of the movement of the daily price, which depends on various factors. To build a predictive model for the analysis of stock data and prediction is an active area of research. However, we found only a few numbers of studies performed on Korean stock market analysis, including both KOSDAQ and KOSPI companies. This study proposed a three-stage approach based on Natural Language Processing and Deep Learning techniques to analyze, comprehends the past and present market scenario, and also predict the future value of a stock. This study involves the application of natural processing techniques and deep learning techniques on around 2500 Korean companies covering KOSDAQ and KOSPI. Firstly, this paper successfully presents the current condition of the stock and overall Korean stock exchange; secondly, it recommends the potential months and weeks for investment, and finally, it predicts the future value of a stock with high accuracy. This paper may pose as a structural framework for developing a complete stock market prediction application.

Keywords: Stock market prediction · Recurrent neural network · LSTM · Natural language processing · KOSDAQ · KOSPI

1 Introduction

A stock market, also known as the share market, is a marketplace combined with share traders in terms of buyers and sellers. Each stock represents a kind of ownership-claim on a company when someone buys a share of its business. Apart from liquid trading, people take advantage of it in the form of investment. The stock market trading and investment are done using either stockbrokers or online trading platforms.

© Springer Nature Switzerland AG 2021
M. Singh et al. (Eds.): IHCI 2020, LNCS 12615, pp. 403–413, 2021.
https://doi.org/10.1007/978-3-030-68449-5_40

Investment requires a mindful strategy and depends on thoughtful prediction of the stock's future price. Stock market prediction determines the future price of a stock based on historical data. It is evident that a close forecast may yield a noteworthy profit to an investor. Although efforts have been made by the stock trading experts to predict, still, it is to be noted that there involves a considerable amount of human errors. For a long time now, computer science and statistics are playing a special role in the analysis of stock markets. Recently, artificial intelligence, specifically machine learning and deep learning are contributing notably.

Literature [3,5,13,15] suggests that an era is going, making efforts to solve the unanswered problems of the stock market analysis and to achieve a robust prediction system. Studies [4,15] involving machine learning applications are quite popular in stock market prediction. Artificial neural networks are also employed to build a recommender system in stock market prediction. Recurrent neural network (R-NN) and convolution neural network (CNN) [2,11] are recently being used besides the classical regression and time-series analysis. We also found some studies [1,7] using the Long-Short Term Memory (LSTM), a type of RNN, to predict the future price of a stock. Alongside the stock market prediction, there are several studies [10] on Natural Language Processing (NLP) that concentrated on finding the clue or traces of the reasons behind market downpour or hike based on tweets and news. Extracting the stock market-related news from different news portal, and then analysis using NLP is a good source of historical information that helps in predicting the future scenario.

The stock market prediction and recommender system based on stock news have not been explored well together. We found very few studies on Korean stock market prediction involving both Korean Securities Dealers Automated Quotations (KOSDAQ) [12] and Korea Composite Stock Price Index (KOSPI) [6]. Thus, we are curious to explore the Korean stock market for developing robust predictions of future prices along with understanding the complexity of the market trends. To the best of our knowledge, this study is the first of its kind to propose a machine learning-based prediction model and NLP based recommender system for both KOSDAQ and KOSPI data. This composite study explores the historical data of the past ten years of approximately 2500 Korean companies. The aims of the paper is to predict the future value of a stock (user's choice) for next one year, alongwith recommending the user whether to buy, sell or hold the stock. This paper also aims to suggest the user by recommending the probable month and time of the month to buy (possibility of low price) or sell (possibility of high price) the particular stock. Firstly, it utilizes the statistical analysis on the stock data to understand the present and past trends; secondly, it uses the NLP to extract the present and past news to understand the hike and fall of stock values at a different time of the year. Finally, it proposes a model using LSTM to predict the future amount. Alongside this, our study also compares the efficacy of our results with other state-of-the-art approaches.

2 Stock Market Prediction and Recommender System

Any stock market prediction process involves a lot of steps or processes. As discussed, in this paper, we are proposing a novel three-stage approach, consisting of our proposed algorithm-based system for the recommendation of potential time of the year to buy/sell the stock, NLP techniques based suggestions on the current condition of the Korean market, and the deep learning-based future price prediction system. The recommender system is based on our proposed approach using a fast and straightforward technique that involves exploring the historical data and analyzing the trend of ups and downs in the stock price in the past ten years. This recommender system will recommend or suggest the user the probable period to buy/sell to achieve a more significant profit while investing.

Stock market prediction, on the other hand, talks about future price prediction. It incorporates the usage of deep learning algorithms to build a model based on historical data for predicting the value of an arbitrary date in the future. We have used the sequential LSTM algorithm to train the model for each of the 2500 Korean companies. The whole process of these three stages was performed for each of the companies (as desired by the user). The detailed experimental setup is described in the Methodology section of this paper. The Fig. 1 shows the complete workflow of our proposed approach. The figure shows a three-stage model, along with user's input and final output.

3 Materials and Methods

3.1 Materials

Dataset: In this study, we have used both KOSPI (Korea Composite Stock Price Index) and KOSDAQ (Korean Securities Dealers Automated Quotations). We have fetched a dataset of around 2500 companies for the past ten years to date. Besides this historical data, we have coded in such a way that we can fetch the most recent data till the day of execution of the script. All the datasets are downloaded from the openly available repository of Yahoo Finance[1]. The dataset of each company contains the information regarding the open, close, high, low price of each stock and total volume for the past ten years.

Tools Used: We have implemented the experiment in an Intel i-7 PC. Natural language processing, machine learning application and visualizations are performed using Python.

4 Methodology

The complete stock market analysis involves all the investigations of past, present and future. Although, in this domain, we only deal with historical time-series

---

[1] https://finance.yahoo.com/.

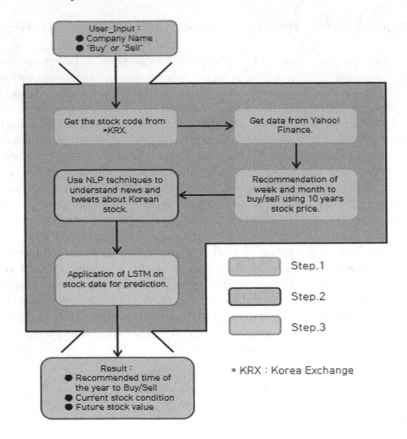

Fig. 1. The workflow of our proposed approach.

data; however, this study has been designed to analyze the past, present, and future based on the past ten years stock data, all news articles, and tweets. Thus, we have divided the experiment into three stages, as follows.

First stage deals entirely with the historical data of the last ten years. All the listed companies (both KOSDAQ and KOSPI) available on the Korea Stock Exchange website was crawled, and then their stock codes were obtained from the list. With the obtained stock code, we have downloaded the data from Yahoo Finance for the past ten years till the last day. This historical data was used for building the prediction model. We have also programmed it in such a way that the recent data can be fetched in real-time while a user searches for it. We have explored the ten years data to find the highest and lowest closing value for each month of a year, and subsequently for each year in the last ten years. We measured the frequency of high prices and low prices in a month for all the years to find a correlation between the month of the year and the stock price. We also performed the same analysis to find the correlation between the days of the month and the stock price. After obtaining the frequency list of the weeks, months and year having high and low value for a stock, we have ranked them accordingly and chose the most occurring months and the week of a month in all the years. This analysis helps us to understand the daily, monthly and yearly trends in the hike or fall of the closing price of a stock. We have selected a date-window of 20 days (+10 days and −10 days) from the day of the lowest and highest closing price to find the reason for the hike or fall in past days. We fetched the news available in Google News during the period (data-window of 20 days), apply NLP to tokenize, and fetched the keywords to find out any clue for the sudden hike or fall. This step will help us in adding knowledge to our understanding of the future price.

The second stage of the proposed approach deals with the data in the present time. In this stage, we have fetched the recent stock data (including the same day). Alongside, Google news, Twitter and web crawling were used to comprehend the present condition of the Korean stock market and the latest status of the desired company. In order to know the situation of the Korean stock market, the contents of the corresponding data were obtained by crawling the links of the news in the stock tab of the Yonhap News homepage[2]. The contents were fetched and converted into the standardized form using Natural Language Toolkit (NLTK). With the application of the NLP package for sentiment analysis, different scores were assigned to each news by scoring whether the article was positive or negative. Summing up the respective scores yielded the sentiment of the overall situation of the Korean stock market. Similarly, we have used Google News to understand the current status of the desired company. Besides the Google News, we have also fetched the Tweets on that company using Python Twitter API and performed the sentiment analysis in the same way to understand the company's situation.

Finally, in the third stage of the study, we deal with future data. In this stage, we have applied advanced machine learning and deep learning algorithms to predict the future price of the stock. In this study, we have used standard Linear Regression [9], the Facebook Prophet model [14], and the widely used LSTM [8] algorithm. In the case of linear regression and FB Prophet model, we have used the default parameters. As we know, LSTM is a type of R-NN that has a memory component and also the hidden layers are needed to preset. LSTM is a compelling model for time-series analysis. The usage of LSTM in stock market prediction proved a beneficial approach to predict an arbitrary number of steps in future dates. It has five essential components allowing the algorithm capable of modeling both long term and short term datasets. The components are cell state (c_t), hidden state (h_t), input gate (i_t), forget gate (f_t) and output gate (o_t). The Fig. 2 given below is a standard representation of a cell in an LSTM model, where we can observe the pathways of the data from input to output, along with controlling the cell state and hidden and forget gate. Here, we have used a sequential model using Keras as it suits appropriate for a straight stack of layers. We have considered the number of epochs as 10 and the number of units as 50. During the compilation of the model, we have used 'Adam' as an

---

[2] https://en.yna.co.kr/.

optimization parameter and 'mean squared error' as the loss function. Along with the validation and test set, we have also tested the data on arbitrary future dates beyond the final date in the test data.

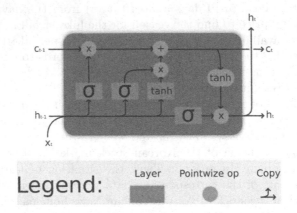

Fig. 2. A standard Long Short-Term Memory (LSTM) Cell (Source: Image by Guillaume Chevalier under CCA 4.0)

We have cross-validated the whole dataset in such a way that the first 60% of the data kept for training, the next 20% kept for validation and the remaining 20% for the test set. We have used the downloaded ten years to build the model. We performed the training and validation operations on the respective dataset using linear regression, FB Prophet, and LSTM, separately. The data value being a continuous variable, we preferred to measure Root Mean Square Error (RMSE) and R^2 value as the metrics for the efficiency of the proposed model.

5   Results

As discussed above, the proposed study is divided into three stages, where each step defines an analysis to develop a recommender system. The first stage of the study results in an investigation of the past data to find out the trend in stock price. It shows the mathematical calculations help to identify the frequently occurring months and weeks of a month for the past ten years, showing the high and low closing value of the stock. This result successfully recommending the user about the probable month(s) of a year and the potential week(s) of a month to sell/buy the desired stock to obtain maximum profit.

The second stage of the proposed algorithm not only shows the current price of the desired stock but also it can successfully report the present condition of the Korean stock market. Alongside this, the results also suggest the present reputation of the selected company. Figure 3 presents a snapshot of a result obtained after Stage-I and Stage-II of our proposed approach. Here, the results

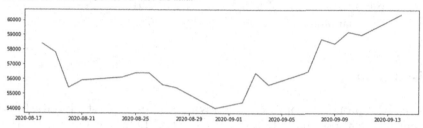

```
Recommendation :
It is recommended to Buy this stock only during the month of 1 or 12 . It is suggested to Buy this Stock during 3week to
get more profit.

Today's price:
삼성전자 (Samsung Electronics)
60400.0 1400.0 2.37% ↑

Suggestions on stock condition :
Samsung 's situation is good.
The situation in the Korean stock market is good.

The trend for the closing price for last one month:
```

Fig. 3. A snapshot of a result obtained after Stage-I and Stage-II of our proposed approach. Here, the results shows the recommendation and suggestions on purchase request of Samsung Electronics stock.

shows the recommendation and suggestions on purchase request of Samsung Electronics stock.

The third stage of the study predicts the future price of the stock. Although the linear regression and FB Prophet models failed to give impressive results, the LSTM model shows promising results with high R^2 value. Although we have implemented the experiment over all the companies data, here in this paper, we are presenting the results of three Korean companies, selected randomly. We have randomly chosen from three different sectors, namely, Samsung Electronics from the Electronics sector, Hyundai Motors from the Automobile manufacturing sector, and Naver from the IT sector. The Fig. 4 shows the original data and the predicted results by LSTM for each of the three companies. The blue line graph denotes the training and testing, whereas the green line graph denotes the validation set. The orange line graph is the predicted results as obtained using LSTM model. It is clear from the plot that the LSTM is giving promising results with the low error between the desired and predicted values. It also the trend in the stock's future closing price. We found promising R^2 values for all the three companies. Table 1 shows the results of the three Korean companies (Samsung Electronics, Naver and Hyundai Motors). The table shows the R^2 value for validation and testing set. It is evident from the numbers in the table, the LSTM shows the most promising results.

Table 1. Results showing the R^2 value for validation and test set while performing the LSTM algorithm on Samsung Electronics, Naver, and Hyundai Motors.

	LSTM	
	R^2 (Validation set)	R^2 (Test set)
Samsung Electronics	0.91	0.96
Naver	0.95	0.91
Hyundai Motor	0.95	0.95

6   Discussions

This three-stage proposed approach is studying the 10-years stock data from 2500 companies covering both KOSDAQ and KOSPI. In the first stage, the proposed algorithm successfully recommends the user with an understanding of the trend of the stock price across the years. Alongside this, with the desired input from the user whether he/she wants to buy or sell a particular stock, our algorithm suggests the potential months and weeks of a month, when the stock may be purchased or sold to acquire maximum profit.

As we can observe from the results of the second stage, the proposed approach can effectively report the present condition of the Korean stock market. With the help of natural language processing and text analytics, we have explored all the news and tweets regarding the same for understanding the sentiment of the public and its present business condition. We have also recommended the same for the individual company as desired by the users. Although we have used a standard sentiment analysis package, we have collected the news, articles, and tweets about the company and the overall KOSDAQ and KOSPI to date. With these texts, we have scored them and subsequently rank them accordingly. By calculating the cumulative sentiment score, we obtained the overall sentiment of the texts.

As hypothesized, the LSTM found promising in the application of stock market prediction. Not only predicting the test data set with high accuracy but also our model can predict the price for arbitrary future dates. Although stock price prediction is a bit complicated task. However, our model is able to predict the Korean stock price and its movement quite efficiently. Using LSTMs to make the stock price predictions many steps ahead of the current date seems to be useful, though it requires further investigation and tuning of the algorithm as the model's hyperparameters are extremely sensitive to the movement of the stock data. We are currently in the process of optimizing the hyperparameters for a robust model. As a part of our future work, we are developing a complete application for stock market prediction with some added features and functionalities.

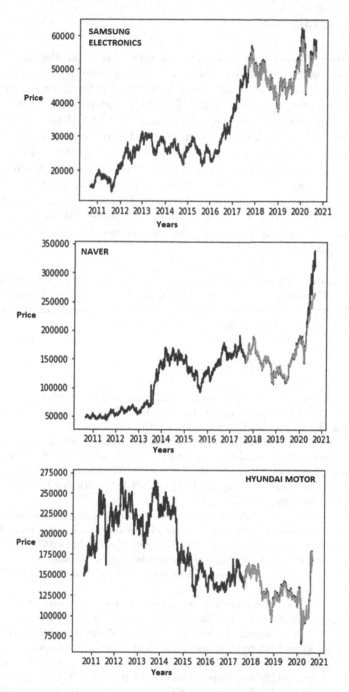

Fig. 4. Results showing the predicted output by LSTM for (a) Samsung Electronics, (b) Naver and (c) Hyundai Motors. The blue line shows the training and testing set, the green line denotes the validation set. The orange line graph is the predicted value obtained on validation and test set using LSTM algorithm. (Color figure online)

7   Conclusions

Although stock market prediction is a complicated task, the involvement of machine learning is inevitable. The key challenge being the nature of ever-changing movement of the daily price; here we propose a three-stage recommendation and prediction model for KOSDAQ and KOSPI stock data. This study is based on Natural Language Processing and LSTM to analyze, comprehends the past and present market scenario, and also predict the future value of a stock. This paper successfully presents the current condition of the stock and overall Korean stock exchange. Alongside, it recommends the potential months and weeks for investment to achieve more profit. Overall, it predicts the future value of a stock with high accuracy. We found LSTM gives the best prediction while comparing with other state-of-the-art techniques. We believe, this paper may pose as a structural framework for developing a complete stock market prediction application.

Acknowledgment. This research was supported by the MISP (Ministry of Science, ICT & Future Planning), Korea, under the National Program for Excellence in SW) supervised by the IITP (Institute for Information & communications Technology Promotion) having number 1711102971.

References

1. Chen, K., Zhou, Y., Dai, F.: A LSTM-based method for stock returns prediction: a case study of china stock market. In: 2015 IEEE International Conference on Big Data (Big Data), pp. 2823–2824. IEEE (2015)
2. Hoseinzade, E., Haratizadeh, S.: CNNpred: CNN-based stock market prediction using a diverse set of variables. Expert Syst. Appl. **129**, 273–285 (2019)
3. Kimoto, T., Asakawa, K., Yoda, M., Takeoka, M.: Stock market prediction system with modular neural networks. In: 1990 IJCNN International Joint Conference on Neural Networks, pp. 1–6. IEEE (1990)
4. Kumar, S., Acharya, S.: Application of machine learning algorithms in stock market prediction: a comparative analysis. In: Handbook of Research on Smart Technology Models for Business and Industry, pp. 153–180. IGI Global (2020)
5. Mizuno, H., Kosaka, M., Yajima, H., Komoda, N.: Application of neural network to technical analysis of stock market prediction. Stud. Inf. Control **7**(3), 111–120 (1998)
6. Na, S.H., Sohn, S.Y.: Forecasting changes in Korea composite stock price index (Kospi) using association rules. Expert Syst. Appl. **38**(7), 9046–9049 (2011)
7. Xu, Y., Chhim, L., Zheng, B., Nojima, Y.: Stacked deep learning structure with bidirectional long-short term memory for stock market prediction. In: Zhang, H., Zhang, Z., Wu, Z., Hao, T. (eds.) NCAA 2020. CCIS, vol. 1265, pp. 447–460. Springer, Singapore (2020). https://doi.org/10.1007/978-981-15-7670-6_37
8. Qu, Y., Zhao, X.: Application of LSTM neural network in forecasting foreign exchange price. In: Journal of Physics: Conference Series, vol. 1237, p. 042036. IOP Publishing (2019)

9. Roy, S.S., Mittal, D., Basu, A., Abraham, A.: Stock market forecasting using LASSO linear regression model. In: Abraham, A., Krömer, P., Snasel, V. (eds.) Afro-European Conference for Industrial Advancement. AISC, vol. 334, pp. 371–381. Springer, Cham (2015). https://doi.org/10.1007/978-3-319-13572-4_31
10. Schumaker, R.P., Chen, H.: Textual analysis of stock market prediction using breaking financial news: the Azfin text system. ACM Trans. Inf. Syst. (TOIS) **27**(2), 1–19 (2009)
11. Selvin, S., Vinayakumar, R., Gopalakrishnan, E., Menon, V.K., Soman, K.: Stock price prediction using LSTM, RNN and CNN-sliding window model. In: 2017 International Conference on Advances in Computing, Communications and Informatics (ICACCI), pp. 1643–1647. IEEE (2017)
12. Song, H.J., Lee, S.J.: A study on the optimal trading frequency pattern and forecasting timing in real time stock trading using deep learning: focused on KOSDAQ. J. Inf. Syst. **27**(3), 123–140 (2018)
13. Tabar, S., Sharma, S., Volkman, D.: A new method for predicting stock market crashes using classification and artificial neural networks. Int. J. Bus. Data Anal. **1**(3), 203–217 (2020)
14. Usher, J., Dondio, P.: BREXIT election: forecasting a conservative party victory through the pound using ARIMA and Facebook's prophet. In: Proceedings of the 10th International Conference on Web Intelligence, Mining and Semantics, pp. 123–128 (2020)
15. Zhang, Y., Wu, L.: Stock market prediction of S&P 500 via combination of improved BCO approach and BP neural network. Expert Syst. Appl. **36**(5), 8849–8854 (2009)

The Commodity Ecology Mobile (CEM) Platform Illustrates Ten Design Points for Achieving a Deep Deliberation in Sustainable Development Goal #12

Mark D. Whitaker[✉]

Department of Technology and Society, Stony Brook University, State University of New York, Korea (SUNY Korea), Incheon, South Korea
mark.whitaker@sunykorea.ac.kr

Abstract. First, this study asks what Information Communication Technology for Sustainable Development (ICT4SD) has to look like in a global decision space? In answering this question, ten important general design issues are suggested as a checklist when making a human computer interface (HCI) to reach cheaply and durably all people, all regions, and all material uses to aid a clean circular economy. Second, a case study recounts how these ten recommended design issues are being integrated in the Commodity Ecology Mobile (CEM) platform, still in development. The CEM is a merged mobile-accessible platform and online database project recognized by the United Nations Academic Impact Office (UNAI) as a good way to really achieve Sustainable Development Goal #12 (Encourage Sustainable Production and Consumption). The CEM design allows all peoples worldwide to debate and to create their own detailed regional sustainable development of a 'circular economy' based on better material choices and better waste flows. This debate occurs within a virtual community platform that integrates simultaneous elements of social media, newswires, archives, crowdsourcing, databases, gamification, voting, and future smart contracts toward business incubation. Therefore, it is important to make design decisions that address the real world of varied technical and social access issues across the world: from stable or intermittent bandwidth and electricity, to respecting local knowledge in different regional agenda settings, to language differences, to age and gender inclusion, to education levels, and how to keep people interested. In conclusion, this paper recounts how to build HCI for deep civil deliberation in global and multi-regional debates, for long periods of time toward sustainable development, while minimizing cost to users or administrators across developed and developing countries alike. Further advice is asked from academics or practitioners for how to design better global-level platforms that can "change the world."

Keywords: Sdgs · SDG#12 · ICT4D · ICT4SD · Smart regions · Sustainable development · Circular economy

"If we are to create a sustainable world—one in which we are accountable to the needs of all future generations and all living creatures—we must recognize that our present

M. Singh et al. (Eds.): IHCI 2020, LNCS 12615, pp. 414–430, 2021.
https://doi.org/10.1007/978-3-030-68449-5_41

forms of agriculture, architecture, engineering, and technology are deeply flawed. To create a sustainable world, we must transform these practices. We must infuse the design of products, buildings, and landscapes with a rich and detailed understanding of ecology. Sustainability needs to be firmly grounded in the nitty-gritty details of design. Policies and pronouncements have their place, but ultimately we must address specific design problems: How can we design our products and manufacturing processes so that materials are completely reclaimed? How can we create wastewater treatment systems that enhance, rather than damage, their surrounding ecosystems? How can we design buildings that produce their own energy and recycle their own wastes? How can we create agricultural systems that are not dependent on subsidies of pesticides, fertilizers, and fossil fuels? Design problems like these bridge conventional scientific and design disciplines. They can be solved only if industrial designers talk to biogeochemists, sanitation engineers to wetland biologists, architects to physicists, and farmers to ecologists. In order to successfully integrate ecology and design, we must mirror nature's deep interconnections in our own epistemology of design. We are still trapped in worn-out mechanical metaphors. It is time to stop designing in the image of the machine and start designing in a way that honors the complexity and diversity of life itself."

-- Sim Van der Ryn and Stuart Cowan,
in *Ecological Design, Tenth Anniversary Edition* (1996)

1   Framing a Global HCI Decision Space and a Checklist of Design Points for Such Projects; Introducing the Decision Space of the Commodity Ecology Mobile (CEM) Platform

This paper describes the decision space for globally accessible Human Computer Interface (HCI) platforms and develops a checklist of ten technology decisions in general toward making more globally accessible HCI platforms for human and sustainable development. This global decision space and checklist is illustrated by a case study of ongoing decisions about software/technology design in a project called the Commodity Ecology Mobile (CEM) platform, a project receiving grants from the Korean National Research Foundation to aid its development. The information and arguments in this paper are an appeal to scholars, practitioners, and entrepreneurs who want to learn or to share similar insights on how to adapt HCI to a global decision space, which includes mostly developing countries and vast inequalities of the quality of access. Data reflected upon in creating this checklist comes from observational research on other successful global platforms that still have limitations particularly on equitable access or resiliency, from long academic teaching in ICT4SD, and from curricular design on our current 'mobile revolution in development' which is how mobile phone networks are changing the world's development via creating 'smart regions' worldwide. Equally, data reflected upon in this checklist comes from this real and rare project designed for this global decision space, the Commodity Ecology Mobile (CEM) platform, which illustrates this project's solutions to the ten ongoing technological problems for a global decision space. Therefore, this is instructive for learning in general how to create better routes for ICT4SD in challenging circumstances. Examples of such circumstances are unstable and variable electricity and bandwidth, weather ruggedness, educational difficulties, gender inclusion, refugee situations, resiliency of access, and business models for such situations. In short, it is argued that at least ten decisions on design principles make a good checklist for global-level human/sustainable development via HCI in general.

Despite globally equalizing digital developmental contexts of the early 21st century [1], there are still many varied and difficult developmental contexts to address to bring integrative sustainable development to all regions of the world with ICT4SD [2, 3]. Therefore, for over twenty years, work has been made on designing and perfecting an easy-to-use, affordable platform that allows people to use their mobile phones to manage resources vital to their livelihood and the environment. It is called the Commodity Ecology Mobile (CEM) Platform. It has been honored three times recently: first, in 2018 with recognition by the United Nations Academic Impact Office (UNAI) as a truly useful way to attempt to achieve Sustainable Development Goal #12 [4] (which is how to develop sustainable production and consumption); second, in 2020 CEM received a multi-year grant from the Korean National Research Foundation to develop the platform; and third, in 2020 CEM received a testimonial from the Chief of Office of the U.N. Academic Impact himself [5].

Three introductory frames help to understand the very wide 'decision space' of global-level HCI projects in general. As an example of this, the CEM platform seems unique when compared to much smaller decision spaces of other SDG projects and sustainability projects. First, the following chart helps to classify and to understand the "decision-making space" of the global-level developmental HCI projects like the CEM platform:

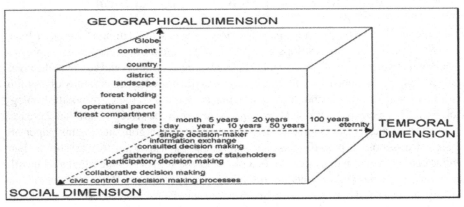

Fig. 1. Introductory Frame: "Three spaces of natural resources decision-making" helps to explain the decision-making space of the Commodity Ecology Mobile (CEM) platform (from Hujala et al., 2009) [6]

You can conceive of all decisions about natural resources, sustainable development, and/or human computer interfaces as *always* involving three dimensions: social, geographical, and temporal. On the one hand, respectively, decisions and platforms are unrequired for the center of the graph in Fig. 1, where a single decision-maker can plant a single tree in a particular place in one day. On the other hand, moving to larger scales, you can see growing requirements for design decisions about platforms and procedures if, for instance, consulted decision-making is used on a forest holding over a five year period. Next, how much more detailed would be the decision making space for global level HCI projects like the Commodity Ecology Mobile platform which is on the outside of all three dimensions: how do you develop platforms for civil control of decision making processes, across the globe, for eternity—or for as long as we live in a mobile phone saturated civilization, whichever comes first?

Second, the frame of this project bridges a disciplinary and cultural divide rarely crossed between the group of pro-technology computer scientists that know a lot about technological design and business incubation yet rarely care about environmental policy, and the other group of environmental policy activists that tend to be anti-technology in policy solutions. Commodity Ecology is a multi-stakeholder idea premised on the idea that these two cultures should come together to work for sustainability via using Information Communication Technologies (ICT) similar to what was discussed by the World Wildlife Fund as long ago as 2002 [7]. Equally, there is commonality with the ongoing use of ICT for Sustainable Development (ICT4SD) [8, 9] (Heeks, 2018; Marolla, 2019), the practice of using ICT for development and services in less developed countries.

Third, another theme is a movement increasingly utilizing HCI potential of deliberative and sustainable 'smart regions' instead of only in smaller green 'smart cities' since mobile ICT networks increasingly saturate our social lives in all world regions simultaneously [10, 11]. With over 5.1 billion mobile phones from 2019 that are more evenly saturated across developed and less developed countries alike, there is a greater "digital parity" from mobile phone ownership than other communication technologies that still have greater "digital divides." Therefore, the whole world is converting itself into 'smart regions' of mutual deliberative real/virtual lives on shared platforms built from aggregate mobile phone network value [12]. This makes HCI choices on mobile-accessible platforms a key requirement for accessing a less developed world's population to enfranchise them in their own development and to link them with the developed world as well. However, there are still other kinds of "digital divides" despite a high "digital parity" in mobile phone ownership, like uneven costs of and access to bandwidth [2] and low electrical access [3]. Both have to be overcome by creative platform decisions to have a global HCI decision space at this time.

To operationalize the social decision space, the meaning of the term "Commodity Ecology" references the most complete example of a circular economy, with 130 categories of world's sustainable materials (commodities) depicted in the Commodity Ecology Wheel diagram in Fig. 2 below. This detailed circular economy and all the social decisions it implies is only one of three frameworks in this global-level HCI platform for making better civic collaborative choices in an 'eternal' global decision space.

How these categories turn into a digital form on the CEM platform is shown in Fig. 3, below. This 'wheel' is converted into a clickable map in which each of the 130 categories lead to group-based posted 'newswires' about good sustainable material ideas and good uses of wastes from each category. Wastes from one category can be good links to other categories if the idea is reducing overall regional pollution to the commons of the air, water, or earth. Increasing economic profit from that waste simultaneously encourages

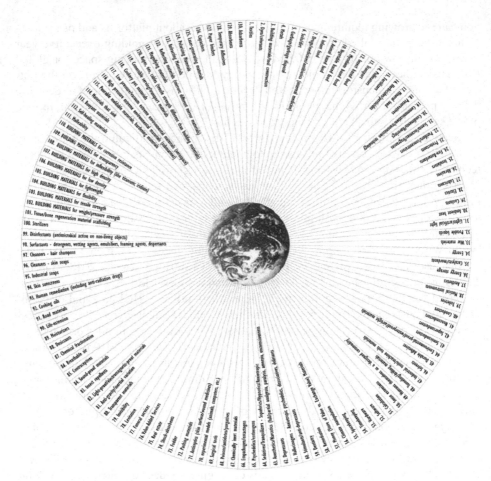

Fig. 2. Commodity Ecology: 130 historically-invented social uses for commodities (Source: com modityecology.blogspot.com)

better sustainable development. In this way, we can reach all material uses using same framework of "Commodity Ecology", in all regions.

However, the second framework is the geographic decision space. This is a global and multi-regional ecoregional map in the CEM platform. Ecoregions are world regions characterized by unique ecological patterns between soil type, plants, animals and micro-climatic conditions. The platform's digitized ecoregional map (in Fig. 4) has 867 stable ecoregions, and it is equally a clickable map like the digitized Commodity Wheel (Fig. 3). When clicking on a unique ecoregion zone in this map, users go to different unique versions of Fig. 3, where they can post and comment. In this way, users of the platform make unique interlacings about what they desire for materials and links in a circular economy in each local ecoregion.

The third framework in this decision space is simultaneously a temporal (eternal and open-ended), geographic (global and multi-regional), and social (civic deliberation)

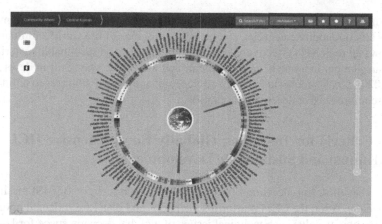

Fig. 3. Commodity Ecology, Digitized: 130 Historically-Invented Uses for Commodities; CEM Platform (Prototype: https://commoditywheel.softlabsgroup.in/.)

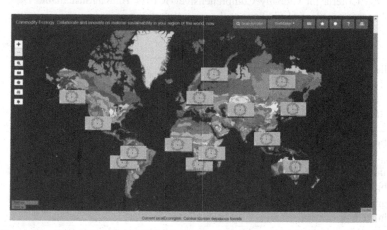

Fig. 4. Ecoregional Map, Digitized: 867 Ecoregions, Each with a Regional Commodity Wheel for Posting about Sustainable Materials/Wastes; CEM Platform (Prototype: https://commodity wheel.softlabsgroup.in/.)

decision space. It is a global participation done by the CEM platform interface being mobile-accessible given over 5.1 billion unique mobile phones rather evenly distributed worldwide from 2019. Half are smartphones already, and over a billion mobile phones are in Sub-Saharan Africa already [1]. Mobile networks allow long-term deliberation globally and are good to connect to HCI platforms. This synergy makes an ideal social, geographic, and temporal network value upon which a common global sustainable value can be erected for the world [12].

In short, the Commodity Ecology Mobile platform is an innovative mobile-accessible platform to effectively manage the world's commodities and reach the United Nations' Sustainable Development Goals. It is a detailed circular economy model that involves deliberating on better sustainable material choices and better profitable waste management per all commodity categories and per all ecoregions of the world simultaneously. This allows all regions to effectively exchange information on sustainable materials and waste management in a quick, easy, efficient, and affordable way. Still in development (2018–2020), what is requested is more advice, aid, and further alliances to build a good globally-comprehensive development platform.

2   Ten Criteria for Designing Globally Comprehensive HCI for Human and Sustainable Development

This checklist of design decisions for global HCI interfaces for ICT4SD is listed in Table 1, below. After, the CEM platform's decisions and plans are reflected upon in ten sections for how to solve such technical issues of a global decision space for HCI.

Table 1. Ten Criteria for Globally Comprehensive HCI for Human/Sustainable Development (*Asterisks represent planned additions to the current prototype.)

Criteria
1. Mobile and Ubiquitous Interfaces
2. Social media use and analysis
3. Language and speech interfaces*
4. Affective and aesthetic interfaces
5. Haptic rendering, haptic input and perception
6. Intelligent visualization tools
7. Pedagogical technologies
8. Interface Design for Accessibility--for Economic Accessibility (Poverty) or Social Disabilities*
9. Decentralized Platforms for more Distributed HCI Experiences*
10. AI in HCI*

This checklist of ten decision points or ten problems in creating global-level HCI interfaces was developed upon four reflections. First, there was a comparative analysis of successful billion-user global HCI platforms (Facebook, Instagram, Amazon, WhatsApp, Google Search, etc.). However, second, these huge platform successes were looked at with a jaundiced eye for concerns about their true human/sustainable developmental potentials or exploitations in their technical platform choices or environmental effects—concerning digital divides, economic centralizations and surveillance capitalism, for instance—and how to learn critically from the largest platforms about solving problems for global-level ICT4SD. Third, ongoing reflection occurred in the process of design decisions of the CEM platform itself in such a wide decision space. Fourth, this checklist is equally the fruit of years of analysis of many good examples of ICT4SD in challenging circumstances from personal teaching and research on the current 'mobile revolution in development' in a course the author designed and had approved into an official university curriculum. So far, this checklist has at least ten criteria required for developing HCI platforms in a global decision-making space that are truly human developmental and truly about sustainable development instead of only about platforms that are corporate or state developmental. Many of the ten criteria are integrated in technical decisions of the CEM platform's design already. Other points await later iterations in planning and funding. The novelty or contribution here is that since there are so few HCI platforms designed for a global decision space in human and sustainable development in ICT4SD, it is worthwhile to show how a humanitarian technology and a truly sustainable developmental HCI can bring global levels of civic design participation into being instead of only bringing unsustainable centralized global passive consumption, monopolization, and short-term profit.

2.1 Mobile and Ubiquitous Interfaces

As said in Sect. 1, we should update an earlier delimited idea of a 'smart city' conceived when ICT was scarce into awareness that development now takes place in an innate 'smart region' since ICT is in abundance due to citizen/consumer mobile ICT networks worldwide [11]. As of 2019, there are an estimated 5.1 billion mobile phones [1] with users rising not only in urban areas though across wider rural areas of the world as well. As said, over half of mobile phones are smartphones already, and there are no zones in the world remaining under 50% mobile phone connectivity now [1, 11]. For the first time in world history, this makes a deep deliberative civic space of 'smart regions' finally possible in all regions and globally as well.

2.2 Social Media Use and Analysis

The social power of the CEM platform relies on user registration to particular ecoregions and global crowdsourcing from those users. This combines the world's local knowledge from actual regional users only, while giving each region a summary analysis of its collective aggregate knowledge and letting them vote upon it. A social media-based and crowdsourced-base of deep civil deliberation on the CEM platform is like a digital version of "rapid rural appraisal." This is a tool in development economics and grass roots development literatures to organize a civic and regional analysis of a food system's vulnerability and abundance in different seasons facilitated by a moderator and with input *only* from the regional population itself in which they make a group map of all issues. Here, this regional self-analysis done on the CEM platform itself by ecoregional users in a platform-based version of this technique. In both cases, groups debate upon a 'map' that they develop themselves showing their whole calendar year of food security and insecurity, month by month—or in this case a 'map' of all commodity uses, links, and waste streams. So, instead of only one category of food being appraised for one year by a civil group, the appraisal is done civically across 130 categories of choices in an open-ended fashion eternally, year after year, in each and every ecoregion of the world. Actual users in particular regions puzzle for the long term how to build a circular economy for themselves based on finding solutions to their own unique regional commons' problems of pollution and material choices for how these innately interact in the long term [13, 14].

First, the CEM platform users can do the following in this social media: (1) they can post as producers or as consumers about current sustainable materials available in any of the regions; (2) they can post about current complaints, future goals, or business incubation desired in a particular region in a particular commodity category; (3) and, they can post about tradable wastes per category to remove wastes from being dumped into the air, water, or earth. (4) Profile pages of posters personalize information and the website experience as well. Users view and communicate on all regional projects and share ideas for linking sustainable materials and wastes in a circular economy, though they can only post in a global level or only in their own ecoregional level. This feeds users in all ecoregions laterally with the latest information on all sustainable materials and consumption across all regions, while allowing only a real geographic community's members to confer and to deliberate on any actual regional discussion and plans.

Then, the CEM platform does the analysis from the data. The digitized 'commodity wheel' diagram automatically updates group data from all social media posts of the website in 130 categories by showing scale of interest or concern per each category of material choices and by showing scale of cross-linked wastes between categories on the same chart. This automatic analysis exists for each ecoregion commodity wheel and for one global commodity wheel of all activity.

2.3 Language and Speech Interfaces for Social Disabilities

It is very important to have multiple language/script interfaces in such a global deci-sion space. Equally important are cross-language translational services so people can learn laterally from each other across regions despite a language barrier. Plus, the CEM platform may help reverse language and culture loss. It is estimated there are over 6,000 languages in the world at present, yet language death and lack of biodiversity is causing many regional languages and cultures to continue to die out—3,000 or half of these will be lost in our generation due to lack of ongoing spoken use in an "ethno-cide—the destruction of a people's way of life" [15]. The CEM platform is designed to encourage multiple regional circular economies, and that in turn can encourage regional language preservation.

Given web browser-based solutions, the CEM platform first will be fitted innately with centralized Google translation services on demand of users. However, relying on such a centralized service is hardly optimal for a world of spotty electrical connections and low bandwidth in many areas despite high mobile phone use per capita and increase worldwide [2, 3]. We ask for feedback on other translation and speech solutions from practitioners of ICT4SD on this issue. For instance, though language translation inter-faces conveniently can be web browser-based, speech interfaces might be organized only on-demand by user-side peripherals used in daily life, and the goal of the platform would simply be to assure compatibility.

A third factor yet to be solved on the CEM platform is how to convert clickable maps into something useful for the blind. I am unaware of applications that design mobile maps for the blind or click maps for the blind, though several companies with external peripherals for mobile phones do translate maps to sounds and touches [16]. These may be useful allies in achieving this future integration on the CEM Platform.

2.4   Affective and Aesthetic Interfaces

In the ancient Greeks' view of the world, physical beauty was connected to truth. In the modern digital world, this remains true today though the beauty that a digital HCI world serves typically is one of well-designed graphs and stunning "visual displays of quantitative information" [17] that allow people to be inspired with data instead of just interpret it. On the CEM platform, drawing on common aesthetic points of billion-user platforms, it is important to keep a clear aesthetic free of words, "chart junk," and full of easily understandable icons. This is argued by information designer Tufte [17]. Tufte's lifelong data design analysis for how to make better graphs and how to make navigation tacit instead of overt is an inspiration to anyone in ICT4SD seeking good examples for how to show data and provide interaction without a distracting interface interposing on the user. Tufte over his 'data design' career became more interested in HCI design instead of only data design lessons from static charts [18]. Similarly, a great deal of beautiful information is organized in the clickable ecoregional map and the clickable commodity wheel diagrams. Both are understandable on sight for what the HCI can do for users without showing detailed instructions. This can only go so far though, since even though non-language and visual icon-based information is a common technique on globally popular websites, the CEM platform is for the literate and numerate.

2.5   Haptic Rendering, Haptic Input and Perception

Equally following Tufte's work, navigation is designed primarily by default to work for mobile users by touch instead of clicking on words, and changing screens is intuitive tunneling through pictures. From the first mass-marketed deeply haptic HCI smartphone by Apple from 2007, the world has been trained on this touch-based navigational standard. However, a flat glass, touch-based smartphone navigation touches something deeper in people since even illiterate chimpanzees, human babies, or other animals sometimes quickly understand the symbolic logic of touch to navigate. Thus globally now, normal pinching and expanding of maps and wheels are already understood on touchscreens, and this HCI on the CEM platform makes it easier to navigate.

2.6   Intelligent Visualization Tools

Salganik argues that intelligent visualization of 'big data' is important more than ever now as more data hardly always means better interpretation and better decision making, and only means faster and cheaper ways of gathering data [19]. The sheer scale and goal of the CEM platform in harnessing a global, eternal, civic deliberation space will provide the only venue in the world for aggregate data about sustainable materials and debate about them. Despite eventually being complete as a platform, the CEM data itself will never really be complete, as our world's only archive of all sustainable materials will be ongoingly updated by people in 867 regional areas by both complaining and dreaming of a better future through the platform. This will be a tremendous amount of data to mine for patterns, and it will require perhaps other kinds of visualization tools added to the platform once 'big data' exists to present it.

2.7 Pedagogical Technologies: Four Kinds of Education Through Gamification

Gamification refers to application of elements of game playing (e.g. points, badges and leaderboards) to encourage engagement with a product, service, or education via external reward motivation [20]. Four kinds of education via gamification are already built into the platform: (1) toward five ethical principles (described below), (2) toward four educational goals of a more holistic education (described below, and required for a circular economy instead of only a specialist education), (3) toward regional- and peer-based voting structures (described below), and (4) toward intentionally merging youth education and adult business education on the same platform for cross-inspiration. All four goals (particularly the fourth point) inspire innovative youth entrepreneurs to contribute early to a sustainable world in their region and to learn more holistically about a circular economy instead of only be a specialist about a few material categories. Adults in turn, by more interactions with the problems or solutions identified by youth, innovate better as well. This child-adult innovation cycle is noted in real world situations of other circular economies in the world [21]. Gamification equally is used to build on innate desires in youth to be socially recognized (on the platform) by peers and adults for good innovative ideas or comments. Since the CEM platform is equally an education service for students and a market service for entrepreneurs and companies, it is a useful single platform for business/industry news to learn about all sustainable material trends in all categories in one place.

First, five ethical principles animate technical solutions of the CEM platform. Issues of gamification are used as well: for age inclusion, for youth education, for regional democratic voting, and for balancing social relations in respect to gender and for balancing developmental feedback of regional inputs into national politics for sustainability.

Second, the CEM platform's gamification achieves four educational goals toward a holistic education: (1) encourage youth and adult participants to develop shared standards of value in holistic knowledge in multiple commodity ecology categories instead of in isolated specialist postings; (2) encourage value of a merged virtual/real geographic community of solidarity between participants; (3) provide incentives virtually to meet people or to trade with people in real life; and (4) encourage learning of comparisons between regions, which accelerates sustainable holistic knowledge via this global learning instead of only a regional learning.

Third, the voting structure of the CEM platform is another form of gamification already integrated. It registers local sentiment in a form of a 'living poll' that recognizes individuals already admired or culturally trusted in social relations and without polarizing influences in a community. This is a non-governmental body. As part of the CEM platform, this Civic Democratic Institution (CDI) is designed to remove seven ethical problems of past democratic voting to get closer to finding ecoregionally-trusted and gender-balanced moderators for a leadership forum of collaborators instead of finding ideologically polarized faction leaders bent on winning elections and crushing enemies. As said above, this voting result is designed to represent and show gender parity in such regional leaders in the group-recognized forum. Due to limitations of space this can only be discussed in brief here. Please contact the author for information on the HCI interface of the CDI already finished.

2.8   Interface Design for Accessibility—From Economic Accessibility (Poverty) to Social Disabilities

Poverty in many countries of the 'bottom billion' is severe. Various 'traps' are related to poverty though not exclusive to poverty [22], and these traps destabilize whole societies due to people and governments being unable or unwilling to plan for the future and build infrastructures for their daily life and for their future life. Two infrastructural issues of the 'bottom billion' zones of the world are crucial problems for the CEM platform that require addressing in the future beyond its current prototype: the issue of missing or subpar electricity connections and the issue of low or very costly bandwidth [2, 3].

First, for a truly good platform that operates in this decision space, it is important to address lack of electricity or unreliable electricity with platform decisions in order to integrate people onto the platform now. For updated data, Routley summarizes this issue, country by country, for the whole world:

Access to electricity is now an afterthought in most parts of the world, so it may come as a surprise to learn that 16% of the world's population—an estimated 1.2 billion people—are still living without this basic necessity. Lack of access to electricity, or "energy poverty", is the ultimate economic hindrance as it prevents people from participating in the modern economy…Where there is an electrical grid, instability is also causing problems. A recent survey found that a majority of Nigerian tech firms face 30 or more power outages per month, and more than half ranked electricity as a "major" or "severe" constraint to doing business [3].

The CEM platform is stillborn unless it can adapt to a lack of distribution of electricity or can adapt to work with irregular power in the poorer areas of the world. However, one bright side is that it is clear that mobile phone networks themselves create desire for people to have more developmental stability in markets for buying more distributed electrical generation, instead of a lack of electrical generation being seen as some 'missing services precursor' that forever stops mobile development in the bottom billion [23]. In other words, mobile-based networks of development are developing themselves. People's strong desire for a developmental platform of tertiary services which a mobile phone civilization brings them can drive the other secondary and primary economy sectors into existence in what this author has called the world's now common 'inverted modernization.' This is contrary to old modernization theory's assumptions that primary and secondary economic issues have to happen in all cases before tertiary services develop. Instead, in many poorer countries the data for this idea fails to exist. Instead, data is beginning to show inverted 'services' first via mobile phones in less developed countries that begin to drive people to more deeply invest in regional electrical generation or primary commodity generation to keep buying such a tertiary services network for their mobile phones. This inverted modernization when "tertiary services come first" is a common feature of mobile development that the author has theorized after much examination of mobile development. Others are coming to the same conclusion despite scanty quantitative data for analysis [23].

Second, inequitable bandwidth and highly varied costs for the same amount of bandwidth around the world is an uneven developmental base that challenges the decision space of the CEM platform [2]. This can be solved in the future by designing interfaces that are as low on bandwidth use as possible or which work in the absence of the internet itself. Mobile applications like this exist already like mapping services in Ghana called Snoocode. Such HCI ideas of a 'no internet app' that store data for daily use or 'wait offline in the mobile phone itself for any later intermittent connection later' would be ideal for ongoing intermittent database retrieval when the internet is unavailable. This would be ideal for arranging later uploading of information similar to internet solutions of ad hoc networks in disaster situations. Both tactics address bandwidth issues and show software-based solutions exist as a way forward. Theoretically, innovations from the global South are called a "reverse innovation" from poorer countries because only the poorest countries have the greatest incentives to develop truly cheaper or more accessible services that the whole word wants instead of these being economical services that only the poorer in the bottom billion want [24]. The CEM platform can learn from this 'reverse innovation. Another solution might be several CEM platforms—one each for high, low, or intermittent bandwidth access yet accessing the same decentralized databases—that may be a way forward.

2.9 Decentralized Platforms in General for More Distributed HCI Experiences: Blockchain, IPFS, and Bluetooth Ad Hoc Networks

Three future developments can make the CEM platform more durable and faster loading for such low bandwidth environments. One technique would be to use decentralized processing power of a Bluetooth-based network of smartphones themselves. This is already taken advantage of in 'non-internet' mobile applications like Firechat that allow smartphones to make their own Internet between users in a particular ad hoc space of smartphones without reliance on an internet or mobile cell phone tower connections. Bluetooth connections between users phones themselves can make ad hoc networks useful for running aspects of the CEM platform ideally in the future on demand in certain user situations. This equally can solve the previous problems mentioned of unstable electricity, uneven bandwidth access, and high bandwidth cost.

Second, the invention of programmer Juan Benet, called the InterPlanetary File System (IPFS), is another way to create a more permanent web and a more distributed web experience for faster loading times and less bandwidth use. This will be required in the future to let local users themselves house torrented instances of the CEM platform on their mobile phones or on altruistic users' desktop computers that locally run parts of the CEM platform on their bandwidth altruistically for other local mobile users. IPFS decentralizes and then distributes big data power. The idea behind IPFS developed from 2014 when Benet was thinking of technical ways to solve numerous problems that plague data sharing today. Only a few months later, IPFS was born as an addendum on top of regular HTML that combined great ideas behind BitTorrent, SFS, and the World Wide Web in the form of a distributed file system that seeks to connect all computing devices with the same system of files. In other words, instead of always having to go through a DNS server, and then back to a single weblink to retrieve a file (which may be expensive and far away in another country or easily censored as a single source), instead IPFS notes where is the most localized true copy of that file downloaded or used, and then 'serves' it from the most localized version since it shares the same hash tag as the original. In some ways, since it is an adaptation of HTML, IPFS is similar to the origins of the Web, but

IPFS is more related to a single BitTorrent swarm exchanging git objects. Today by 2020, IPFS is live with thousands of nodes running across the world, and several popular web streaming services like d.tube, a 'decentralized YouTube,' cannot be censored because of its distributed file system. IPFS has a big community of open-source contributors, and it provides a simple interface similar to a hypertext protocol for the World Wide Web. Benet solved the 'third party' problems of data retrieval, just as the unknown inventor "Satoshi" and his Blockchain solved 'third party' problems in online transactions via distributed ledgers.

Third, Blockchain is another addendum to a future CEM platform. Blockchain technology is being considered as an ethical option for data management due to compelling features such as being a fault-tolerant distributed ledger, having a chronological data record that is irreversible, auditable, untamperable, and cryptographically sealed in information blocks, near real-time data updates, consensus-based transactions, and the ability to authorize private smart contracts and policy-based access to facilitate data protection without a third party's surveillance or interference [25]. The ethical point here is Blockchain solutions encourage trust in numbers like totals and exchanges where trust was hard to create in the past like in anonymous online virtual communities, in online voting, and in online digital authentication or in online financial transactions without clear visibility of parties. Blockchain is an ethical principle because it is a strong security against rigging a process, a transaction, or a sense of privacy. It creates a kind of pseudo-anonymity (of secret users yet publicly shared, transparent, and group-verified data transactions) impossible to even conceive of before. First invented and applied in the online cryptocurrency Bitcoin which solved the double spending problem without recourse to the regular solution of a public (corruptible) third-party knowledge privy to all private transactions, Blockchain moved beyond finance when it was realized Blockchains were cost-effective distributed ledgers without high cost or without potential compromise from a biased central monitoring clearinghouse. This respects the ethics of anonymity and privacy in users' data while giving deep ethical trust in public exchanges simultaneously. Three uses of Blockchain exist in future CEM platform uses since the technology can: (1) facilitate trust in gamification/leaderboard totals against rigging of points by hackers or even administrators; (2) facilitate processes of voting in the Civic Democratic Institution (CDI) by tracking user nominations, candidate voting, and leaderboard result phases to a trustworthy unrigged result; and (3) facilitate sustainable product/material journey and local business growth by tracking origin and transfer of products/wastes purchases and supporting transparent bidding process. The product/material journey envisioned is similar to product/material journey in supply-chain management where the material, information and services flow from a producer to consumer.

Blockchain on the CEM platform can assure security and lack of centralized monitoring, control, or censorship of the detailed knowledge that different regions are creating. In short, Blockchain would be used later for facilitating smart contracts by the platform's users between each other as they truck and trade sustainable materials or marketable wastes per category, facilitated as a trade on the CEM platform itself.

In conclusion, whether by Bluetooth, IPFS, or Blockchain, all three different approaches to decentralized data are the future of the CEM platform if it is to operate successfully in this global-level decision space instead of only be a developed world platform. The platform and interface should serve the world's neediest who have spotty electrical connections, low and costly bandwidth, and likely a fear of tyrannical governments—yet who can be users with securities from tyranny via Bluetooth-based mobile phone nodes among themselves, IPFS-enhanced websites unable to be shut down, and Blockchain-based databases unable to track users' transactions. This would make the CEM platform less likely to be censored or spied upon by corrupt governments or other hackers. Safety would encourage sustainable democratic regional development by civic groups in particular regions.

2.10 AI in HCI

Other future ideas for the CEM platform are to have more on-demand use of artificial intelligence (AI) like other large billion-user sites that mine their own 'big data' to enhance user experiences. An AI-enhanced CEM platform could show users useful tailored comparisons between different big data matrices of different Commodity Wheels of different ecoregions that interest them. Perhaps such comparisons will help users understand the best routes to take to link up different commodity categories and their wastes in the beginning by telling us which kinds of early trellises of interactions in a circular economy are more successful, profitable, or durable over time.

3 Conclusion

What is the impact for academics and practitioners? The impact of this paper helps to start our brainstorming about ICT4SD in such a global decision space because much about ICT decision spaces before were only very localized and limited in application. Instead, the implications of thinking on this global holistic level are that we can think global and local at the same time with good HCI design, and we can adapt some information from the world's largest successful web services toward more humanitarian technology solutions like the CEM platform. In this paper, the ongoing work on the Commodity Ecology Mobile platform was described towards realizing Goal #12 of the SDGs, "Responsible Consumption and Production." The prototype is already available online, and it provides many services mentioned above in the ten criteria already to empower members to become an inclusive, sustainable, 'smart region' for themselves. The team works to extend the existing platform to use other technologies to facilitate sustainable product/material journey, better data resiliency, and privacy. Equally, the importance of this paper is the sharing of a common future holistic vision. The best way to develop a better common future is to invent it. This better future vision is one of better democratic ecological design in which people, professions, and industries talk to people they rarely talk to in their daily life. This encourages ongoing democratic risk assessment between producers and consumers in 'smart regions.' The CEM platform catalyzes a virtual community to enhance a real geographic community and to start a globally cosmopolitan debate at the same time. Our vision of the future is Commodity Ecology operating in every ecoregion of the world, allowing every region to be a 'smart region.' This shared sustainable vision of the future is already in the mobile phone in the palm of your hand.

References

1. We Are Social/Hootsuite: Digital 2019: global internet use accelerates (2019). https://weares ocial.com/blog/2019/01/digital-2019-global-internet-use-accelerates. Accessed 15 Sept 2020
2. Ang, C.: What does 1GB of mobile data cost in every country? Visual Capitalist (2020). https://www.visualcapitalist.com/cost-of-mobile-data-worldwide/. Accessed 15 Sept 2020
3. Routley, N.: Mapped: the 1.2 billion people without access to electricity. Visual Capitalist (2019). https://www.visualcapitalist.com/mapped-billion-people-without-access-to-electr icity/. Accessed 15 Sept 2020
4. United Nations Academic Impact (UNAI): #SDGsinAcademia: Goal 12 (2018). https://academicimpact.un.org/content/sdgsinacademia-goal-12. https://academicimpact.un.org/con tent/commodity-ecology-initiative-facilitate-sustainable-development. Accessed 15 Sept 2020
5. Damodaran, R.: Why we care (2020). https://academicimpact.un.org/content/why-we-care-20-august-2020. Accessed 30 Oct 2020
6. Hujala, T., Kainulainen, T., Leskinen, P.: Psychological aspects of decision making in natural resource management. In: Murphy, D., Longo, D. (eds.) Encyclopedia of Psychology of Decision Making, pp. 9–14 (2009)
7. Pamlin, D., (ed.): Sustainability at the Speed of Light: Opportunities and Challenges for Tomorrow's Society. WWF Sweden (2002)
8. Heeks, R.: Information and Communication Technology for Development. Routledge, Abingdon (2018)
9. Marolla, C.: Information and Communication Technology for Sustainable Development. CRC Press, Boca Raton (2019)
10. Morandi, C., Rolando, A., Di Vita, S.: From Smart City to Smart Region: Digital Services for an Internet of Places. Springer, Heidelberg (2016). https://doi.org/10.1007/978-3-319-173 38-2
11. Whitaker, M., Pawar, P.: Commodity ecology: from smart cities to smart regions via a blockchain-based virtual community platform for ecological design in choosing all materials and wastes. In: Singh, D. , Rajput, N.S. (eds.) Blockchain Technology for Smart Cities. BT, pp. 77–97. Springer, Singapore (2020). https://doi.org/10.1007/978-981-15-2205-5_4
12. Benkler, Y.: The Wealth of Networks: How Social Production Transforms Markets and Freedom. Yale University Press, London (2007)
13. Ostrom, E.: How inexorable is tragedy of the commons? Institutional arrangements for changing the structure of social dilemmas. In: Workshop in Political Theory and Policy Analysis, Indiana University, Bloomington, Distinguished Faculty Research Lecture, Indiana University, 3 April (1986)
14. Ostrom, E.: Governing the Commons: The Evolution of Institutions for Collective Action (Canto Classics), Reissue Ed. Cambridge University Press, Cambridge (1990, 2015)
15. Davis, W.: Dreams of Endangered Cultures Ted.com (2003). https://www.ted.com/talks/wade_davis_dreams_from_endangered_cultures#t-1112455. Accessed 15 Sept 2020
16. Darell, R.: Braille map iphone accessory guides the blind through touch. BitRebels (2012). https://bitrebels.com/technology/braille-map-iphone-accessory-concept/. Video and article
17. Tufte, E.: The Visual Display of Quantitative Information, 2nd Ed. Graphics Press. Edward R. Tufte: Beautiful Evidence, First Edition. Graphics Press (2006)

18. Tufte, E.: Video: Edward Tufte critiques the iPhone interface (2008). https://www.edward tufte.com/bboard/q-and-a-fetch-msg?msg_id=00036T. [discussion of museum websites as well]
19. Salganik, M.: Bit by Bit: Social Research in the Digital Age. Princeton University Press, Princeton (2019)
20. Clint, F.: Points, badges, and leaderboards in gamification. Study.com, 19 May 2017. https://study.com/academy/lesson/points-badges-leaderboards-in-gamification.html
21. Weisman, A.: Gaviotas: A Village to Reinvent the World. Chelsea Green Publishing Company, Hartford (1999)
22. Collier, P.: The Bottom Billion: Why the Poorest Countries Are Failing and What Can Be Done About It. Oxford University Press, Oxford (2007)
23. Park, G.: Rethinking mobile diffusion: explanatory analysis of factors affecting mobile diffusion in African Countries in the Sub-Saharan region. Paper for the Completion of "'Part A' Exam" in the Graduate School of Stony Brook University. Department of Technology and Society. The State University of New York, Korea (2019)
24. Govindarajan, V., Trimble, C., et al.: Reverse Innovation: Create Far From Home, Win Everywhere. Harvard Business Review Press, Cambridge (2012)
25. Shafagh, H., et al.: Towards blockchain-based auditable storage and sharing of IoT data. In: Proceedings of the 2017 on Cloud Computing Security Workshop. ACM (2017)

The Design and Development of School Visitor Information Management System: Malaysia Perspective

Check-Yee Law[1]([✉]) [iD], Yong-Wee Sek[2] [iD], Choo-Chuan Tay[2] [iD], Wei-Ann Lim[1] [iD], and Tze-Hui Liew[1] [iD]

[1] Multimedia University, Melaka, Malaysia
{cylaw,thliew}@mmu.edu.my, denisweiann98@gmail.com
[2] Universiti Teknikal Malaysia Melaka, Melaka, Malaysia
{ywsek,tay}@utem.edu.my

Abstract. Visitors especially unwelcome visitors may pose possible risks to school children at school. This leads school administrators to reinforce their school visitor management procedures. Current visitor registration practice in Malaysia still lack of a convenient and robust system to manage school visitor registration process and visitor record tracking. This paper reports the design and development of a new school management system with the integration of Malaysian identity card (MyKad) with extra security and data protection features known as EasyVZ school visitor management system. This system can manage either pre-registration or walk-in visitors as well as local and nonlocal visitors. A complete system testing has been conducted and no major issue is reported. With EasyVZ, the school visitor information can accurately be captured, easily be tracked and data can securely be kept. It helps to improve the work of school administrative officers and school security guards in handling school visitors. Privacy and sensitive personal data of all visitors are protected besides ensuring school children safety at school.

Keywords: Visitor information management system · School visitor · MyKad · Smart card reader · Security · School guard · School safety

1 Introduction

Nowadays, teachers and school administrators are far more security-conscious and clearly aware of the risks posed to school children from unwelcome visitors [1, 2]. This leads schools to reinforce their school visitor management procedures. One of the initiatives is the introduction of school visitor management system. There are many software platforms available in the market to facilitate visitor registration process. However, each platform has its own strengths and weaknesses. Also, the use of these systems would be different from one country to another due to different policies and practices. Therefore, users might have different needs and requirements. Existing visitor management systems available are unable to cater wide range of school requirements especially

© Springer Nature Switzerland AG 2021
M. Singh et al. (Eds.): IHCI 2020, LNCS 12615, pp. 431–444, 2021.
https://doi.org/10.1007/978-3-030-68449-5_42

from the perspective of developed and developing countries due to different policies and practices.

As a developing country like Malaysia, the school visitor management policy has its own unique practice and procedures. The current procedure and process being practiced by Malaysian schools are mostly in paper form or even worse by using verbal form. These ways can be easy and convenient at time but could pose some safety issues. For example, in the event of an emergency caused by an unwelcome visitor, there is no accurate record to be tracked or no proper person to be accounted for. Also, with the current practices, data can be fabricated or the purpose of visits can be a lie in order to pass the security checking. Besides that, records being written down on paper are hard to be tracked and can be easily lost or sensitive data can easily be disclosed. A complete and robust system that can eliminate the entire possible problems associated with the current practices is needed.

The objective of this research is to design and develop a new school management system with the integration of Malaysian identity card (MyKad) known as EasyVZ school visitor management system. This system involves three types of users: the Security guard, the Admin personnel, and the Visitor. It can manage either pre-registration or walk-in visitors. It is carefully designed to provide users with easy to access, easy to navigate and interactive graphical user interface.

This paper presents the research work of designing and developing EasyVZ school visitor management system. Section 2 provides the concepts underpinned the development of the EasyVZ system that includes the review of school safety, visitor management systems, MyKad, smart card reader and hardware connection. Section 3 discusses the prototype design and development of EasyVZ system. Some screenshots of the prototype interface are provided to delineate the functionality of the prototype system. Section 4 presents some functional testing processes in EasyVZ system. Section 5 presents the system limitation. We conclude the paper in Sect. 6 with future work.

2 Literature Review

2.1 School Safety

Accepting and welcoming visitors is a routine of any organization especially school. However, simply accept any visitor without a proper procedure can cause serious safety issue at school. School safety is deemed as one of the important school environment factor to improve school climate [3]. The concerns of school safety are crucial as feeling unsafe at school can lead to some negative effects such as lower school performance and achievement, lower physical and mental well-being, and inclined to risk-taking behaviors among students [4].

In some developed countries such as United States (US) and Canada, various technological devices and artificial intelligence surveillance technologies have been introduced to improve school safety. The use of security cameras, advanced cameras, video surveillance, video intercoms, and body scanners are some examples of the technological devices used to monitor school safety [5]. Some schools use biometrics and artificial intelligence (AI) technology to recognize faces, detect weapons, and track individuals' whereabouts in schools in order to have a greater control over the school environment

[6, 7]. Radio frequency identification (RFID) wireless technology has also been used to replace the manual process of capturing file or information that allows the schools or universities to keep track of student movement in the premises.

The application of technologies to enhance school safety is still very lack in Malaysia. At present, the procedure to capture the presence of a visitor in most Malaysian schools is in paper form or even worse, in verbal form. Such procedure would pose safety issues to the school children.

2.2   Visitor Information Management Systems

Various visitor information management systems have been developed to manage visitors for visiting private and public premises such as Vizitor, Visitor Management, Visitor Book, and Secured Pass. Each system has its own key features to support various user needs and requirement.

The Vizitor application allows the management team or users to handle visitor records easily. It allows visitors to fill up the information required through any device that runs the application system. The system then notifies the host when a guest checks in. It also has Send Invites feature that enables the host to invite guests and provide visit details to guests prior to their visit. In this way, the host can easily track down visitors [8].

The Indovisitor visitor management application can be used by the office to replace the visitor book in paper form, substitute guest marriage, events guest books, seminars, meetings, training and other activities that require visitors to register themselves. It also allows users to pre-register themselves. Key features in Indovisitor management application include digital guest book, auto notifications (e-mail, short message system (SMS)), digital signature, and print or share visitor card, scan QR code, take photo, and perform facial visual identification [9].

Visitor Book is basically the exact transform of paper form into a digital one. The working concept is still the same where the user still has to fill up all the details and a visitor record will be shown instantly. The record is not stored in the database but just for the purpose of displaying. After fields such as name, age, purpose of visits, address, identity, and others information are filled up, the photo of the visitor is taken and displayed on the dashboard [10].

Secured Pass is another application designed to ease the management personnel of handling their visitors. Secured Pass is a visitor management system bundled with a strong verification system to ensure visitors who have authenticated as "right visitor" to access to the facility quickly and efficiently. Its key features include walk-in management and pre-registered invites validation, self-registration kiosk, and tracknig visitor records. It is also integrated with vehicle parking management and tracking application, human resource interview management module, face resonation, and government identification (ID) validation [11].

Table 1 compares and contrasts the features of these four visitor management systems and the proposed system called EasyVZ school visitor management system.

Table 1. Comparison of system features

Features	Apps				
	Vizitor	Indovisitor	Visitor Book	Secured Pass	Proposed System (EasyVZ)
Free-Access	Yes	No	Yes	No	Yes
Pre-registration	No	Yes	No	Yes	Yes
QR code storage	No	Yes	No	No	Yes, pre-registration
Encryption of data	No	No	No	Yes	Yes
Viewing visitor summary	Yes	No	No	Yes, via website	Yes
Smart card scanner	No	No	No	Yes	Yes

Various visitor management systems have been proposed to manage visitor registration process including school visitors. These systems, however, have its own weaknesses, for example, many of the systems do not capture visitors' information through using smart card. This can be a major problem as information obtains through different channels could pose to various risk such as fake or inaccurate information, and insecure or leak of user sensitive data. To overcome these problems, a school visitor management system using a MyKad and a smart card reader has been proposed to manage the registration process of school visitors. The proposed system not only can eliminate the inaccuracy of the information, but it also can provide detailed information about the visitors and ease visitor record tracking.

2.3 MyKad

MyKad refers to Malaysian personal Identification Card (IC). It is a mandatory identity card to be possessed by a Malaysian who is 12 years old or above. Some personal information contained in it includes a card holder's fingerprint minutiae, full name, address, race, citizenship, religion, birth date, gender, IC number, driving license and passport information [12]. MyKad is developed under Malaysian Government Multi-Purpose Smart Card Project. It has also been integrated with several applications and contains information about a card holder's health status, "Touch and Go" transit application and the Public Key Infrastructure (PKI) feature for authentication, encryption, digital signatures and online transactions purposes [12]. Figure 1 shows a sample of MyKad.

Fig. 1. A sample of MyKad

2.4 MyKad Smart Card Reader

In this project, a MyKad smart card reader is required to establish the electronic connection between MyKad and the application system. With this connection, the information in the MyKad chip can be extracted. Some set of commands called MyKad command set is used to retrieve the information and sent to the intended application system. Some personal information that can be obtained includes a card holder's name, birth date, identity card number, photograph, driving license and passport. There is a variety of MyKad readers available in the market. The MyKad reader model used in this project is ACS ACR38U-A4 (Fig. 2) produced by Advanced Card Systems Ltd. (ACS). The card authentication of this model is processed through Secure Access Module (SAM) interface in which authentication process is done via card to reader and vice versa to enhance the security in the smart card application [13].

Fig. 2. MyKad reader (Model: ACS ACR38U-A4)

2.5 Hardware Connection

Figure 3 illustrates the hardware connection of the proposed system and the way information is retrieved from MyKad. When a MyKad is inserted into a MyKad reader, the program that has been coded to extract the information from the chips will retrieve the information from MyKad and send it to the application system for data processing and storage purpose. A laptop or computer equipped with a webcam is required to host the application system and take the photo of a visitor. A database management system is

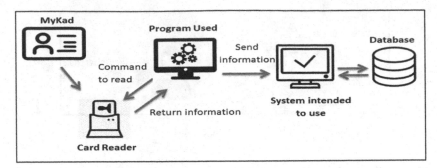

Fig. 3. Hardware connection of EasyVZ school visitor management system.

needed to control and manage data capturing, processing, updating and storing. Figure 4 illustrates the connection between the application system and the database server. The connection is done through PHP web services. JavaScript Object Notation (JSON) is used to store and send data from a server to a web page in a lightweight format.

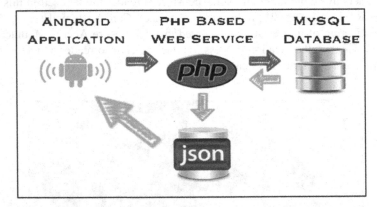

Fig. 4. The connection between the application system and the database server.

3   Prototype Design

EasyVZ visitor registration system with MyKad is a web app cum mobile app system. It allows a school security guard to easily and accurately obtain a visitor's details from MyKad by using a smart card reader. The prototype design has two main modules that include the Security guard and the Admin users. Figure 5 shows the interface of selecting the user types in EasyVZ system.

When Security is clicked, the Security main page interface will be shown as in Fig. 6. This interface allows a security guard to have a direct interaction with the visitors. Three types of registrations are available, i.e. for a local visitor, nonlocal visitor and pre-registration visitor with a QR code access. The View Visitor interface provides the

Fig. 5. Interface for selecting user type

number of visitors remaining in the school premises. The Check Out interface allows the security to check-out a visitor. The main page is equipped with an analog clock to show the time to the security guard. The red button at the right top corner of the interface will turn green once a Mykad is inserted into the smart card reader.

Fig. 6. Security guard interface

The flowchart as shown in Fig. 7 describes the flow of a security guard as a user. The security guard will select the feature based on whether a visitor is a pre-registered visitor and whether the visitor is a local or nonlocal visitor. To ease the process of registration and save up time, a visitor can apply for visitor pass in advance through EasyVZ application. Once the application is approved by the Admin, the visitor will be given a QR code as a visitor pass. The visitor can then present the QR code to the security guard. A webcam will be used to scan the QR code presented by the visitor. If a valid QR code is prompted, the visitor's information that has been submitted in advance

438   C.-Y. Law et al.

to the EasyVZ system will be displayed. The data will then be sent to the database. The visitor's photo will be taken before the information is saved in the database. For a local visitor, registration can be done by inserting MyKad to the smart card reader. The information retrieved from MyKad will then be displayed and sent to the database. For a nonlocal visitor, the security guard would need to key-in visitor information and take a photo of the visitor by using the webcam in EasyVZ system.

Figure 8 shows that a QR code has been granted to a visitor as a visitor pass. Figure 9 shows the interface of capturing a pre-registered visitor's information and photo once a valid QR is scanned. Figure 10 shows the interface of retrieving a visitor's information

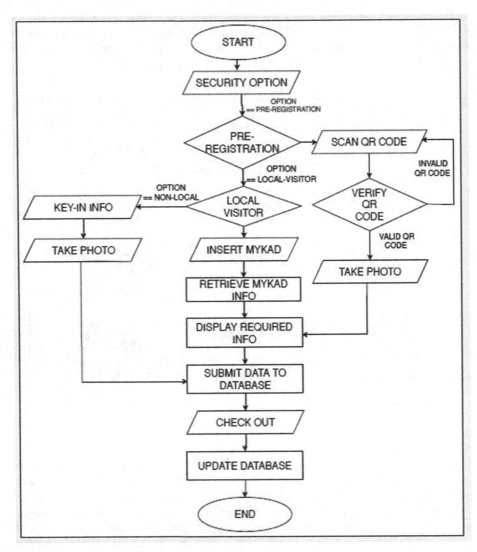

Fig. 7. Flowchart of a Security guard as a user

Fig. 8. A sample of QR code presented by a pre-registered visitor

Fig. 9. Interface of capturing a pre-registered visitor's information using QR code

Fig. 10. The interface of retrieving a local visitor's information from MyKad

from MyKad by using a smart card reader. The status of the reader will be shown on top, i.e., "OFF" indicates no MyKad is inserted into the smart card reader and "ON" indicates a MyKad is inserted into the reader. When a MyKad is inserted into the smart card reader, a visitor's data will be displayed when the download icon (↓) is clicked. The tick icon (✓) is clicked for saving the data to the database. The brush icon (●) is used to clear all the information in the field including the photo. To protect user information, some data screenshotted from EasyVZ system have been revised or redacted to protect user personal data for the presentation of this paper.

Figure 11 shows the interface of registering a nonlocal visitor's information by keying-in personal data and taking a photo of the visitor by using a webcam. The information being captured includes name, passport number, age, citizenship, gender and purpose. The tick icon (✓) is clicked in order to submit the record to the database. The record can be cleared by clicking the brush icon (●).

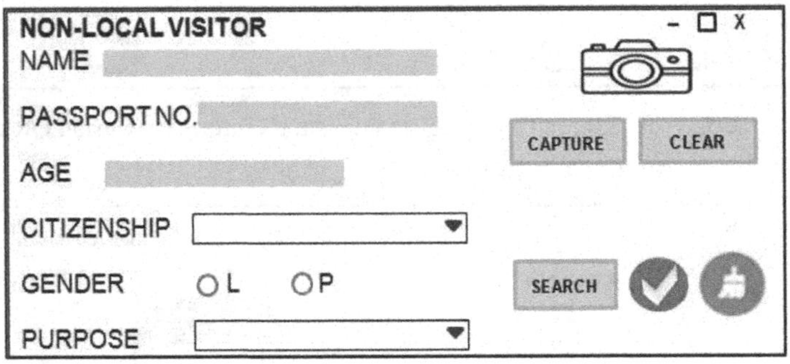

Fig. 11. The interface of recording information of a nonlocal visitor

Referring back to Fig. 5, when Admin is clicked, the Admin main page interface will be shown as in Fig. 12.

Fig. 12. The interface of Admin main page

Figure 13 shows the process flow of the Admin as a user. Three key features for Admin include managing pre-registration process, viewing visitor record and visitor statistics. For managing pre-registration process, the Admin would be able to receive visitor application, view the status of pre-registration, and approve or reject visitor pre-registration application. Figure 14 presents the interface of all pre-registration records that can be viewed by the Admin. Upon approval from Admin, EasyVZ system will generate a QR code for a pre-registered visitor. The visitor's particular and the QR code will then be stored in EasyVZ host server. For the Viewing Visitor feature, the Admin would be able to view visitor records for the day. Filter function is available for Admin to view some specific records such as filtering by visitor name, purpose and visiting date. The Admin can also export the visitor record in Excel file format for filing purpose. An example of viewing visitor record interface for the Admin is as shown in Fig. 15. For Viewing Visitor Statistics feature, the Admin is able to view visitor by a specific month or by purpose. Statistics about visitor by category will be shown in percentage. Figure 16 shows the screenshot of Viewing Statistics interface.

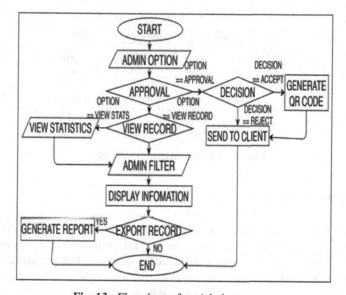

Fig. 13. Flowchart of an Admin as a user

Approval								
Approved ▼								
Pending		PASSPORT	AGE	CITIZENSHIPS	GENDER	PURPOSE	DATE	ACCOUNT
Approved		950610-04-5353	25	16, Jalan Bb 5, T...	L	Visitor	2020-02-21	Dennis
Rejected	Lim Chee Boon	950610-04-5353	25	16, Jalan Bb 5, T...	L	Cleaner	2020-02-22	Dennis
	Lim Chee Boon	950610-04-5353	25	16, Jalan Bb 5, T...	L	Vendor	2020-02-20	Dennis
	Lim Chee Boon	950610-04-5353	25	16, Jalan Bb 5, T...	L	Guardian	2020-02-21	Dennis
•								

Fig. 14. The interface showing all pre-registration records

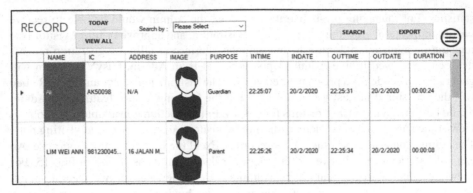

Fig. 15. Viewing Visitor interface

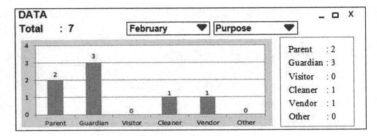

Fig. 16. Viewing Statistics interface

4 Prototype Testing

System testing is conducted at the host side. This testing is mainly focused on functionality testing to ensure EasyVZ system functions properly and fulfills the expected results. A complete system testing has been conducted including testing for Admin, Security Guard and Visitor features, software navigation, and user sign up and login. It is important to note that all functional testing for these parts fulfill the functional requirements and work well. For the presentation purpose of this paper, only the functional testing for Security Guard is shown. Table 2 shows the results.

Table 2. Prototype testing results

Test cases	Expected result	Outcome
Capturing MyKad information		
a) Retrieve MyKad information	a) MyKad information is shown	Successful
b) Submission of data to database	b) Data are inserted into database	Successful
c) Clear button to clear information	c) The field in the form is cleared	Successful
d) Capture a visitor's photo	d) A visitor's photo is shown	Successful
Scanning QR code		
a) Scan QR code	a) Decode the QR code	Successful
b) Get the details of QR code	b) Information of QR code is shown	Successful
c) Invalid QR code is shown	c) Display invalid message	Successful
Checking Out a visitor		
a) Check out the selected name	a) The selected name is checked out and successful message is shown	Successful
b) Check out without selecting a name	b) A message prompted for name	Successful

5 Limitation

Every school needs to equip with a smart card reader and has stable internet connection in order to ensure the system works properly and efficiently. Furthermore, nonlocal visitors without a valid MyKad can only register by using passport in which the information has to be keyed in by the security guard in the EasyVZ system.

6 Conclusion and Future Research

The proposed EasyVZ school visitor information management system is able to eliminate problems and mitigate issues encountered in manual visitor registration. With the extra security feature using MyKad to register through EasyVZ, school administrators are able to obtain a visitor's information in secured, fast and accurate manners. In addition, data manipulation can easily and quickly be performed through searching, filtering and report generation. This helps to improve the work of school administrative officers and school security guards in handling school visitors. The system also can prevent an unwelcome visitor for providing fake information. Privacy and sensitive personal data of all visitors are protected besides ensuring school children safety at school. To further enhance the security feature of EasyVZ system, future work can be done by integrating face recognition or iris recognition function. For nonlocal visitor, Optical Character Recognition (OCR) technology can be integrated to provide passport scanning and recognition.

References

1. Musu-Gillette, L., et al.: Indicators of school crime and safety: 2017 (2018)
2. Perumean-Chaney, S.E., Sutton, L.M.: Students and perceived school safety: the impact of school security measures. Am. J. Crim. Justice **38**, 570–588 (2013). https://doi.org/10.1007/s12103-012-9182-2
3. Voight, A., Nation, M.: Practices for improving secondary school climate: A systematic review of the research literature. Am. J. Community Psychol. **58**, 174–191 (2016). https://doi.org/10.1002/ajcp.12074
4. Lenzi, M., Sharkey, J., Furlong, M.J., Mayworm, A., Hunnicutt, K., Vieno, A.: School sense of community, teacher support, and students' school safety perceptions. Am. J. Community Psychol. **60**(3–4), 527–537 (2017). https://doi.org/10.1002/ajcp.12174
5. Weinstein, M.: School surveillance: The students' rights implications of artificial intelligence as K-12 school security. North Carolina Law Rev. **98**(2), 438–480 (2020)
6. Locker, M.: Schools can now get free facial recognition software to track students (2018). https://www.fastcompany.com/90205116/schools-can-now-get-free-facial-recognition-software-to-track-students. Accessed 21 Oct 2020
7. Zimmerman, E.: Company offers free facial recognition software to boost school security. EdTech Magazine (2018). https://edtechmagazine.com/k12/article/2018/08/company-offers-free-facial-recognition-software-boost-school-security. Accessed 21 Oct 2020
8. Vizitor. Vizitor: Touchless visitor management system, make your workplace safe and secure (2020). https://www.vizitorapp.com/. Accessed 30 Sept 2020
9. Indovisitor: Perfect visitor management, best for your company (2018). https://indovisitor.com/. Accessed 30 Sept 2020
10. Google: Visitor book: Visitor management system (2020). https://play.google.com/store/apps/details?id=com.tech.gahlot.visitorbook. Accessed 30 Sept 2020
11. Google. Secured pass: Visitor management system (2020). https://play.google.com/store/apps/details?id=com.automatebuddy.securitytab&hl=uz. Accessed 30 Sept 2020
12. Malaysian_National_Registration_Department: MyKad sebagai kad pengenalan. Jabatan Pendaftaran Negara/Home/Aplikasi Utama 2020 (2020). https://www.jpn.gov.my/informasi mykad/aplikasi-utama/. Accessed 30 Sept 2020
13. Advanced_Card_Systems_Ltd: ACS ACR38U-A4 Smart card reader technical specifications V2.03, Advanced Card Systems Ltd: Hong Kong

The Impact of the Measurement Process in Intelligent System of Data Gathering Strategies

Mario José Diván[1] ⓘ and Madhusudan Singh[2](✉) ⓘ

[1] Data Science Research Group, National University of La Pampa, 6300 Santa Rosa, Argentina
mjdivan@eco.unlpam.edu.ar
[2] School of Technology Studies, Endicott College of International Studies, Woosong University, Daejeon, South Korea
msingh@wsu.ac.kr

Abstract. An intelligent system is a technology that has opened diverse ranges of possibilities due to its availability, value, and accessibility. Such an intelligence resides in the possibility to measure and know the current state of entities or environments. Thus, the measurement process constitutes a key asset for many aspects related to the intelligence of systems (e.g., decision-making) and the relationship with the data gathering strategies. A strategy for formalizing monitoring projects based on entity states and scenarios is introduced. It integrates the visualization pipeline to align the visual communication to measurement requirements. From the methodological point of view, an extension of a measurement and evaluation framework which supports the modeling of entity states and scenarios is considered. The framework allows agreeing on concepts required to formalize a measurement project. Thus, a specialization of the Goal-oriented Context-aware Measurement and Evaluation strategy is introduced using the business process model to describe how scenarios and entity states are articulated jointly with their transition models. Also, the visualization pipeline is integrated into the new strategy to articulate the information need that gives origin to the measurement project jointly with the visual communication strategy. A synthesized case as a proof-of-concept is introduced. In this way, a monitoring strategy aware of scenarios, entity states, and the visualization pipeline into a measurement project is reached. Thus, traceability about each visual perspective is aligned with measurement points of view.

Keywords: Measurement process · Intelligent system · Data gathering · Strategy · Monitoring

1 Introduction

The measurement process represents a key asset in the intelligent systems because it allows quantifying the behavior of entities under monitoring but also about the related environments. It fosters a context-aware decision-making process, where the system is capable of characterizing each monitored aspect to make decisions [1].

© Springer Nature Switzerland AG 2021
M. Singh et al. (Eds.): IHCI 2020, LNCS 12615, pp. 445–457, 2021.
https://doi.org/10.1007/978-3-030-68449-5_43

A measurement process should be consistently defined, extensible in front of new requirements, feasible, repeatable, and the measures comparable [2]. C-INCAMI (Context-Information Need, Concept Model, Attribute, Metric, and Indicator) is a multi-purpose measurement and evaluation framework based on an ontology that defines terms, concepts, and the relationships needed to implement a measurement process [3]. An extension incorporating the possibility to model scenarios through which an entity under monitoring could develop different activities jointly with the entity states were defined in [4]. The GOCAME (Goal-oriented context-aware Measurement and Evaluation) strategy [5] uses the C-INCAMI framework to formalize a measurement project definition aligned with an information need. However, once the measures have been obtained, the strategy does not provide a reference to the visualization strategy. There, the visualization guidelines [6] provide different recommendations for developing visualization schemas based on the target audience and information needs. However, it is pending an articulation between the transition from the data gathering strategy to a visual prototype.

The contributions could be synthesized as follows a) A strategy named *GOCAME-ESVI (The Entity States, Scenarios, and Visualization Pipeline)* is introduced as a specialization of GOCAME. It allows considering the entity states, scenarios, and visualization guidelines from the measurement project definition, giving traceability from how the measure is obtained to how the indicators are visually represented, b) A synthesized case of study based on news of Corona Virus Disease -19 (COVID-19) published in online newspapers of Santa Rosa (La Pampa, Argentina) between March 1 and April 2 of 2020 is illustrated to describe a basic application.

This article is organized as follows. Section 2 describes some related works referred to measurement strategies used to monitor COVID-19 news. Section 3 describes GOCAME-ESVI. Section 4 synthetically introduces the study case and describes some results. Finally, some conclusions and future works are presented.

2   Related Works

The chosen study case refers to how to measure the news polarity based on COVID-19 news. For that reason, here is introduced different approaches used to monitor the news polarity for supporting different decision-making processes.

De O. Carosia et al. [7] propose the use of Perceptron multilayer-based neural networks for studying the relationship between text and sentiments expressed in tweets. From such a relationship, the person's decision about to invest or not in the Brazilian stock market is analyzed. From the processed news, the authors propose a metric to describe the predominant sentiment. As a difference with this proposal, the measurement process is not formally described. For that reason, it makes fuzzy how the need for information and aims are aligned with the data collection strategy. Adnan and Shamsudin [8] introduce an analysis based on unstructured data written in English obtained from a News portal. A tag-based analysis was performed around the MH17 flight accident. It is not specified how the transition from unstructured text to a quantitative perspective is made. Our strategy focuses on providing traceability from the information need up to the content visualization, even when the analyzed content is text and not a sensor. Abd-Alrazaq et al. [9] introduced a COVID-19 tweets analysis between February 20 and March 15 of 2020. Data and metadata were extracted (e.g., number of likes, etc.). A unigram, bigram, and frequency analysis allowed arriving a news classification.

Table 1. An example of simple decision criteria for the news polarity level.

Article	Aim	Approach	Year
De O. Carosia et al. [7]	It analyzes the influence of tweets on decisions of persons in the Brazilian stock market	Neural networks based on multilayer perceptron	2019
Adnan and Shamsudin [8]	It implements a concordance analysis based on the frequency of words used in real-world contexts	Based on corpus	2019
Abd-Alrazaq et al. [9]	It identifies main topics about the COVID-19 pandemic from tweets	Word frequencies based on unigrams and bigrams are employed. The Python's tweepy library with PostgreSQL is proposed jointly with the LDA (Latent Dirichlet Allocation) model	2020
Zavarrone et al. [10]	It identifies the main topics of the COVID-19 pandemic using tweets	It is composed of a sentiment analysis jointly with co-occurrence network analysis and LDA for integrated visualization of the communication strategies	2020

Zavarrone et al. [10] describe the media monitoring related to COVID-19 news written in the Italian language. The idea is related to detect the level of polarity associated with each media based on their published news. As a difference, our proposal is supported by a formal measurement project definition that gathers the aim, requirements, metrics, measurements, and visualization schemas. Table 1 describes a comparative perspective on the aim and approach of each alternative.

3   GOCAME-ESVI

In C-INCAMI, the entity category represents the concept under monitoring. It is analyzed from different points of view named attributes. Each attribute is quantified through a metric that describes its scale, unit, method, instrument, value domain, etc. Each entity category is immersed in a context that interacts with it. By analogy, the context could be described through context properties that could be quantified through metrics. Metrics define how to obtain the numerical value (i.e., a measure) but not how to interpret it.

Indicators are responsible for interpreting each measure using decision criteria [11]. However, attributes characterizing the entity category could configure different states, while context properties could describe different scenarios for the context [4]. This implies that indicators could have different decision criteria according to the current scenario and entity states. Thus, decision criteria should be adapted to the current scenario and entity state to increase the precision and accuracy of decision criteria in contrast to the monitored entity before a decision needs to be made.

Munzner visualization guidelines [6] describe four-level design abstractions for developing different visualization schemas, based on the target audience and information need to communicate. These abstractions describe a) An aim or information need that the visualization strategy needs to satisfy (i.e., the problem characterization), b) A way to translate from the aim to the User-expected perspectives (i.e., Data abstractions and associated tasks), c) A design contemplating data, aim, and visualization techniques to use (i.e., visual interacting and coding), and 4) An implementation articulating the different perspectives around the aim.

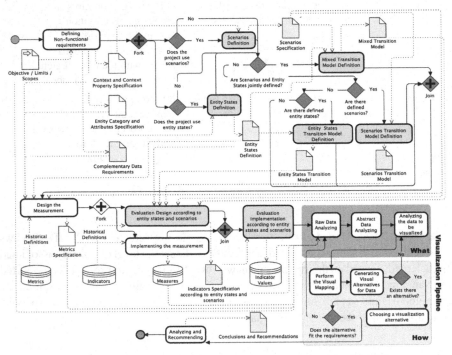

Fig. 1. A high-level description of GOCAME-ESVI using BPMN.

In this way, GOCAME-ESVI is proposed as a GOCAME's specialization able to manage entity states, scenarios, and the Munzner visualization pipelines (See Fig. 1). The capacity of formally specifying and structuring measurement projects supporting entity states and scenarios is articulated with the problem characterization required by Munzner to advance in the visualization itself. This allows providing traceability to the visualized data and aligns them with a specific monitoring need. Figure 1 uses a Business Process Model and Notation (BPMN) diagram for describing GOCAME-ESVI. Symbols with white background belong to the original version of GOCAME and Munzner respectively, highlighting with a color background the new elements.

The process starts with the specification of the aim, limits, and scope of the measurement process. After that, the entity category is defined jointly with their attributes, context, context properties, and the data complementary requirements when it is required (e.g., a geographic position). Depending on the indicated aim, it is possible that not always is required the context specification, entity states, or scenarios. Scenarios are defined as a combination of context properties, while the entity states are defined as a combination of attributes from the entity under monitoring.

Based on the specification of entity states and/or scenarios, the next activity is related to the transition model definition. There, expected transitions among scenarios, entity states, or a combination of both are modeled. Thus, the following activity consists of the metric definition, an instant where how each attribute or context property is quantified is defined (also, it could be enriched from previous experience from other projects). Thus, metric implementation refers to run the method to quantify and collect measures effectively, that later will be analyzed using indicators. Parallelly to the implementation of the measurement process, the evaluation is designed. There, the decision criteria are specified according to the different variations of expected measures, considering the definition of entity states and scenarios for its interpretation. Once the evaluation has been detailed, the indicators are computed to provide data for visualization. Thus, the Munzner's problem characterization required for the visualization guideline is articulated with a) The objective, limits, and scopes related to the measurement project, b) The metrics specifications, c) Measures generated according to the metric specification, d) The indicator specifications to implement the assessment, e) The indicator values got based on its definition. From there, the visual mapping, prototypes, and visual strategies are iteratively generated up to satisfy the measurement and evaluation requirements indicated by the user. Once they have been implemented (e.g., using software such as Tableau, Qlik Sense, etc.), it is obtained enough feedback for analyzing and recommending different actions tend to reach the defined aim in the measurement project.

4   A Case of Study About Online News Polarity

GOCAME-ESVI could be applied in situations where the origin is a sensor providing a numerical value, but also in a case of study based on published news related to COVID-19 in online newspapers from La Pampa (Argentina) is synthetically described. The underlying idea is to describe how from unstructured text is possible to monitor the news polarity, bringing the text written in Spanish into a quantitative representation. In such a sense, the Spanish dictionary of affections developed by Gravano and Dell' Amerlina is employed [12]. Independently of the measurement process and its formalization, each word in the dictionary contains a level of polarity. Because of that, the choice of the dictionary is critical, affecting how the global polarity is measured. A brief example of this aspect is described in Fig. 3.

Following the guidelines of the Extended C-INCAMI framework and GOCAME-ESVI, the objective of the measurement project is *to monitor the news polarity published in online newspapers from La Pampa* (Argentina) between March 1 and April 2 of 2020. The main newspapers were considered for this analysis (i.e., La Arena -LA-, Diario Textual -DT-, La Reforma -RF-, El Diario -ED-, and InfoPico -IP-). The entity category is limited to the *news with a previous editorial process*. The news analysis is carried out from the following perspectives 1) Terms-based polarity, 2) News-based polarity, 3) Media-based polarity, and 4) Topic-based polarity. Over 217 news related to local news about COVID-19 were analyzed to schematize this proof-of-concept.

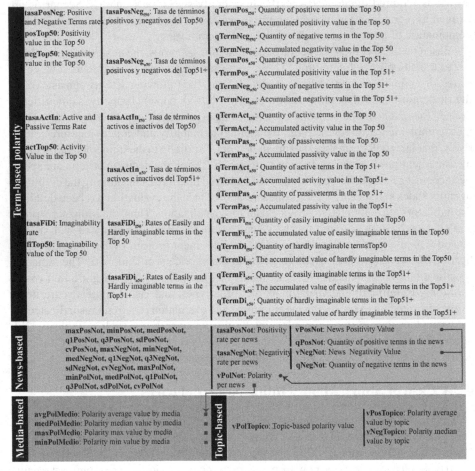

Fig. 2. Synthesis of metrics based on defined attributes for COVID-19-based news monitoring.

Figure 2 synthetically describes the defined metrics based on each specified attribute for the entity category (i.e., the outcome from the activity *Design the Measurement*, See Fig. 1). From right to left, it is possible to read the IDs and names for direct metrics (i.e., the most basics), while the metrics on the left are computed from the direct metrics. Before implementing the metrics, all the news were processed as follows 1) They were loaded in memory to create the corpus; 2) The text was preprocessed to remove punctuation signs, special characters, numbers, indicating the composed words (e.g., the name for the province of La Pampa was indicated as la_pampa) and converting all the text to lowercase; 3) The stop words are removed using a Spanish list contained in the tm-0.7–7 package; 4) The docterm matrix is generated containing each news as a row, while each column represents a given term. Each intersection represents the frequency of a given term in a given document; 5) The affinity dictionary is loaded using the term as key. Thus, given a term in the docterm matrix, it is possible to search for the polarity (P), Imaginability (I), and Activity (A) values. The three perspectives in the dictionary are quantified from 1 to 3, being 2 neutral, close to 3 high, and close to 1 low. For example, an affinity lower than 2 would indicate negativity, while an affinity upper than 2 would represent positivity. Analogously with the imaginability and the activity.

That is to say, a value of activity upper than 2 would represent a certain level of activity, while a value lower than 2 would represent a certain level of inactivity.

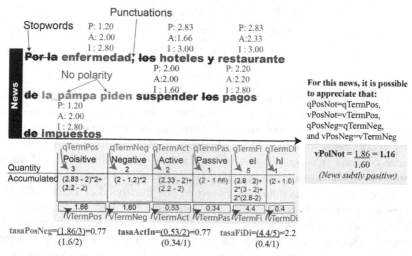

Fig. 3. An example of polarity calculation per news.

Figure 3 introduces an example to describe the calculation of polarity per news using an unstructured text. The example indicates a piece of news written in Spanish. The English meaning is "*Restaurants and hotels from La Pampa request suspend the tax payments because of the disease*". Once stop words and punctuations are removed, the list composed of (Polarity, Activity, Imaginability) is identified using the affinity dictionary for each word. Figure 3 indicates with red color the negative terms, green color the positive terms, blue color the neutral, and gray color the missing terms in the dictionary. Below the news, it is shown how the direct metrics qtermPos, qTermNeg, qTermAct, qTermPas, qTermFi, qTermDi, vtermPos, vTermNeg, vTermAct, vTermPas, vTermFi, and vTermDi are calculated (They were introduced in Fig. 2). The indirect metrics were calculated from the direct metrics. For example, a value of 0.77 for tasaPosNeg means that negative polarity terms average used in the news is upper than the average weight of the positive.

Here, vPosNot and vNegNot are equal to vTermPos and vTermNeg respectively. Thus, the news polarity could be defined as the division between vPosNot and vNegNot. However, the metric will give a value (i.e., 1.16 in Fig. 3), but nothing is said about its interpretation. At this point, it is where the indicator definition becomes meaningful. Because an aspect related to the extension, this case does not define entity states and scenarios for news. However, the indicator definition (exemplified in Table 2) could be fitted to each combination of entity states and scenarios.

Table 2. An example of simple decision criteria for the news polarity level.

Interval	Polarity level
$(-\infty; 0.95)$	Negative
$[0.95; 1.05)$	Neutral
$[1.05; 2.00)$	Subtly positive
$[2.00; 5.00)$	Moderately positive
$[5.00; 10.00)$	Positive
$[10.00; +\infty)$	Highly positive

From the metrics related to the terms, it is calculated the metrics related to media and news (i.e., the news could be viewed as a bag of words. Thus, the media could be analyzed as a set of news coming from a common source). On the other hand, the topic modeling is addressed from the LDA (Latent Dirichlet Allocation) perspective, where each piece of news is a combination of multinomial distributions [13]. There, each document (i.e., news) is expressed as a random combination of latent topics, where each topic is characterized by a combination of terms. The likelihood of a term belongs to a given topic is known as a beta probability, while the theta distribution refers to the topics' distribution for a given piece of news. On the one hand, the characteristic terms for a topic are obtained ordering in a descendant way the terms according to the beta probability. On the other hand, given two topics, it is possible to find the discriminating terms using their beta probabilities according to the following expression $log_2(beta^{t1}_{topic1}/beta^{t2}_{topic2})$ to make symmetric the difference. Thus, when the probability of the term 2 in topic 2 $(beta^{t2}_{topic2})$ is the double of the probability of the term 1 in the topic 1 $(beta^{t1}_{topic1})$, the logarithm tends to be 1, while that when the $beta^{t1}_{topic1}$ is the double of $beta^{t2}_{topic2}$, the logarithm tends to be -1.

The use of the LDA model requires to define the number of topics jointly with an alfa parameter. Here, the model perplexity was obtained through a simulation where topics were varied from 2 to 60 (See Fig. 4). Also, the VEM (*Variational Expectation-Maximization*) method jointly with cross-validation with k-folds of 5 was used.

The text processing and implementation of the measurement project following the GOCAME-ESVI strategy was carried out using R Studio (1.2.5033) and R (3.6.2) [14]. The used libraries were ggplot2 (3.3.0) for graphics generating [15], dplyr (0.8.4) [16], tidyr (1.0.2) [17], and tidyverse (1.3.0) for the data processing [18]. SnowballC (0.7.0) was used for stemming [19], tm (0.7–7) was applied in text mining [20, 21], wordcloud2

(0.2.1) to generate word clouds [22], topicmodels (0.2–11) for LDA modeling [13]. The affinity dictionary, scripts, and the set of news employed in this case are freely available at GitHub (github.com/mjdivan/IHCI20K).

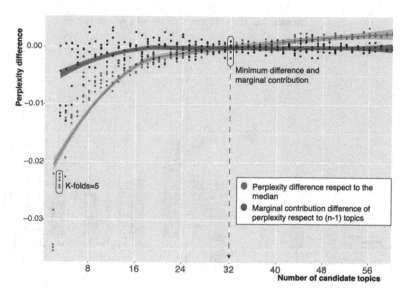

Fig. 4. Topics estimation using the absolute minimum difference with respect to the perplexity median of the model.

Figure 5 describes the heat map related to the news polarity level indicator according to the definition introduced in Table 2 organized by week.

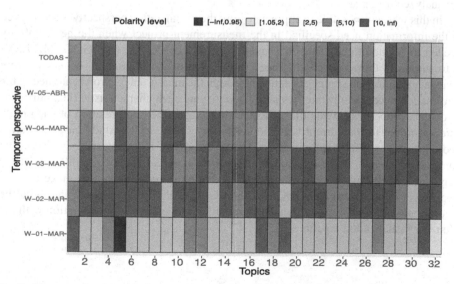

Fig. 5. Heat map for the polarity level indicator organized by week for COVID-19 news based on the LDA topics.

Figure 6 shows the evolution per week organized by media (i.e., online newspaper). As it was previously introduced, the DT, ED, IP, LA, and RF acronyms refer to the online newspapers considered this case of study.

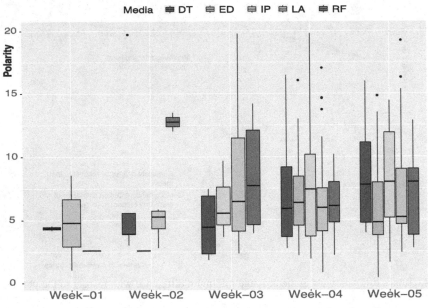

Fig. 6. Boxplots for the vPolNot metric organized by week and media.

Figure 7 shows the evolution of the news polarity median organized by day inside the analyzed timespan.

In this way, it is possible to appreciate how the visualization perspectives are aligned to the information need specified in the measurement project when the news polarity needed to be monitored. Thus, the underlying idea associated with GOCAME-ESVI is to start with those aspects that need to be monitored for an entity category. From there, a quantification strategy is defined through the measurement project aligned to the information need. Finally, metrics allow quantifying each characteristic (i.e., attributes or context properties), while the indicators make easy the interpretation of each measure according to the current scenario and entity states. In this study case, scenarios and entity states were not defined due to the extension. However, the news could be contextualized (e.g., it is published online, paper, etc.) and could be in different states (e.g., such as editing, recently published, etc.). The improvements that make the difference between the original GOCAME and GOCAME-ESVI are associated with the incorporation of scenarios, entity states, transition models, and the integration with the Munzner visualization pipelines.

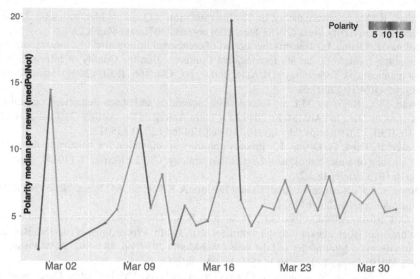

Fig. 7. The evolution of the median of news in the analyzed period.

5 Conclusions

The GOCAME-ESVI strategy was introduced as a specialization of GOCAME based on an extended C-INCAMI framework. The extended framework introduces concepts such as entity states, scenarios, entity states transition model, scenarios transition model, allowing articulating them about the indicator definition for better interpretation. The GOCAME strategy was modified to incorporate and adapt specific activities for such concepts, while that Munzner visualization guidelines were integrated. Thus, it is possible the traceability from the measurement project definition to the visualization strategy. On the one hand, the extended measurement framework provides a way to specify the requirements (describing how the measures are obtained and interpreted), integrating them into the problem description required by the visualization guidelines. On the other hand, the guidelines provide a visual way to describe the quantitative approaches aligned to a given aim. A study case was introduced to describe the GOCAME-ESVI application, using a set of COVID-19 news published in the online newspapers of La Pampa (Argentina). As future work, the impact of scenarios and entity states transition models on the decision-making process based on real-time data gathering strategies will be analyzed.

Acknowledgements. This research partially is supported by the project Res.CD 312/18 at the NULP.

References

1. Yang, M., et al.: Measurements and cluster-based modeling of vehicle-to-vehicle channels with large vehicle obstructions. IEEE Trans. Wirel. Commun. **19**, 5860–5874 (2020). https://doi.org/10.1109/TWC.2020.2997808

2. Omnès, R.: Measurement theory. In: The Interpretation of Quantum Mechanics, pp. 324–377. Princeton University Press (2018). https://doi.org/10.2307/j.ctv346qpb.13
3. Molina, H., Olsina, L.: Towards the support of contextual information to a measurement and evaluation framework. In: 6th International Conference on the Quality of Information and Communications Technology (QUATIC 2007), pp. 154–166. IEEE (2007). https://doi.org/10.1109/QUATIC.2007.21
4. Divan, M.J., Reynoso, M.L.S.: Incorporating scenarios and states definitions on real-time entity monitoring in PAbMM. In: 2019 XLV Latin American Computing Conference (CLEI), p. 10. IEEE (2019). https://doi.org/10.1109/CLEI47609.2019.235072
5. Becker, P., Papa, F., Olsina, L.: Process ontology specification for enhancing the process compliance of a measurement and evaluation strategy. CLEI Electron. J. (2015). https://doi.org/10.19153/cleiej.18.1.2
6. Munzner, T.: Visualization Analysis and Design. A K Peters/CRC Press, New York (2014). https://doi.org/10.1201/b17511
7. de O. Carosia, A.E., Coelho, G.P., da Silva, A.E.A.: The influence of tweets and news on the brazilian stock market through sentiment analysis. In: Proceedings of the 25th Brazillian Symposium on Multimedia and the Web - WebMedia 2019, pp. 385–392. ACM Press, New York (2019). https://doi.org/10.1145/3323503.3349564
8. Adnan, W.N.W.M., Shamsudin, S.: Corpus-based analysis of MH17 online Dutch news articles. In: Proceedings of the 2019 3rd International Conference on Education and Multimedia Technology, pp. 350–354. Association for Computing Machinery, New York (2019). https://doi.org/10.1145/3345120.3345191
9. Abd-Alrazaq, A., Alhuwail, D., Househ, M., Hamdi, M., Shah, Z.: Top concerns of tweeters during the COVID-19 pandemic: infoveillance study. J. Med. Internet Res. 22, e19016 (2020). https://doi.org/10.2196/19016
10. Zavarrone, E., Grassia, M.G., Marino, M., Cataldo, R., Mazza, R., Canestrari, N.: CO.ME.T.A. -- covid-19 media textual analysis. A dashboard for media monitoring (2020)
11. Rossi, G., Schwabe, D.: Modeling and Implementing web applications with OOHDM. In: Rossi, G., Pastor, O., Schwabe, D., Olsina, L. (eds.) Web Engineering: Modelling and Implementing Web Applications, pp. 109–155. Springer , London (2008). https://doi.org/10.1007/978-1-84628-923-1_6
12. Dell' Amerlina Ríos, M., Gravano, A.: Spanish DAL: a Spanish dictionary of affect in language. In: Proceedings of the 4th Workshop on Computational Approaches to Subjectivity, Sentiment and Social Media Analysis, pp. 21–28. Association for Computational Linguistics, Atlanta (2013)
13. Grün, B., Hornik, K.: Topicmodels: an R package for fitting topic models. J. Stat. Softw. (2011). https://doi.org/10.18637/jss.v040.i13
14. R Core Team: R: A Language and Environment for Statistical Computing (2019). https://www.r-project.org/
15. Wickham, H.: ggplot2: Elegant Graphics for Data Analysis. Springer, New York (2016). https://doi.org/10.1007/978-3-319-24277-4
16. Wickham, H., François, R., Henry, L., Müller, K.: dplyr: a grammar of data manipulation (2020). https://cran.r-project.org/package=dplyr
17. Wickham, H., Henry, L.: tidyr: tidy messy data (2020). https://cran.r-project.org/package=tidyr
18. Wickham, H., et al.: Welcome to the tidyverse. J. Open Source Softw. 4, 1686 (2019). https://doi.org/10.21105/joss.01686
19. Bouchet-Valat, M.: SnowballC: snowball stemmers based on the C "libstemmer" UTF-8 Library (2020). https://cran.r-project.org/package=SnowballC
20. Feinerer, I., Hornik, K.: tm: text mining package (2019). https://cran.r-project.org/package=tm

21. Feinerer, I., Hornik, K., Meyer, D.: Text mining infrastructure in R. J. Stat. Softw. **25**, 1–54 (2008)
22. Lang, D., Chien, G.: wordcloud2: create word cloud by "htmlwidget," https://cran.r-project.org/package=wordcloud2 (2018)

Detecting Arson and Stone Pelting in Extreme Violence: A Deep Learning Based Identification Approach

Gaurav Tripathi[1]([⊠]), Kuldeep Singh[2], and Dinesh Kumar Vishwakarma[3]

[1] Department of ECE, Delhi Technological University, Delhi 110042, India
gauravtripathy@gmail.com
[2] Department of Electronics and Communication Engineering, MNIT, Jaipur 302017, India
kuldeep.ece@mnit.ac.in
[3] Department of IT, Delhi Technological University, Delhi 110042, India
dinesh@dtu.ac.in

Abstract. Violence is an extreme activity that presents a clear danger to human lives, human properties and governing authorities . Violence emancipates from strong protests to extreme disturbing activities by the mob. Violence unfolds itself from low level violence to extreme violence. This journey has two major steps. Stone Pelting and Arson activities are the two most ferocious activities by mobs from which classify the violence as building from low level violence to extreme violence. Arson is a spontaneous activity which is executed by protestors for showing extreme emotional dissent to the governing authorities. Lead by mob, arson is a dangerous activity and constitute most ferocious form of violence. Stone pelting is again an extreme case of mob fury against governing authorities. Pelting refers to throwing number of things at someone or something very quickly. Stone pelting is thus the most feared form of crowd violence that needs to be tackled on priority. Arson and stone pelting activities generate a fear in crowd and endangers life of humans and public and private property. The paper presents an application oriented deep learning framework using transfer learning approach for identification of arson and stone pelting in the images and videos. Cities which are classified as sensitive can opt for arson and stone pelting identification scheme for the protection of people and properties. We present a 2D ConvNets based transfer learning model for classifying extreme violence of arson and stone pelting. For a proof of concept, a small dataset is curated containing arson images, stone pelting and normal images.

Keywords: Arson activity · Stone pelting · Convolutional neural network · Deep learning · Transfer learning

1 Introduction

The world systems have converged towards surveillance systems. With rise of urban cities, smartphones and CCTV cameras, getting inputs from images and videos have

© Springer Nature Switzerland AG 2021
M. Singh et al. (Eds.): IHCI 2020, LNCS 12615, pp. 458–468, 2021.
https://doi.org/10.1007/978-3-030-68449-5_44

become lot easier. Crowd surveillance has become absolute essential to maintain peace and order in urban areas. Automatic deciphering of evolving situation needs to be monitored for detecting extreme violence. Normally protests are normal in any country. There can be various reasons for such an action of public ranging from personal, ethical, religious or even political reasons. Detection of key events in violence arising from such protests needs to be closely monitored. Arson and stone pelting are such activities which signals change in the attitude of crowd. It also signifies transformations of gathering in to extreme violence. Dissenting crowds go on rampage and they start behaving violently. Violence slowly converges in to extreme nature as soon as the arson and stone pelting are detected. There are always situations when there is a lack of sync between government and public on some matters. These conflicts give rise to peaceful as well as violent protests. On some matter there are high chances that these protests transform themselves in to extreme violence. Our focus is on arson and stone pelting activities that needs to be detected. Arson is a form of showing dissent using symbols, voice and other means to the concerned authorities. Arson is usually a solemn declaration of opinion and usually denotes dissent [1]. It's an act of challenge against an idea or a course of action usually executed by governing authorities. Similarly stone pelting refers to a criminal assault in the form of stone throwing by mob who pelt, bombard or throw stones on police forces deployed for crowd control in some protest areas [2]. Sample figure describe what exactly is stone pelting and arson images in case of violence (Figs. 1 and 2) .

Fig. 1. Stone pelting sample images

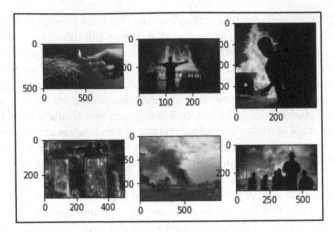

Fig. 2. Sample arson images

Dissatisfaction with any government decisions, laws, religious diktat, and so on makes the public to go all out on streets for showing dissent. Large number of people who have the same emotions of dissent against governing authorities also fall under the purview of arsons and stone pelting. Hence it becomes a strong medium to attract attention from all over the world and subsequently put pressure to the authorities to withdraw the notice, diktat, law and so on. Arson and stone pelting have become the single most effective of strategy for bringing the opinions in the public form. Above images give us an idea about arsons. Since there is a large gathering of crowds for a unifying cause, it is important to detect such activities using surveillance so that governing authorities must be aware of the arson development.

1.1 Motivation

There have been previous attempts to study arson. There have been qualitative models for the arson's domain [3, 4]. The advent of Internet and smartphones has heated the debate whether they add fuel to more arsons [5, 6]. Urbanization of cities has forced people to migrate in big cities in search of opportunities. The law and order present a huge challenge in such multi-cultural society. Naturally disagreements and dissents would be observed in such places. The focus is on cities and safety of its infrastructure. Most of the city places are under CCTV surveillance. Important public place, railways stations, airport and prominent market places are under surveillance. This input feed can be utilized for arson activity classification. The process of arson activities is taken by the group of dissenter or protestor which have a personal or community grudge against another community or governing authorities. They believe that arson activities can seek the attention of concerned parties to the matter in their hands. Usually arson activity starts from s group of people which gathers at the same place and starts with protests. The protests take the center stage when a small group of people starts gathering at a common place and then add strength to their numbers. They display their intentions by showing placards, posters and banners and adding emotional strength to it. Similarly, on this line stone pelting has emerged as a strong violent form of option exercised by dissenters all over the world. Stone pelting has been widely reported in India, Israel [7]. Security forces sustains grievance injury in these stone pelting activities. This experiment uses MobileNet [8] for generating a model that classifies arson activities from normal activities. The model was initiated using the concept of transfer learning from a model trained specifically on the ImageNet Large Visual Recognition Challenge dataset [9]. For stone pelting classification, VGG Net has been used [10] which is a very popular pretrained model used for transfer learning. The papers conceptualize a real time solution to detect arson activity and stone pelting in any input images or videos. It becomes important to classify arson activity and stone pelting and subsequently monitor them to immediately stop dissenters from destroying more properties and control the damage. Arson activity amounts to extreme inhuman form of cruelty especially in a country like India where poor people earn their livelihood on streets by exhibiting their daily products. Stone pelting in one of the Indian states have resulted in death of Indian security forces [11]. Hence the problem is critical and needs to have solution via automatic surveillance.

1.2 Research Contributions

Following are the research contributions of this paper.

- A review of current state-of-the-art approaches for the crowd related violent activities.
- Violence start from low level and spiral in to high level. The indicator for high level violence is arson and stone pelting. These two activities have the potential to cause great harm to human lives and destruction of private and government properties.
- Curating image datasets for arsons and normal activity images from Internet. Curating image datasets for stone pelting and normal activity images from Internet.
- A novel application of transfer learning approach using VGG16 to predict the arson activity classification of any image/video is proposed.
- A novel application of transfer learning using Inception V3 to predict the stone pelting activity classification of any image/video is proposed.
- Finally, the performance evaluation of the model is done using standard parameters.

1.3 Organization

Section 2 presents a brief introduction of transfer learning. Section 3 presents the existing work in the area of arson classification. Section 4 presents examine the overall framework of the arson classification architecture. Section 5 presents experiments and results of the arson classification. Section 6 ends the paper with discussion and key takeaways of the results.

2 Transfer Learning

Convolution Neural Networks (ConvNets) are an excellent tool for various computer vision applications [12–14]. Convolutional neural networks (CNNs) are tools for analysis of visual information [15]. An insight in to CNN has been aptly discussed in [16]. ConvNets have been successful as a feature extractor. There are two popular approaches for CNN based applications. Training the dataset from scratch is one approach and the other is to use transfer learning. The other approach and a popular one is to use the pretrained model which has been trained on a very large dataset. We have curated a small dataset of arson and normal activity images for using with pretrained models. Transfer learning uses pre-trained models and makes small changes in the architecture [17]. In deep learning, since large amount of data is needed to achieve good results; transfer learning is popular to avoid expensive training of new deep neural networks [18]. Hence usage of transfer learning has increased in number due to availability of many pretrained models such as AlexNet, VGG16, VGG19, and InceptionV3 for training curated dataset.

3   State-Of-The-Art

Arsons and violence in mass gatherings and their expected impacts on crowds and society is definitely a topic of interest for crowd safety and smooth functioning of governing authorities. From psychology to journalism, there have been attempts to understand the system. Recent attempts have been made in the fields of computer vision by utilizing social images of mass gatherings. One of the attempts was to use twitter images [19] for classifying facial attributes using prominent cues such as race, age and gender to identify the support base for U.S. politicians [20]. Analysis of images of prominent personalities of the countries [21] shared on social media has also been studied for carrying out the vision of governments and authorities to play their agendas. Public opinion about scheme of things of government and authorities have also been studied in journalism [22]. These analysis presents the importance of computer vision-based analysis in the field of multimedia. These images enhance the idea of perceived dissent among the public. These images/videos can be seen in the light of different topic such as full of creativity [23], interesting images [24] or promoting adult content [25]. Some previous work in the classification domain pertaining to social event detection has been done in the past [26, 27]. Most of these studies were focused on event classification using clustering as a key element, this paper is more focused on classifying arsons from any given images. There have been various studies of violence detection in images post mass gatherings in cities. These violence detection methods are based on ConvNets and its variation in 2D ConvNets and 3D ConvNets. The final aim of the arson is to make their voice heard to the higher governing authorities. Many a times arsons transform itself in to violent arsons. These arsons become violent arsons. There have been many instances of utilizing ConvNets in deciphering violent activity recognition in the input images and videos. ConvNets stream using 3D CNN using different classifiers have been tested for violence detection [16]. A unique method using Bidirectional Convolutional LSTM has been explained demonstrated for violence detection [28]. A unique method of flow detection for violence detection is described in [29]. Another method includes extraction of visual cues for violence scene detection (VSD) in the input images and videos [30]. These cues help the model to identify violent and non-violent activities. The paper is deeply inspired by the convolution methods of violence detection. A recent arson activity detection method introduced its own datasets and came out with the concept of perceived violence detection [31]. The paper is specifically targeting classification of arsons as compared to non-arson images. The use of 2D ConvNets using transfer learning approach gives us the platform for a deep learning model which can be used for arson classification in crowd related images.

4 Proposed Approach

In this section, we implement the proposed methods for the classification of employee performance. The paper focusses on the pretrained network which needs retraining on the custom dataset. Our novel framework exploits the transfer learning approach in which we have tried to maximize the existing capability of using pre-trained ConvNets and produce discriminant feature extractor. This is deployed on the new riot dataset that has been introduced in this paper. Post feature extraction we use standard classification approach to classify the images in to riot and non-riot images. Our curated dataset consists of sequences of images. We have used most popular VGG16 [10] architecture as it was one of the best performing architecture in ILSVRC challenge 2014.It was also a runner up in the classification task with top-5 classification error of 7.32%. VGG16 was also the winner of localization task with 25.32% localization error. The second approach is using Inception-V3 [32]. This model is trained on the ImageNet datasets. We utilize Inception V3 model by retraining the last layer. The number of output nodes in the last layer is equal to the number of the classification categories which are identification of stone pelting and arson separately from normal behavior. The simple philosophy is to curate the respective datasets, choose the best pre trained model as per domain preference (VGG16 and Inception V3 in this case), freeze all the previous layers of these models, train the final dense layers and classify the problem as per class definition.

The background platform used is Keras and Tensorflow. The model can be extended for mobile application as well as embedded applications too [33]. We can clearly infer that the pretrained model has used the weights of ImageNet which is generic for the image processing domain. The previous layers trainability is set to false. The top layers are used especially the dense layers for the classification of the activities which in our case is binary. It is important to note that the last dense layer is used for classification among two groups of datasets, namely arson and normal activities. A proper arson dataset was curated using Bing image downloader [34]. Our dataset contains 5000 images which have been collected using Bing image downloader from popular search engine. The dataset consists of 5000 images belonging to 2 classes. Generic approach is to use the class division in to two categories. These classes are annotated as arson and normal categories in one category and stone pelting and normal in second category. Each arson image usually consists of fires, burning vehicles, torched houses, throwing fire and so on. There can be multiple approaches to detect arson as a whole by classifying all the above-mentioned criteria (Figs. 4 and 5).

Below is the Inception V3 architecture diagram used for classification of stone pelting:

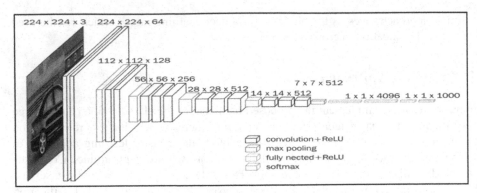

Fig. 3. Pretrained VGG16 architecture used in arson classification [35]

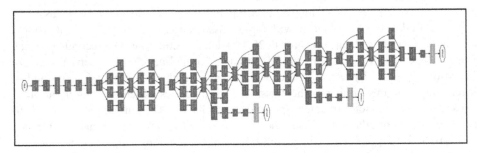

Fig. 4. Inception V3 adapted from [36].

5   Experiment and Results

The concept is proven by utilizing the dataset that is curated using popular search engines and Bing image downloader. The primary aim is to detect arsons in any given image. To justify the arsons activity detection and stone pelting detection, we have chosen normal crowd activity such as walking, strolling, celebrating and many more as normal datasets. Between these two activities, the model is initiated for training on this dataset. For the implementation of proposed arson detection in crowded places, we used standard Keras application with Python on Ubuntu operating system and experiments were conducted on a standard machine (Intel Core i5,8 GB RAM) equipped with NVIDIA TITAN X GPU. The training was performed to generate a model file. Since the dataset is small the number of epochs is set to 30. The model is described in Fig. 3. The model uses Keras [37] version 2.3.1 and the underlying environment is that of Tensorflow. We have used the metric accuracy for image classification evaluation. Accuracy is defined as the proportion of tweets that has been correctly classified among all image content. Accuracy is a very intuitive metric. It is computed using below equation.

$$\text{Accuracy} = \frac{\text{TP} + \text{TN}}{\text{TP} + \text{TN} + \text{FP} + \text{FN}} \times 100\% \tag{1}$$

Our model achieved 97.25% accuracy on the validation set.

Fig. 5. Arson and celebration classification results

Figure 7 clearly brings out the results of our approach for successful modelling of arson classification. The model accuracy and loss are depicted in the figure below (Fig. 6):

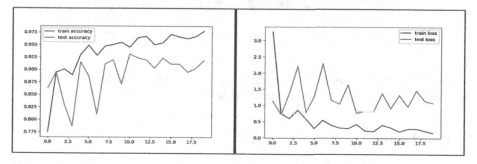

Fig. 6. Training accuracy and loss diagram for arson identification

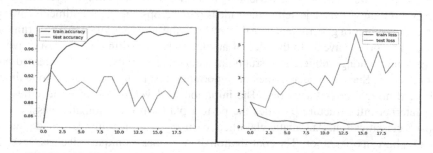

Fig. 7. Training accuracy and loss diagram using inception V3 of stone pelting

The model achieved the accuracy of 97.25% in just 20 epochs. This is the power of transfer learning. After curating a dataset of similar nature as that of pretrained model, and then training and then train the model in a smaller number of epochs.

The results of stone pelting are as follows:

Resulting images of stone pelting model is as follows (Fig. 8):

Fig. 8. Stone pelting results identified using inception V3

The Inception V3 model achieved more than 98% of the accuracy on the curated stone pelting dataset. The results are depicted below:

6 Conclusion

Arson is seen everywhere as the amount of protests has increased everywhere in the world . Dissent and protests are normal in a country like India where democracy is the main stay of the governance . Stone Pelting is a common phenomenon in one of the India's state where mob shows their extreme dissent by resorting to stone pelting. Psychologically when a community gets dejected in their ideas or develops negative sentiments towards other communities of government then they resort to the arson activities. All governments around the world is averse to the idea of arson and stone pelting like activities which creates panic among public and disrupts the economic activity of the concerned area or community. Surveillance activities of important cities is important so as to maintain peace and normal crowd activities. The initiation of development of smart cities and huge Internet infrastructure provides the perfect platform for automatic classification of arson classification. Automatic classification using deep learning has made the job easier for human orientation. The contribution of this paper is unique as it classifies stone pelting and arsons from normal activities. It is one of the most unique works that has been attempted to research in to journey between dissent and violence . By giving two models of different pre trained model, it has been established that identification of these heinous activities is possible and it shall become boon for security agencies. They can detect these activities using surveillance and can take evasive actions. In the future this model can be integrated with any Android technology for more generalized solution.

References

1. https://www.merriam-webster.com/dictionary/protest,   https://www.merriam-webster.com/dictionary/protest. Accessed 6 July 2020
2. https://en.wikipedia.org/wiki/Stone_pelting_in_Kashmir (2020). https://en.wikipedia.org/wiki/Stone_pelting_in_Kashmir. Accessed 29 Sep 2020
3. Liu, Z., Luo, P., Wang, X., Tang, X.: Deep learning face attributes in the wild. In: Proceedings of the IEEE International Conference on Computer Vision, pp. 3730–3738 (2015)
4. Steinert-Threlkeld, Z.C., Mocanu, D., Vespignani, A., Fowler, J.: Online social networks and offline protest. EPJ Data Sci. **4**(1), 1–9 (2015). https://doi.org/10.1140/epjds/s13688-015-0056-y
5. Leetaru, K., Wang, S., Cao, G., Padmanabhan, A., Shook, E.: Mapping the global Twitter heartbeat: the geography of Twitter. First Monday **18**(5) (2013)
6. Redi, M., O'Hare, N., Schifanella, R., Trevisiol, M., Jaimes, A.: 6 seconds of sound and vision: creativity in micro-videos. In: Proceedings of the IEEE Conference on Computer Vision and Pattern Recognition, pp. 4272–4279 (2014)
7. https://indianexpress.com/article/world/israeli-troops-killed-two-stone-pelters-in-west-bank-palestinian-officials-4746966/. Accessed 29 Sep 2020
8. Howard, A.G., et al.: Mobilenets: efficient convolutional neural networks for mobile vision applications. arXiv preprint arXiv:1704.04861 (2017)
9. Russakovsky, O., et al.: Imagenet large scale visual recognition challenge. Int. J. Comput. Vision **115**(3), 211–252 (2015)
10. Simonyan, K., Zisserman, A.: Very deep convolutional networks for large-scale image recognition. arXiv preprint arXiv:1409.1556 (2014)
11. https://en.wikipedia.org/wiki/Stone_pelting_in_Kashmir#:~:text=On%2025%20October%202018%2C%20an,22%20year%20old%20from%20Uttarakhand. Accessed 29 Sep 2020
12. Krizhevsky, A., Sutskever, I., Geoffrey, E.H.: Imagenet classification with deep convolutional neural networks. In: Advances in Neural Information Processing Systems (NIPS 2012), vol. 25 (2012)
13. Simonyan, K., Zisserman, A.: Two-stream convolutional networks for action recognition in videos. In: Advances Neural Information Processing Systems (2014)
14. Girshick, R., Donahue, J., Darrell, T., Malik, J.: Rich feature hierarchies for accurate object detection and semantic segmentation. In: CVPR (2014)
15. LeCun, Y., Kavukcuoglu, K., Farabet, C.: Convolutional networks and applications in vision. In: Proceedings of 2010 IEEE International Symposium on Circuits and Systems (2010)
16. Perez, M., Kot, A.C., Rocha, A.: Detection of real-world fights in surveillance videos. In: International Conference on Acoustics, Speech and Signal Processing (ICASSP) (2019)
17. Torrey, L., Shavlik, J.: Transfer learning. In: Handbook of Research on Machine Learning Applications and Trends: Algorithms, Methods, and Techniques, pp. 242–264. IGI global (2010)
18. Gupta, D., Jain, S., Shaikh, F., Singh, G.: Transfer learning \& the art of using pre-trained Models in deep learning. Anal. Vidhya (2017)
19. Tufekci, Z.: Big questions for social media big data: representativeness, validity and other methodological pitfalls. In: Eighth International AAAI Conference on Weblogs and Social Media (2014)

20. Little, A.T.: Communication technology and protest. J. Polit. **78**(1), 152–166 (2016)
21. Yang, G.: Achieving emotions in collective action: emotional processes and movement mobilization in the 1989 Chinese student movement. Sociol. Q. **41**, 593–614 (2000)
22. Isola, P., Xiao, J., Torralba, A., Oliva, A.: What makes an image memorable? In: CVPR 2011, pp. 145–152. IEEE (2011)
23. Petkos, G., Papadopoulos, S., Schinas, E., Kompatsiaris, Y.: Graph-based multimodal clustering for social event detection in large collections of images. In: Gurrin, C., Hopfgartner, F., Hurst, W., Johansen, H., Lee, H., O'Connor, N. (eds.) MMM 2014. LNCS, vol. 8325, pp. 146–158. Springer, Cham (2014). https://doi.org/10.1007/978-3-319-04114-8_13
24. González-Bailón, S., Borge-Holthoefer, J., Moreno, Y.: Broadcasters and hidden influentials in online protest diffusion. Am. Behav. Sci. **57**(7), 943–965 (2013)
25. Fisher, D.R.: Studying Large-Scale Protest: Understanding Mobilization and Participation at the People's Climate, March (2014). http://www.sindark.com/phd/thesis/sources/PCM_Pre liminaryResults.pdf
26. Parikh, D., Grauman, K.: Relative attributes. In: 2011 International Conference on Computer Vision, pp. 503–510. IEEE (2011)
27. Petkos, G., Papadopoulos, S., Kompatsiaris, Y.: Social event detection using multimodal clustering and integrating supervisory signals. In: Proceedings of the 2nd ACM International Conference on Multimedia Retrieval, pp. 1–8 (2012)
28. Hanson, A., PNVR, K., Krishnagopal, S., Davis, L.: Bidirectional convolutional LSTM for the detection of violence in videos. In: Leal-Taixé, L., Roth, S. (eds.) ECCV 2018. LNCS, vol. 11130, pp. 280–295. Springer, Cham (2019). https://doi.org/10.1007/978-3-030-11012-3_24
29. Sumon, S.A., Shahria, M.T., Goni, M.R., Hasan, N., Almarufuzzaman, A., Rahman, R.M.: Violent crowd flow detection using DEEP learning. In: Asian Conference on Intelligent Information and Database Systems (2019)
30. Mu, G., Cao, H., Jin, Q.: Violent scene detection using convolutional neural networks and deep audio feature. In: Chinese Conference on Pattern Recognition (2016)
31. Won, D., Steinert-Threlkeld, Z.C., Joo, J.: Protest activity detection and perceived violence estimation from social media images. In: Proceedings of the 25th ACM International Conference on Multimedia, pp. 786–794 (2017)
32. Szegedy, C., Vanhoucke, V., Ioffe, S., Shlens, J., Wojna, Z.: Rethinking the inception architecture for computer vision. In: Proceedings of the IEEE Conference on Computer Vision and Pattern Recognition, pp. 2818–2826 (2016)
33. Harvey, M.: Creating insanely fast image classifiers with MobileNet in TensorFlow. Hacker Noon (2017)
34. https://github.com/ostrolucky/Bulk-Bing-Image-downloader. Accessed 06 July 2020
35. https://neurohive.io/en/popular-networks/vgg16/, https://neurohive.io/en/popular-networks/vgg16/. Accessed 05 Sep 2020
36. https://alexisbcook.github.io/2017/using-transfer-learning-to-classify-images-with-keras/. Accessed 05 Sep 2020
37. Chollet, F.: o. "Keras," GitHub (2015)

A Scientometric Review of Digital Economy for Intelligent Human-Computer Interaction Research

Han-Teng Liao, Chung-Lien Pan(✉), and Jieqi Huang

Higher Education Impact Assessment Center, Sun Yat-Sen University Nanfang College, Guangzhou, China
peter5612@gmail.com

Abstract. As the economy goes digital, Human-Computer Interaction (HCI) professionals have been helping companies to better understand the usability, experience, and thus profitability issues, suggesting the contribution of HCI professionals to the digital economy. However, there is no comprehensive review or theory-driven work that comes from the research area of the digital economy itself. This exploratory study, based on a scientometric analysis of digital economy literature, aims to outline the possibilities and application areas for future research and policy development for HCI research and its intelligent applications. By identifying and analyzing top key authors from 2,778 articles and their more than 100,000 citations, collected from the Web of Science database, the study reveals a dense network with a few clusters of concepts and research work.

Keywords: Human-Computer Interaction · Digital economy · Platform economy · Artificial intelligence

1 Introduction

As the economy goes digital, Human-Computer Interaction (HCI) professionals have been helping companies to better understand the usability, experience, and thus profitability issues, suggesting the contribution of HCI professionals to the digital economy [1]. Indeed, the main UK funding body, the Engineering and Physical Sciences Research Council (EPSRC) as one of the 100+ research areas, aims for an HCI research portfolio "that contributes high-quality work relevant to the Digital Economy Theme and development of the Internet of Things", with strategies to support people interaction with complex and intelligent systems, under EPSCR's "Future Intelligent Technologies cross-ICT priority, and enable interaction with and action from data, under EPSCR's "Data Enabled Decision Making cross-ICT priority" [2]. Thus, data-driven intelligent HCI design research must keep renewing the ongoing exchanges between the HCI and artificial intelligence (AI) communities so as to advance better design generation and evaluation of AI practices for better interaction design [3], especially on the topic of the digital economy. Managing the digital economy effectively means that digital technologies are applied effectively for various economic transactions, processes, activities, and interactions, and thus HCI professionals should participate in its design.

© Springer Nature Switzerland AG 2021
M. Singh et al. (Eds.): IHCI 2020, LNCS 12615, pp. 469–480, 2021.
https://doi.org/10.1007/978-3-030-68449-5_45

Previous research has explored how HCI research can be applied to AI systems design and vice versa [3], with several conferences such as the ACM International Conference on Intelligent User Interfaces (ACM ICI), the International Conference on Intelligent Human-Computer Interaction (IHCI), etc. However, to the best our knowledge, there is no comprehensive review or theory-driven work that comes from the research area of the digital economy itself. A better understanding of digital economy literature, as a broader theoretical lens, is expected to be useful to elevate the level of considerations higher, from the lower level of computer systems and the context of use, to the higher level of social and organizational dynamics. With the aim to broaden the pool of ideas and capabilities of intelligent HCI design generation and evaluation, it is important to provide a systematic review of digital economy literature to outline the ways in which HCI design can better shape the worldwide network of livelihood activities, commercial transactions and everyday interactions enabled and empowered with the use of information and communications technologies (ICT).

2   Research Methodology and Data

With the purpose to systematically collect the relevant literature on related concepts on the digital economy, the study has used snowballing process until no more significant and directly relevant author keywords were found in the pool of articles.

Focusing on key authors and the associated concepts, the overall research design is based on techniques such as co-word, co-citation and bibliographic analyses. We have used the following advanced query string in Clarivate Analytics' Web of Science (WoS), on September 1, 2020:

TS = "information economy" OR TS = ("digital economy" OR "digital market" OR "digital capitalism" OR "digital eco*systems") OR TS = "Data is the new oil" OR TS = ("data economy" OR "data innovation*" OR "data market" OR "data capitalism" OR "data eco*system" OR "data-driven economy" OR "data-driven innovation*" OR "data-driven economy market" OR "data-driven economy capitalism" OR "data-driven eco*system") OR TS = ("platform economy" OR "platform market" OR "platform capitalism" OR "platform eco*systems")

Aiming to cover the broad concepts related to the digital economy, 2,778 articles were collected from the WoS Core Collection of SCI-EXPANDED, SSCI, A&HCI, and ESCI index. Their citations, more than 100,000 of them, were also parsed, cleaned, and analyzed using Python scripts. Both the article and citation data sets are then processed by VOSviewer and Python data science packages for analysis and insights.

With the aim to identify the main ideas of the digital economy literature, the key authors and their main keywords are extracted from the citations. Also, via iterative processes of examining author keywords, a thesaurus is built. Such a thesaurus is helpful in organizing keywords into sets of concepts. Special attention is paid to the literature on the ways in which big data, artificial intelligence, etc. can be used as pathways for "digital public goods" and "digital inclusion", both of which are part of the main policy agenda by the UN for building an inclusive digital economy and society [4].

3 Findings

3.1 Annual Trends and Periodization

Analyzing how research has evolved over time, Fig. 1 shows the piecewise linear regression results of the number of research articles. Much faster growth begins after 2015, as evidenced by the steeper upward tilt of regression lines. Closer examination of periodized data shows that during the first period, the main disciplines are Economics and Management, with main topics of "information economy", "digital economy" and "Internet". The trend shifted towards the topics such as "digital economy", "platform economy", "digitalization", "platform capitalism" and "big data", with more diverse disciplines such as Communication, Business, Information Science, Computer Science, etc.

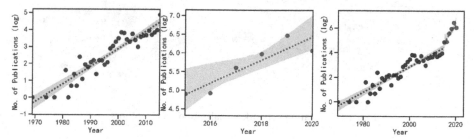

Fig. 1. Annual publications and time periods: piecewise linear regression results

In particular, the rise of "platform economy", "platform capitalism" and "big data" suggests the central role of data-driven and platform-based characteristics of more recent phenomena of digital economy, and the following discussions on key authors and their concepts should provide more details.

3.2 Key Authors: Based on Co-citations and Their Network

To explore the digital economy research fronts [5] based on key authors and main topics, we first examine where current research fronts of the digital economy have been *citing* at the level of authors. Co-citation networks, showing the shared links among citing and cited articles, can be used to observe similarities among areas of knowledge, and bodies of work. Co-citation networks, at the level of authors, not only show the number of citations an author (represented as a node) has received, but also show how often a pair of authors have been cited in the literature (represented as a link or edge), thereby providing meaningful insights into how related two authors are in the overall structure of co-citation networks.

Figure 2 shows two separate clusters at the left and right, with four other clusters connecting them in the middle. The red cluster at the right mainly discusses "platform markets", whereas the green cluster at the left discusses topics such as "informational capitalism", "digital labor", "post-industrial society", etc. The clusters of cited authors also help in organizing key authors and key ideas, as shown in Table 1.

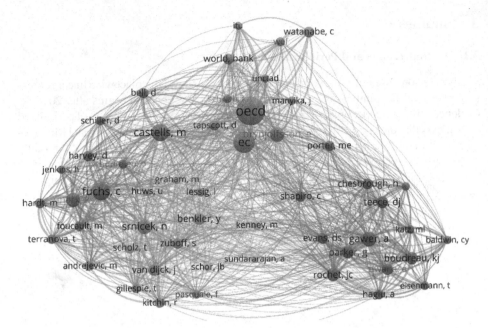

Fig. 2. Co-citation network visualization: cited authors.

3.3 Key Ideas and Implications for Intelligent HCI

The clustering outcomes based on co-citation networks (see Fig. 2) help to organize key authors as shown in the first two columns in Table 1. The paper will first outline the key authors and corresponding concepts based on the clustering outcomes. Then it will summarize the key ideas as different conceptual perspectives, followed by brief discussions on the implications for intelligent HCI research and design.

Key Authors and Concepts. Based on the affinity feature of co-citation networks [5], the cluster membership should help to organize key authors and therefore concepts.

The green cluster at the left side of Fig. 2 contains authors such as Christian Fuchs [6], Manuel Castells [7], and Henry Jenkins [8], with key concepts such as "information capitalism" [6], "network society" [7] and "convergence culture" [8]. Such work appears to focus on the political economy and socio-economic conditions of Internet society, from the disciplines of communication, sociology, and culture. The cluster contains work regarding the social and cultural aspects of change brought by information and communication technologies, and thus provide authoritative work on our continuous efforts in understanding the impact of digital technologies on economic, social, and cultural fronts, and the different impact patterns for different people groups.

The red cluster at the right side of Fig. 2 include authors such as Annabelle Gawer [9], Jean-Charles Rochet [10], and Henry William Chesbrough [11], with key concepts such as "technological platforms" [9], "platform competition" [10], and "open innovation" [11]. Such work appears to focus on the perspectives of markets and firms on platforms and innovations, mostly from the disciplines of economics and management.

Table 1. Key ideas and authors on discussing digital economy

Cluster	Authors	Concepts	Ideas
Green	Fuchs, C.	informational capitalism, digital capitalism, labor	New notions of class, labor, and knowledge labor, especially on user personal data and usage behavior
	Castells, M.	network society, Internet Galaxy	New forms of society and political economy with Internet becoming the essential communication and information medium
	Jenkins, H.	convergence culture	New power relationship on how media producers interact with consumers
Red	Gawer, A.	technological platforms	New concepts and hypotheses regarding platforms as evolving organizations, within firms, supply chains and ecosystems
	Rochet, J.C.	platform competition, two-sided markets	A new model of platform competition for most markets with network effects, especially in software, media, and the Internet industries where platforms must get both sides on board
	Chesbrough, H.	open innovation	A new paradigm for understanding industrial innovation as the purposive curation of inflows and outflows of knowledge for both the expansion of innovation and markets
Blue	OECD	tax challenges	An analysis of new challenges on taxation when digital economy becomes the economy itself
	Tapscott, D.	digital revolution	A new generation of computer, Internet, and digital technologies, with new ways in which individuals and society interact

(*continued*)

Table 1. (*continued*)

Cluster	Authors	Concepts	Ideas
Yellow	Srnicek, N.	platform capitalism	A new form of business and capitalism new business forms, by providing software or hardware foundation for others to operate on
	van Dijck, J.	platform society, cultural connectivity on social media	A new analytical framework to understand online platforms and social media through cultural connectivity and public values
Purple	Huws, U.	digital labor, gig economy	A new form of employment where "on-demand" work is managed by online platforms, challenging labor standards and protections
	Graham, M.	digital labor, livelihoods, gig economy	A new summary and analysis of digital labor in gig economy with four key concerns: "bargaining power, economic inclusion, intermediated value chains, and upgrading"
	Schor, J.B.	sharing economy	A new distinction between for-profit and non-profit platforms, suggesting the need for democratization of the ownership and governance of the platforms
Cyan Blue	Benkler, Y.	wealth of networks, commons-based peer production	A new framework of "networked information economy" where the rise of effective non-market, non-proprietary forms of social production becomes viable and critical

(*continued*)

The blue cluster at the top side of Fig. 2 includes authors such as OECD [12], European Commissions, and Don Tapscott [13], with key concepts such as "digital economy" and "digital revolution" [13]. Such work appears to be published earlier. The OECD report on tax challenges has signaled a strong interest in taxation when the digital economy becomes the economy itself. Don Tapscott's work on digital economy in 1996 [13] focuses on the new ways in which-individuals and society interact.

The yellow cluster at the center of Fig. 2 includes more recent authors on "platform capitalism" [14] and "platform society" [15]. Such work appears to provide a synthesis

Table 1. (*continued*)

Cluster	Authors	Concepts	Ideas
	Lessig, L.	coded architecture, software architecture	A new regulatory or even legal regime that increasingly regulates our lives through software architecture

of the green cluster of internet society and culture and the red cluster of platforms. Similarly, the purple cluster features the work on digital labor and gig economy [9, 16, 17] and better sharing economy platforms [18], providing a similar synthesis that focuses more on the labor issues on platforms.

Finally, the cyan blue cluster features Yochai Benkler's notion of "networked information economy" [19] and Lawrence Lessig's notion of coded architecture or software architecture [20]. Such work appears to provide a technology and law perspective suggesting new social production is made possible via software codes.

Key Ideas: Structuring and Arranging Interactions in Digital Economy. Based on the above discussions on key authors and concepts, this section further synthesizes the corresponding ideas, so as to see how interactions are structured and arranged in the digital economy, thereby exploring the ways in which we can make such interactions more intelligent.

Platform Economy Perspective. As defined by [21], a platform enables value-creating interactions between producers and consumers, thereby harnessing the power of network effects to scale such value-creation through successful platform design.

Thus, the platform perspective first highlights the importance of technological platforms that shape interactions among platform participants. Better and more advanced interactions and their design should maximize the network effects from the lens of the platform economy. Also, looking beyond the firm organizations, "technological platforms" can be seen as evolving new forms of organizations, within firms, supply chains and ecosystems [9], with exemplar case studies in software, media and the Internet industries that demand new forms of competition and collaboration in a two-side market [10], and also a new innovation paradigm that regulates the inflows and outflows of knowledge to expand both the innovation and marketing outcomes, preferably via "open innovation" [11].

For intelligent HCI, the platform perspective of the digital economy provides several insights. First, the design of platform itself can be defined as the design of "core interactions" between producers and consumers [21] in a two-sided market [10], involving key components such as the participants, the value unit, and the filtering mechanisms. Then, intelligent interventions can be introduced or even designed with data-driven mechanisms to recruit new participants, generate new value units, and facilitate better recommendation systems for filtering. Finally, the market success in the competition should be evaluated with more advanced and smarter machine-learning models that can capture the complex dynamics of eco-systems.

Digital Labor and Capitalism Perspective. As proposed by [6], new notions of class, labor and knowledge labor are needed to understand the use of user personal data and usage behavior data, which indeed constitutes the main power source to drive a platform.

Thus, the digital labor perspective first emphasizes the importance of human investment and participation in online platforms. As already empirically studied by [16, 17, 22], workers' key concerns on bargaining power, economic inclusion, intermediated value chains, and capability-upgrading must be addressed, especially in the gig economy where platforms dominate the production and consumption of labor. Also, in making sure the fruits of labor fairly distributed to avoid the unjust accumulation of capital, the digital labor and capitalism perspective highlights the responsibility of platform or interaction designers to arrange on-demand work, so as to comply with labor standards and protections, or even more proactively, to improve the livelihoods and well-beings of users and workers. Finally, the notion of "technology for good" or "AI for good" [23, 24] should be introduced, combining with more just sharing economy platform design [18], so as to achieve better democratization of the ownership and governance of the platforms through design.

Platform Governance Perspective. Based on the cross-disciplinary work by legal scholars such as Yochai Benkler's [19] and Lawrence Lessig's [20], the coded nature of software and software developers in producing new governance or even "law" within the platforms highlight the rule-making power of interaction and platform designers in shaping the platform dynamics.

Thus, the platform governance perspective demonstrates the rule-making and normative culture of platform and interaction design. Such an idea can be found in the latest wording by the United Nations on digital collaboration to "promote open-source software, open data, open artificial intelligence models, open standards and open content that adhere to privacy and other applicable international and domestic laws, standards and best practices and do no harm" [4]. In addition, the same document also argues for the norm of explainable AI that "autonomous intelligent systems should be designed in ways that enable their decisions to be explained and humans to be accountable for their use". Hence, the platform governance perspective provides important insights for intelligent HCI to make sure that the issues of explainability and accountability of technological governance are addressed alongside the issues of usability and user experience.

Table 2 summarizes the three perspectives of the digital economy, and the implications for HCI and intelligent HCI, as discussed above.

Overall, the platform perspectives appear to dominate the Digital Economy literature perspectives, with one perspective ("Platform Economy") focusing more on the market implications and the other ("Platform Governance") perspective focusing more on the governance implications.

Nevertheless, the perspective of "Digital Labor and Capitalism" remains important, pointing to the continuous efforts in understanding the advancement of digital technologies, which is expected to digitally transform our fabric of economic, social, and cultural lives, may or may not be welcomed and accepted by different people groups. From the collected literature dataset, critical work such as a systematic theoretical discussion of risk, innovation, and democracy in the digital economy [25], a geospatial critique of Malaysia's Multimedia development with the associated financial and social costs [26],

Table 2. Digital economy perspectives for intelligent HCI

Perspectives	HCI	Intelligent HCI
Platform economy	successful design of core interactions between producers and consumers to harness network effects	data-driven mechanisms to model the complex dynamics of platforms for smarter firms, supply chains and ecosystems
Digital labor and capitalism	fair design and distribution of labor-based value creation (including user data and user behavior data)	fair design and distribution of labor-based value creation (including user data and user behavior data-driven intelligent models) and collective intelligence
Platform governance	explainable and accountable of interaction design	explainable and accountable of intelligent interaction design

and an education reform proposal that integrates the digital, technology and citizenship into a notion of "radical digital citizenship" [27], etc., point to the need to design and implement the digital economy that works for more people groups, or in other words, more equal or universal design of the digital economy. The HCI professionals can benefit from such a body of literature that questions the digital remaking of interpersonal and solitary life, the socio-technical threats of AI on employment, etc. [25].

Such continuous concerns for equality in the network society or the digital economy from the "Digital Labor and Capitalism" perspective do have implications for both "Platform Economy" and "Platform Governance" perspectives, especially in prioritizing policy and design interventions when it comes to platforms. For instance, whether and how platforms of sharing economy increase inequality [28, 29], reshape global labor management practices [30, 31], etc., requires continuous empirical work that may in the near future incorporate more data and design interventions contributed by HCI professionals and researchers. Indeed, as reported and studied by Wood et al. [30], these mechanisms of controlling labor on gig economy platforms can result in overwork, sleep deprivation, low pay, isolation, and irregular hours, and such mechanisms are increasingly implemented by algorithmic control that regulates the platform operations. Such control mechanisms are indeed increasingly implemented by the work of intelligent HCI professionals and researchers. Given the important legacy of "Digital Labor and Capitalism" perspective of the digital economy, it is then also important to keep the legacy in mind when pursuing the newer perspectives of "Platform Economy" and "Platform Governance".

4 Conclusion

Overall, facing a wide variety of new phenomena due to the impacts of digital and ICTs, researchers and policy-makers have so far come up with several notions and conceptual frameworks to understand and address the opportunities and challenges regarding rethinking and redesigning our economic and social activities. Questions of digital labor, digital literacy, digital divides, digital attention, etc., must address the well-being and sustainable livelihoods of individuals on earth, arguably the most important aspect of the 2030 agenda for sustainable development [4].

Based on the findings, the paper concludes with key ideas and implications of the digital economy literature for intelligent HCI, as detailed above in Table 1, Table 2 and Sect. 3.3. The "Platform Economy" perspective focuses more on the market implications, the "Platform Governance" perspective more on the governance implications, and the "Digital Labor and Capitalism" perspective more on the wider impact of technologies on important equality and justice issues. For HCI professionals, it means that when interaction design becomes central to the design of platforms, which is central to the dynamics and mechanisms of the digital economy, wider system thinking is required to consider different people groups and stakeholders.

Some future research directions surrounding the digital economy for intelligent HCI can be derived from the discussions above. The platform economy perspective highlights the importance of data-driven mechanisms to model the complex dynamics of platforms for smarter firms, supply chains, and ecosystems. The digital labor and capitalism perspective highlights the significance of fair design and distribution of labor-based value creation, including user data and user behavior data-driven models and collective intelligence. The platform governance perspective of the digital economy indicates the priority for developing explainable and accountable for intelligent interaction design. To sum up, as the life and workplace have become more digitized, networked, and even automated, the important notions of individual and social well-being require better integration of, on one hand, the digital labor and digital work, and on the other hand, the digital and increasingly data-driven life of consumption and meaning-making.

Acknowledgment. The research is funded by a project of Smart App Design Innovation Research in the Age of New Business, Arts and Engineering Disciplines (2019GXJK186), under the 2019 Guangdong Education Grants, China.

References

1. Lohse, G.J.L.: Usability and profits in the digital economy. In: McDonald, S., Waern, Y., Cockton, G. (eds.) People and Computers XIV—Usability or Else!, pp. 3–15. Springer, London (2000). https://doi.org/10.1007/978-1-4471-0515-2_1
2. EPSRC: Research areas: Human-computer interaction: Human-computer interaction, https://epsrc.ukri.org/research/ourportfolio/researchareas/hci/#2. Accessed 27 Sept 2020
3. Blandford, A.: Intelligent interaction design: the role of human-computer interaction research in the design of intelligent systems. Expert Syst. **18**, 3–18 (2001). https://doi.org/10.1111/1468-0394.00151
4. United Nations General Assembly: Road map for digital cooperation: implementation of the recommendations of the High-level Panel on Digital Cooperation (2020)
5. Garfield, E.: Research fronts. Current Comments (1994)
6. Fuchs, C.: Labor in informational capitalism and on the internet. Inf. Soc. **26**, 179–196 (2010). https://doi.org/10.1080/01972241003712215

7. Castells, M.: The Internet Galaxy: Reflections on the Internet, Business, and Society. Oxford University Press, Oxford (2003)
8. Jenkins, H.: Convergence Culture: Where Old and New Media Collide. New York University Press, New York (2008)
9. Gawer, A.: Bridging differing perspectives on technological platforms: toward an integrative framework. Res. Policy **43**, 1239–1249 (2014). https://doi.org/10.1016/j.respol.2014.03.006
10. Rochet, J.-C., Tirole, J.: Platform competition in two-sided markets. J. Eur. Econ. Assoc. **1**, 990–1029 (2003). https://doi.org/10.1162/154247603322493212
11. Chesbrough, H.W.: A new paradigm for understanding industrial innovation. In: Chesbrough, H.W., Van Haverbeke, W., West, J. (eds.) Open Innovation: Researching a New Paradigm. Oxford Univ. Press, Oxford (2006)
12. OECD: Addressing the Tax Challenges of the Digital Economy, Action 1 - 2015 Final Report. OECD (2015). https://doi.org/10.1787/9789264241046-en
13. Tapscott, D.: The Digital Economy: Promise and Peril in the Age of Networked Intelligence. McGraw-Hill, New York (1996)
14. Srnicek, N., De Sutter, L.: Platform Capitalism. Polity, Cambridge, Malden (2017)
15. van Dijck, J., Poell, T., de Waal, M.: The Platform Society. Oxford University Press, New York (2018)
16. Huws, U.: Labor in the Global Digital Economy: The Cybertariat Comes of Age. Monthly Review Press, New York (2014)
17. Graham, M., Hjorth, I., Lehdonvirta, V.: Digital labour and development: impacts of global digital labour platforms and the gig economy on worker livelihoods. Transf. Eur. Rev. Labour Res. **23**, 135–162 (2017). https://doi.org/10.1177/1024258916687250
18. Schor, J.: Debating the sharing economy. J. Self-Gov. Manag. Econ. **4**, 7 (2016). https://doi.org/10.22381/JSME4320161
19. Benkler, Y.: The Wealth of Networks: How Social Production Transforms Markets and Freedom. Yale University Press, New Haven, London (2006)
20. Lessig, L.: Code and Other Laws of Cyberspace. Basic Books, New York (1999)
21. Parker, G.G., Alstyne, M.W.V., Choudary, S.P.: Platform Revolution: How Networked Markets Are Transforming the Economy and How to Make Them Work for You. W. W. Norton & Company, New York (2016)
22. Huws, U., Spencer, N., Syrdal, D.S., Holts, K.: Work in the European gig economy : research results from the UK, Sweden, Germany, Austria, The Netherlands, Switzerland and Italy (2017)
23. Taddeo, M., Floridi, L.: How AI can be a force for good. Science **361**, 751–752 (2018). https://doi.org/10.1126/science.aat5991
24. ITU: AI for Good Global Summit, Geneva, Switzerland. https://aiforgood.itu.int/. Accessed 11 Nov 2019
25. Curran, D.: Risk, innovation, and democracy in the digital economy. Eur. J. Soc. Theory **21**, 207–226 (2018). https://doi.org/10.1177/1368431017710907
26. Bunnell, T.: Multimedia utopia? A geographical critique of high-tech development in Malaysia's multimedia super corridor. Antipode **34**, 265–295 (2002). https://doi.org/10.1111/1467-8330.00238
27. Emejulu, A., McGregor, C.: Towards a radical digital citizenship in digital education. Crit. Stud. Educ. **60**, 131–147 (2019). https://doi.org/10.1080/17508487.2016.1234494
28. Schor, J.B., Attwood-Charles, W.: The "sharing" economy: labor, inequality, and social connection on for-profit platforms. Sociol. Compass **11**, e12493 (2017). https://doi.org/10.1111/soc4.12493
29. Schor, J.B.: Does the sharing economy increase inequality within the eighty percent?: findings from a qualitative study of platform providers. Cambridge J. Regions Econ. Soc. **10**, 263–279 (2017). https://doi.org/10.1093/cjres/rsw047

30. Wood, A.J., Graham, M., Lehdonvirta, V., Hjorth, I.: Good gig, bad gig: autonomy and algorithmic control in the global gig economy. Work Employ Soc. **33**, 56–75 (2019)
31. Lehdonvirta, V., Kässi, O., Hjorth, I., Barnard, H., Graham, M.: The global platform economy: a new offshoring institution enabling emerging-economy microproviders. J. Manage. **45**, 567–599 (2019). https://doi.org/10.1177/0149206318786781

eGovernance for Citizen Awareness and Corruption Mitigation

A. B. Sagar[1(✉)], M. Nagamani[1], Rambabu Banothu[2], K. Ramesh Babu[1], Venkateswara Rao Juturi[3], and Preethi Kothari[4]

[1] HCU, Hyderabad, India
bablusagar@gmail.com
[2] IIIT Hyderabad, Hyderabad, Telangana, India
[3] American International Group Inc., New York, USA
[4] National Informatics Center, New Delhi, India

Abstract. 'Corruption' can be defined as the exploitation of delegated authority for illegitimate profits. We consider the kind of corruption happening on several levels in the country in regards to infrastructure and public works. Corruption can be identified by studying the issue from both sides—the government and the public. We look at how ICT, with the help of citizens, can help in identifying misuse and corruption. Corruption is a big menace in India. Transparency and accountability are generally improved when we enable support and citizen involvement in addition to closer collaboration between government and citizens. We dont claim that our model is not as such a complete remedy to fight corruption, and it can also be misused by fraudulent officials. Practically, impact depends on the aptness of ICT for local circumstances and requirements and in utilising technology. There are several corruption prone public services and citizens can be for effectively monitoring and mitigating corruption as citizens would generally like to participate in activities that would help in mitigating corruption at their provincial level. But there are several challenges and issues involved in providing such a platform and implement an effective workflow and dataflow, and also avoid misuse or abuse of the system. The proposed model discusses how ICT empowers the citizens with information to combat corruption and see that the abuse of the system is minimized by making information transparent to the citizens and letting them report on discrepancies.

Keywords: Corruption · Egovernance · Smart city

1 Introduction

1.1 Corruption and ICT

'Corruption' can be defined as the exploitation of delegated authority for illegitimate profits. We consider the kind of corruption happening on different levels in the country in regards to infrastructure and public works. Corruption can be

© Springer Nature Switzerland AG 2021
M. Singh et al. (Eds.): IHCI 2020, LNCS 12615, pp. 481–487, 2021.
https://doi.org/10.1007/978-3-030-68449-5_46

identified by studying the issue from both sides—the government and the public. We look at how ICT, with the help of citizens, can help in identifying misuse and corruption.

There is a general consent that ICTs have the power to make a major impact on combatting corruption [2]. Searching for role of ICTs in combatting corruption has been of much interest in the research community. There is enough ICT to enable a successful national integrity system [6,9,10,12,13]), and ICT is being considered by governments and civil societies as significant tool to help transparency and accountability and lessen corruption. ICT applications that have spawned a lot of interest, particularly among the media and anti-corruption specialists, are the crowd-based corruption-reporting applications. In the recent past, several corruption-reporting applications were publicised in several countries with the goal of drawing consideration to the scale, scope and geographic spread of corruption. These applications have taken benefit of the speedily increasing Internet and mobile technology in the developed world to offer an answer to bribery [8]. There are three main mechanisms through which these applications assist in the fight against corruption [14]): First, they provide an stress-free access to information These tools dissuade corruption through the power of visibility and public shaming. Second, crowd reporting helps break the silence around the incidence of corruption. When others realise that many other fellow citizens are acting upon and reporting the cases, others too join the efforts. Third, these corruption-reporting applications can also help trigger collective action. When the concealment of secrecy is removed, citizens become courageous and tend to speak up and show that a majority of people do not commend of or endure these corrupt practices. Corruption is a big menace in India. India has been ranked at the 80th spot among 180 countries and territories in the Corruption Perception Index (CPI) created by Transparency International [1].

2   Challenges and Research Problem

There are numerous corruption-prone public services and citizens can be for effectively monitoring and mitigating corruption as citizens would generally like to participate in activities that would help in mitigating corruption at their provincial level. But there are many other challenges and issues that come up when providing such a platform and create an effective workflow and dataflow, and also avoid misuse or abuse of the system. In spite of the academic agreement that ICTs offer a great prospect to help fight corruption, the success of bribery-reporting platforms has not yet been comprehensively studied. This section summarizes some of the most common challenges and downsides. Firstly, to be able to recognize and report of any corrupted events happening in the projects in their provinces, citizens require awareness. For example, if we consider public works, for citizens to be able to identify and report corruption in the public works such as roads, water tanks, parks, bus shelters, auditoriums, drainage systems, health centers, etc., they first need to know funds that were sanctioned by the government for the said projects, conditions of sanction, duration, etc. Common basic knowledge of the citizens about the funds for the projects would be enough in the most cases to bring several discrepancies to light. Some of the proposed systems had some downsides which resulted in some systems failing and others not capable of enduring for long. One complaint is that online reporting applications are usually used by a lesser sector of the population, i.e. young, technologically advanced individuals, and are often not accessible to the people who face most of the effects of bribery, i.e. the poor and people in rural areas without access to the internet [11]. Therefore, the analysis of the reports obtained will most possibly be skewed towards the experiences of fewer percentage of population. But this is an old situation. In present days, mobile users are present in all parts of the country. As per 2019 data, Internet usage in India has exceeded half a billion people for first time [3–5]. Another objection is that these platforms are forgotten soon by the public. The high number of reporting platforms that fail to survive once the media attention and initial publicity fades away attests that engaging citizens is no easy task. Being featured in the media often generates a spike in incoming reports, but once the platform stops being featured, citizens seem to forget about it.

3   Proposed Model

he proposed model discusses how ICT empowers the citizens with information to combat corruption and see that the abuse of the system is minimized by making information transparent to the citizens and letting them report on discrepancies. Disclosure of information is made between public office holders and citizens so that the latter can find it easier to monitor and report for discrepancies. The model works by providing transparency, which can help reduce the room for corruption. It also detects potential cases of corruption through the identification of outliers, under performance and other anomalies. This model involves citizens for monitoring the efficiency and integrity of usage of finances (Fig. 1).

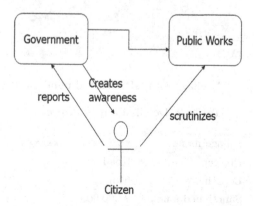

Fig. 1. Basic working principle for citizen awareness and citizen reporting

Suppose the map for a village is represented by the Fig. 2. The government has several public works and constructions in that village, for example, water tank, roads, library, sports complex, auditorium, bus shelters, etc. Each work is marked by a clickable GPS icon as per the data available with the government. The markings would represent both the works that are completed and are to be initiated. This information is open to all and all residents of that village will have access to that map. As residents of a village will be well acquainted with the locations of the markings, they will easily be able identify them and will be of great interest to them.

Selecting each marking would provide essential details about the project. For example, clicking on one marking could reveal information such as (Tables 1 and 2):

Table 1. Detail of a water tank project

Village name:	xxxxxxxxxxxxxxxxx
Project	Water Tank
Capacity	50,000 litres
Sanctioned amount	5,00,000 /-
Sanctioned date	xx/xx/xxxx

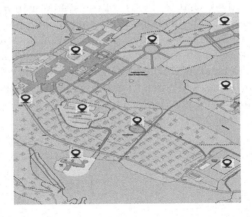

Fig. 2. A village with GPS marked public works

Table 2. Detail of a road project

Village name:	xxxxxxxxxxxxxxxx
Project	Road
Length	5 km
Sanctioned amount	25,00,000 /-
Sanctioned date	xx/xx/xxxx

Confirmation Layer. A confirmation layer exists between the government and the citizen consumed information. In this layer, the public office holders will confirm the information. This is essential because it makes the information authentic and also excludes the case where the public office holder denies any such furbished information. For this, an extra field is added to the information displayed regarding a project. This layer will limit the scope for denial of funds or knowledge of it. Not confirmed by the officer can be regarded as an anomaly and to be treated as a possible case for corruption (Table 3).

Table 3. An extra field of confirmation added as part of confirmation layer

Village name:	xxxxxxxxxxxxxxxxx
Project	Water tank
Capacity	50,000 litres
Sanctioned amount	5,00,000 /–
Sanctioned date	xx/xx/xxxx
Is info confirmed by officer?	Yes

3.1   Reporting

The citizens have an option to report any corruption or discrepancies that they find in relation to any project. Suppose they identify a project named road, extending over a five kilometers, the amount sanctioned is 5 million rupees but they discover that the quality of the road is cheap and that it would not have cost more than one million rupees. In such a case, the citizen reports to government saying that he/she found a discrepancy or a possibility of corruption of about four million rupees regarding the project road (Fig. 3).

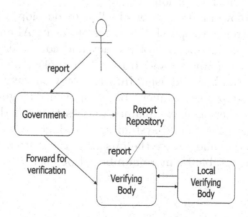

Fig. 3. Workflow of reporting

3.2   Addressed Issues

Personal vengeance: This is overcome by providing not detail of individuals, but only of projects.

3.3   Quality and Reliability of Results

How does the system meet the non-functional quality factors such as reliability? Regardless of the number of reports got through a corruption-reporting platform, it is always necessary to note that the data gathered through them needs to be treated with care. A known weakness of these platforms is that they do not paint an objective picture of petty corruption where they are being used as online reporting applications are mostly used by a definite segment of the population, i.e. young, technologically advanced individuals, and are often not commonly available to the masses who face the effects of corruption, i.e. the poor and rural populace who do not have access to the internet [11]. Because of this, the conclusions of the reports obtained will most probably be skewed towards the experiences of only few individuals. There are, however, a small number of elements that can help make a corruption crowdsourcing platform more operational. As theorized by [15], the following methods could help in the success of crowd-reporting platforms. His proposals include: emphasizing action possibilities on what to do to help reduce corruption, etc.

4   Future Work

To find means to stop the system from being abused. For example, there might be a project handled by a person who might hire people to provide positive feedback so that it would not come under suspicion. Again, there might be some other person who might hire people to give negative remark so that a well executed project might come under suspicion.

In addition to above system we would like to develop a query based spoken dialogue system refers to speech based system using AI and ML techniques, which helps the people irrespective of literacy and computer illiterates to save themselves from the corruption and bribe. The spoken dialogue system is a cloud based system, can be build using Information Communication Technology (ICT) systems, by adapting the query based spoken dialogue system, the citizens are able to register their complaints/queries through their voice in their native language. The query based system spoken dialogue system can increase the transparency, detect and reduce the misuse of the system and as it is speech based system it is adaptable by more number of people.

References

1. https://economictimes.indiatimes.com/news/politics-and-nation/india-ranked-80th-in-corruption-perception-index/articleshow/73560064.cms. Accessed 20 Sept 2020

2. Chne, M.: Use of mobile phones to detect and deter corruption, U4 expert answer. Chr. Michelsen Institute, Bergen (2012). http://www.u4.no/publications/use-of-mobile-phones-to-detect-and-deter-corruption/
3. https://economictimes.indiatimes.com/tech/internet/internet-users-in-india-to-reach-627-million-in-2019-report/articleshow/68288868.cms. Accessed 20 Sept 2020
4. https://digitalequality.in/smart-phones-as-educational-tools-a-reality-check-from-rural-india/. Accessed 20 Sept 2020, Accessed 1 Aug 2020
5. https://www.statista.com/statistics/262966/number-of-internet-users-in-selected-countries/. Accessed 20 Sept 2020, Accessed 21 Aug 2020
6. Bertot, J.C., Jaeger, P.T., Grimes, J.M.: Using ICTs to create a culture of transparency: e-government and social media as openness and anti-corruption tools for societies. Govern. Inf. Q. **27**, 264–271 (2010)
7. Bailard, C., Baker, R., Hindman, M., Livingston, S., Meier, P.: Mapping the maps: a meta-level analysis of Ushahidi and Crowdmap. Internews Center for Innovation and Learning, Washington, DC (2012)
8. Crawford, C.: Crowdsourced anticorruption reporting, 2.0. The Global Anti-Corruption Blog: Law, Social Science, and Policy (2014). https://globalanticorruptionblog.com/2014/12/29/crowdsourced-anticorruption-reporting-2-0/
9. Elbahnasawy, N.G.: E-government, internet adoption, and corruption: an empirical investigation. World Dev. **57**, 114–126 (2014)
10. Gurin, J.: Open governments, open data: a new lever for transparency, citizen engagement, and economic growth. SAIS Rev. Int. Affair. **34**(1), 71–82 (2014)
11. IACC 2012: New technologies against Petty Corruption: Tactics and Lessons. IACC 2012 Conference Paper. http://15iacc.org/wp-content/uploads/New_Technologies_Against_Petty_Corruption.pdf
12. Starke, C., Naab, T.K., Scherer, H.: Free to expose corruption: the impact of media freedom, internet access and governmental online service delivery on corruption. Int. J. Commun. **10**, 21 (2016)
13. Sturges, P.: Corruption, transparency and a role for ICT? Int. J. Inf. Ethics **2**(11), 1–9 (2004)
14. Zinnbauer, D.: False dawn, window dressing or taking integrity to the next level? Governments Using ICTs for integrity and accountability – some thoughts on an emerging research and advocacy agenda (2012)
15. Zinnbauer, D.: Crowdsourced corruption reporting: what petrified forests, street music, bath towels, and the taxman can tell us about the prospects for its future. Policy Internet **7**(1), 1–24 (2015)

Towards a Responsible Intelligent HCI for Journalism: A Systematic Review of Digital Journalism

Yujin Zhou[1] and Zixian Zhou[2]

[1] GDT for Green and Digital Transformation, Guangzhou, China
[2] Sun Yat-Sen University Nanfang College, Guangzhou, China
zixianzhou@qq.com

Abstract. Previous research has been exploring the user experiences, interface design, and production processes of journalistic products, and it is important to continue such discussions on how better Human-Computer Interaction (HCI) research and practices improve the production and consumption of news. Specifically, because of the development of Big Data and AI digital technologies, it is now vital to investigate how such technologies can be integrated into the HCI design for better user experience. With the aim to systematically examine digital journalism research fronts for better HCI, a scientometric study has been conducted by analyzing 2156 articles on digital journalism, collected from the Web of Science (WoS) database, so as to critically evaluate the extant literature investigating such relationship between HCI and digital journalism for future research directions. By analyzing major disciplines, core publications, keywords, and key authors' ideas, this study found that the multidisciplinary knowledge surrounding digital journalism, especially top disciplines such as communication, computer science, information science & library science can be divided into four main categories: journalism and framing; interactivity; analytics and metrics; big data and transparency. The main contribution of this paper is to provide a better overall picture on the relationship between journalism and interaction design, thereby helping advancing both fields with preliminary but important findings.

Keywords: Human-Computer Interaction · Digital journalism · Scientometric · Big data · Computer science

1 Introduction

In the context of modern society, data has become the key to the development of various industries that are actively seeking digital transformation [1]. Take digital journalism as an example, it is a new mode in journalism that mainly enhances users' interactive reading experience by means of visual presentation and guides public opinion by tracking social hot spots from a unique perspective. The interactive map and virtual reality technology give readers more sense of presence through human-computer interaction (HCI), making them interact with the news while reading it, so as to obtain better effects

© Springer Nature Switzerland AG 2021
M. Singh et al. (Eds.): IHCI 2020, LNCS 12615, pp. 488–498, 2021.
https://doi.org/10.1007/978-3-030-68449-5_47

of communication [2]. Because of a large amount of information and misinformation around us, digital journalism has gradually developed into an important interdisciplinary field of research and practices, helping to cope with the shift in people's demand for news consumption in the era of big data. Indeed, news, as sizable proportion of the digital content world, along with other content such as advertisements and chats, has become the major data sources for generating algorithms, associated algorithm property rights, etc. for marketing, advertising, and campaigns [3]. Indeed, as consumption of wider content becomes central to market success, data ownership and technological competitiveness, news texts and patterns of consumption have become major sources for the consumer sentiment analysis [4], for tracking tourism and leisure marketing outcomes [5], etc. It appears that the advancement of intelligent HCI will require better understanding of how news content, especially online, have been used to better understand and model the content consumption patterns, which are expected to be shaped by better interaction designs.

Also, news organizations and practices increasingly have to consider the relationship between journalism and interaction designs. For example, the value of digital journalism has been highlighted during the covid-19 epidemic, for instance, the importance of verifying and checking information in the overall process of knowledge production, management and sharing [6], and such processes often require adequate interaction design and can be automated. And many researchers believe that digital journalism should be research focus in 2020, with the need to generate better understanding of the characteristics, applications and challenges in conducting digital journalism during the pandemic [7]. Indeed, as online news sometimes depend on social media to gain online audience using more data and better interaction design, online news organizations increasing have to consider several interaction platforms such as search engines, news aggregators, or social networking sites [8].

Despite the need to understand the relationship between journalism and interaction designs, there is no comprehensive and systematic review on how such relationship can be better understood, not only to improve the digital journalism practices, but also the HCI knowledge and practices. With the goal to better understand the relationship between journalism and interaction in our digital lives, this article presents a scientometric review of the digital journalism literature for the HCI, with a focus on the use of Big Data and Artificial Intelligence (AI). A better understanding should help both researchers and professionals to develop better interaction design, and also better consumption and production of news content.

2 Research Methodology and Data

To fill the gap, the paper conducted a bibliometric analysis based on articles collected from the Web of Science (WoS) database. The "Advanced Search" query below shows how our data query design tries to cover all relevant papers related to digital journalism.

TS = ("digital journalism" OR "online news" OR "online newspaper" OR "digital journalistic products" OR "digital news data journalism" OR "data news" OR "data journalistic products" OR "data-driven journalism" OR "data-driven news" OR "data journalist*" OR "online journalist*" OR "digital journalist*" OR "network journalist*" OR "internet journalist*").

In August 2020, the query above gave us 2156 records of bibliographic data include results indexed by SCI-EXPANDED, SSCI, A&HCI, and ESCI. By building a thesaurus that treat terms such as "online news", "data-driven journalism", etc. as the same concept of "data journalism", we tried to overcome the analytical challenges of different terms in author keywords, then the research conducted various bibliometric analysis using VOSviewer.

3 Research Findings

3.1 Annual Trends and Periodization

First, annual trends and time periods show how research has evolved over the years, reflecting changes in researchers' interests, as shown in Fig. 1. The results of piecewise linear regression show that 1997 was a turning point in the development of digital journalism. Since then, the research on digital journalism has increased year by year.

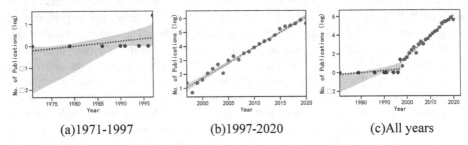

$\qquad$(a)1971-1997$\qquad\qquad$(b)1997-2020$\qquad\qquad$(c)All years

Fig. 1. Annual publications trend and time periods identified by piecewise linear regression.

3.2 Main Cluster Sources and Disciplines

Second, since bibliographic coupling networks can reveal research fronts and discipline foundation [9], we visualize the top publication sources and use Python for data analysis of subject division, aiming to grasp current situation of interdisciplinary integration in digital journalism research.

Figure 2 indicates where such work has been published and similar based on their bibliography (i.e. bibliographic coupling). The red cluster shows the top publication sources such as *(4) New Media & Society, (6) Journalism & Mass Communication Quarterly, (9) Computers in Human Behavior*, etc. The green cluster shows mostly communication and social studies journal such as *(1) Digital Journalism, (5) Journalism Practice, (19-1) Convergence-the International*, etc. The blue cluster interspersed between the red cluster and the green cluster mainly consists of the journals of communication, such as *(21-4) Discourse & Communication, (19-2) Social Media + Society, etc.* The yellow cluster shows the main journal, *(17-2) International Journal of Press-Politics.*

Table 1 summarizes the respective disciplines (based on WoS research areas), showing the detailed information of top publication sources. And, according to the data analysis results, Fig. 3 shows the distribution and relationships of disciplines.

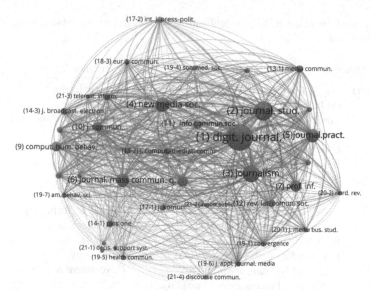

Fig. 2. Top publication sources: a bibliographic coupling relationship network visualization

On the whole, the disciplines integration on digital journalism research can be divided into three main categories-communication and sociology, communication and science, communication and government & law. In addition to the basic discipline-communication, computer science and information science & library science are the dominate subjects, which has a little difference from the results based on "What is digital journalism" by Steensen [10], demonstrating the growing importance of computer science.

3.3 The Dominant Object of Study in Digital Journalism Studies

Third, Fig. 4 shows how keyword data can be visualized based on the interconnection of author keywords resulting in a keyword co-occurrence network. Such "interactivity" often refers to the availability of user comments, user-generated content, and therefore the possibilities of participatory journalism and citizen journalism. At the left, the emerging concepts of framing and exposure points to the fact that social media platforms such as twitter and Facebook can shape how news is exposed, framed and thus consumed. Also, at the top, the rise of big data enables both network analysis and data journalism.

3.4 Main Co-authors Countries

From the data points in the visual chart (Fig. 5), it can be seen that the contribution rate of American scholars in digital news is the highest, followed by the UK and Spain, all of which are developed countries. On the contrary, France, Taiwan, Brazil and Ireland have less influence on it. It shows that economically developed countries pay more attention to digital journalism.

Table 1. Detailed information of top publication sources based on bibliographic coupling relationship network

Clusters	Journals	Disciplines
#1	(1) Digital Journalism, (2) Journalism Studies, (5) Journalism Practice, (13-1) Media and Communication, (19-1) Convergence-The International, (7) Profesional De La Informacion, (20-1) Journal of Media Business Studies, (18-2) Media Culture & Society	Communication* Sociology Information Science & Library Science Business & Economics
#2	(4) New Media & Society, (6) Journalism & Mass Communication Quarterly (19-5) American Behavioral Scientist, (9) Computers in Human Behavior, (21-6) Decision Sup-port Systems, (21-5) IEEE Access, (14-2) plos one, (21-1) Information Processing & Management, (14-1) Political Communication	Communication* Science & Technology Computer science Social Sciences Psychology Government & Law
#3	(19-2) Social Media + Society, (21-7) Javnost-The Public, (11) Information Communication & Society, (21-4) Discourse & Communication	Communication*
#4	(17-2) International Journal of Press-Politics	Communication* Government & Law

Moreover, from the perspective of the number of links, the United States, as the core node, is active in exchanging and cooperating with other countries, among which China, South Korea and Norway are related to it mostly. It also reveals that the Contribution of the United States to digital journalism lies not only in its own research, but also in expanding its influence to the East and the West. In terms of the intensity of the connection, the Digital journalism communication in Europe is dominated by the UK and Spain, and the two countries are more closely connected, compared with other Countries in Europe.

Therefore, what can be concluded is that there is an imbalance in the development of digital journalism in Europe. However, the imbalance is not like the market in which the two sides compete against each other and strive to be the first. Instead, a combination of strong and strong is adopted.

3.5 Main Organizations

Table 2 shows the top 6 organizations in the field of digital journalism, including the number of documents they have published and the number of citations. From the results, the international organizations focused on digital journalism research are mostly universities, which have been the core strength. What's more, all of the top organizations come from developed countries, revealing their excellent research ability.

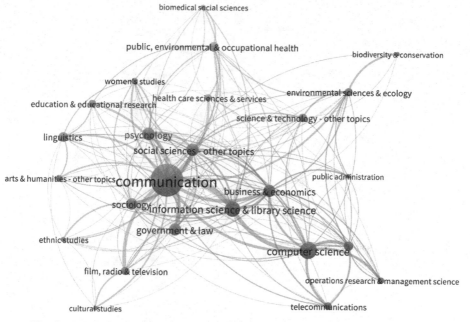

Fig. 3. Top disciplines: a bibliographic coupling relationship network visualization

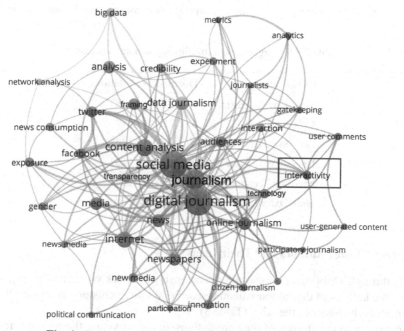

Fig. 4. Top author keywords: a co-word network visualization

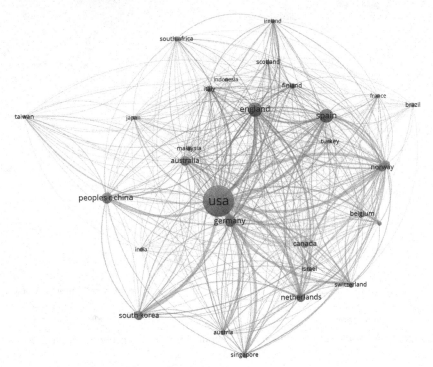

Fig. 5. Top countries: a bibliographic coupling relationship network visualization

Table 2. Top organizations of digital journalism research

Organization	Documents	Citations
Univ Texas - Austin	44	1087
Univ Amsterdam	43	577
Ohio State Univ	31	1401
Univ of Wisconsin	30	1151
Nanyang Technol Univ	27	501
Univ Oxford	26	550

3.6 Main Cluster Authors and Their Ideas

Fourth, through a bibliographic coupling relationship network visualization (Fig. 6), we divided the authors of digital journalism research into three clusters, and listed the core authors in each cluster in the table (Table 3).

According to the studies of the core authors in recent years, the research focus of digital journalism can roughly be summarized, giving us some enlightenment for the future development.

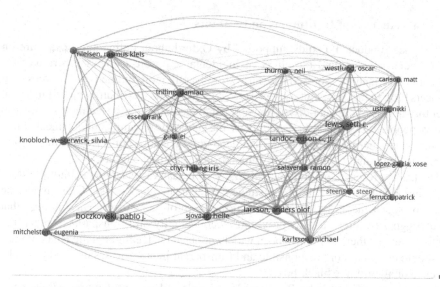

Fig. 6. Top authors: a bibliographic coupling relationship network visualization

Table 3. Detailed information of core author

Clusters	Core authors
#1	Lewis, SC, Tandoc, EC, Larsson, AO, Westlund, O, Karlsson, M, Steensen, S
#2	Knobloch-Westerwick, Nielsen, RK, Sjovaag, H, Trilling, D
#3	Boczkowski, PJ, Mitchelstein, E

The authors of Cluster 1 mainly conduct macro research on the basis of the current background, such as discussing the motivation of digital journalism sharing based on the rise of social media [11] and reflecting on the transformation of digital journalism in the context of covid-19 in order to cope with the future uncertainty [12].

Issues in digital journalism are the main focus of the authors in Cluster 2. For example, Knobloch-Westerwick suggested that digital journalism fosters greater confirmation bias than traditional media [13], and Nielsen, RK developed a way of measuring how polarized news audience behavior is at the national level [14].

The authors in the third cluster focus on feminism which is a small incision in the study of digital journalism, theorizing about the role of technological change in the dynamics of gender inequality and reflecting on how the relative scarcity of women has resulted [15].

The above research results represent the direction of digital journalism research at the present stage, and at the same time, they also give us some future inspiration, that is committing to solving the problems brought by digital journalism and promoting digital journalism development in a more diversified direction.

3.7   Cross-Analysis of Institutions and Authors

Nielsen once conducted a study supported by Oxford about online news exposure on social media, which found the effect of incidental exposure to news is stronger for younger people, and the exposed users use significantly more online news sources than non-users [16]. In fact, incidental exposure to news is a news distribution mode emerging under the support of big data-driven and algorithm recommendation technology, so as to improve people interaction experience. Thus, the automatic recommendation mechanism in human-computer interaction is reasonable, and what can be considered is that digital journalism community apply it to the interaction design in the future.

The leading research institution, University of Texas-Austin conducts multiple studies on digital journalism. One of the studies implemented by Lewis, SC examined the impact of social media on journalists journalism professionalism. The findings that microblogging features such as providing accountability and transparency regarding how they conduct their work have been widely adopted [17], enlighten the HCI community should design a more responsible and transparent interactive system, reducing the existence of algorithmic black-box.

Also, an article written by Boczkowski, PJ reflected on how factors such as news topics, the format of the news article, and the type of digital source interact with gender as a structuring mechanism of media representations [15]. For HCI community, the design of human-computer interaction should adhere to the principle of equality, dilute the influence of gender on people's acceptance of news, and try to make people expose to more comprehensive news issues.

4   Conclusion

The main contribution of this paper is to provide a better overall picture on the relationship between journalism and interaction design, thereby helping advancing both fields with preliminary but important findings. The scientometric analysis has shown not only the overall picture of the current digital journalism literature, but also the key concepts that may connect with the HCI research and practices: journalism and framing; interactivity; analytics and metrics; big data and transparency. In short, the data-driven efforts in understanding and harnessing the dominant interaction activities, especially with the development of automation and algorithm-based decision-making, within and surrounding social media defines the main line of inquiry between the digital journalism and the HCI. Such a main line of inquiry explains why the current digital journalism literature consists of the top three disciplines (communication, computer science, and information science & library science).

From the perspective of country analysis of co-authors and organizations, universities in developed countries are the leading force in the development of digital journalism. In the future, the development of digital journalism also needs the input of advanced technology enterprises or think tanks in the society, and developing countries should actively cooperate with developed countries to enhance their own digital transformation capabilities. Also, the analysis of top authors content shows the future direction of digital journalism studies, such as a more nuanced understanding of the journalism sector and towards more diversity. Thus, it is extremely vital to cultivate students' interdisciplinary ability in order that they can facilitate the interactive development of digital journalism and improve the efficiency of big data analysis.

Although preliminary, the series of visualization maps and tables offer both digital journalism researchers and practitioners on the one hand, and the HCI professionals and researchers to navigate and explore multiple possibilities for collaboration. Indeed, the relationship among different sources, disciplines, object of study, co-authors' countries, organizations, and key ideas. For instance, as sentiment analysis becomes dominant in terms of available data and applications, it becomes important to ask whether and how emotional mobilization may be designed to guide or regulate such consumption and production of content, including news. Researchers and practitioners can use the findings of the research so as to find the key stakeholders for collaboration.

One is to make digital journalism more open, with users' comments widely heard and accepted. Another one is to study and apply new technology actively in sentiment analysis.

Interaction design and HCI professionals can benefit from learning from the digital journalism community in the following ways. Firstly, maybe they can use algorithmic recommendation techniques appropriately, making people be incidental exposed to some other news when reading the selective news. Secondly, human-computer interaction design should adhere to the principle of equality and dilute the influence of gender on news acceptance by disclosing design details partly and overcoming the discrimination defects inherent in data and algorithms.

In short, future work can potentially include in-depth analysis or expanded data collection of neighboring disciplines, such as information science, computer science and sociology. Though our findings are preliminary, the systematic approach can be followed and improved by any future work on the topic.

References

1. Digital Journalism: Analytical advances through open science. https://think.taylorandfrancis.com/special_issues/analytical-advances-open-science-analysis-and-reporting/. Accessed 21 Sept 2020
2. Lewis, S.C., Guzman, A.L., Schmidt, T.R.: Automation, journalism, and human-machine communication: rethinking roles and relationships of humans and machines in news. Digit. Journal. **7**, 409–427 (2019). https://doi.org/10.1080/21670811.2019.1577147
3. Kleinnijenhuis, J.: News, ads, chats, and property rights over algorithms. MaC **6**, 77–82 (2018). https://doi.org/10.17645/mac.v6i3.1601.
4. Bai, X.: Predicting consumer sentiments from online text. Decis. Support Syst. **50**, 732–742 (2011). https://doi.org/10.1016/j.dss.2010.08.024
5. Swart, K., Linley, M., Bob, U.: The media impact of South Africa's historical hosting of Africa's first mega-event: sport and leisure consumption patterns. Int. J. Hist. Sport **30**, 1976–1993 (2013)
6. Deng, H., Fang, J.: Data News Research in COVID-19 -- From the perspective of knowledge Production. Chin. Journal. **4**, 105–108 (2020)
7. Zhou, X.: Features and applications of data journalism during coVID-19 outbreaks. Guide News Res. **7**, 11–100 (2020)
8. Hong, S.: Online news on Twitter: newspapers' social media adoption and their online readership. Inf. Econ. Policy **24**, 69–74 (2012). https://doi.org/10.1016/j.infoecopol.2012.01.004

9. Persson, O.: The Intellectual base and research fronts of JASIS 1986–1990. J. Am. Soc. Inf. Sci. **45**, 31–38 (1994)

10. Steensen, S., Westlund, O.: What is Digital Journalism Studies? Routledge, London, New York (2020)

11. Tandoc, E.C., Huang, A., Duffy, A., Ling, R., Kim, N.: To share is to receive: news as social currency for social media reciprocity. J. Appl. Journal. Media Stud. **9**, 3–20 (2020). https://doi.org/10.1386/ajms_00008_1

12. Ekström, M., Lewis, S.C., Westlund, O.: Epistemologies of digital journalism and the study of misinformation. New Med. Soc. **22**, 205–212 (2020). https://doi.org/10.1177/1461444481 9856914

13. Pearson, G.D.H., Knobloch-Westerwick, S.: Is the confirmation bias bubble larger online? Pre-election confirmation bias in selective exposure to online versus print political information. Mass Commun. Soc. **22**, 466–486 (2019). https://doi.org/10.1080/15205436.2019.1599956

14. Fletcher, R., Cornia, A., Nielsen, R.K.: How polarized are online and offline news audiences? A comparative analysis of twelve countries. Int. J. Press-Polit. **25**, 169–195 (2020)

15. Mitchelstein, E., Boczkowski, P.J., Andelsman, V., Etenberg, P., Weinstein, M., Bombau, T.: Whose voices are heard? The byline gender gap on Argentine news sites. Journalism **21**, 307–326 (2020). https://doi.org/10.1177/1464884919848183

16. Mitchelstein, E., Boczkowski, P.J.: Online news consumption research: an assessment of past work and an agenda for the future. New Media Soc. **12**, 1085–1102 (2010). https://doi.org/10.1177/1461444809350193

17. Lasorsa, D.L., Lewis, S.C., Holton, A.E.: Normalizing Twitter: journalism practice in an emerging communication space. Journal. Stud. **13**, 19–36 (2012). https://doi.org/10.1080/1461670X.2011.571825

A Systematic Review of Social Media for Intelligent Human-Computer Interaction Research: Why Smart Social Media is Not Enough

Han-Teng Liao[1] , Zixian Zhou[1](✉) , and Yujin Zhou[2]

[1] Higher Education Impact Assessment Center, Sun Yat-Sen University Nanfang College, Guangzhou, China
zixianzhou@qq.com
[2] GDT for Green and Digital Transformation, Guangzhou, China

Abstract. As social media shapes human behavior and social interactions, especially with the help of Big Data and artificial intelligence, it becomes an important site for policy and design interventions. Since no systematic review on social media research for intelligent HCI has been conducted, the article presents exploratory findings on a scientometric analysis of the literature at the intersections of social media and AI. By identifying and discussing the main and emerging disciplines and the related keywords from 2,443 articles along with more than 18,000 citations, the findings show that while Twitter and Facebook have been the main platforms for study, Chinese social media platforms emerge as new sites of research with the COVID-19. Also, sentiment analysis appears to be the most prominent research practices, with implications on the issues of privacy, misinformation, depression, and mental health). Four key dimensions of social media are summarized as foundations for the proposed research agenda for intelligent HCI that is not only smart, but also fair and inclusive.

Keywords: Human-Computer Interaction · Social media · Artificial intelligence · Interaction design · Service design · Socio-technical systems

1 Introduction

As social networking sites (SNSs), or social media, allow their users to curate their individual profiles, interact with friends and strangers, and receive news and information, the behavior-changing outcomes has been tested and examined, for example on the topics of addictions and mental health [1]. For instance, misinformation has been spread, with the help of Big Data and artificial intelligence (AI), mostly via social media. Recently the United Nations leaders have made it as one of the top priorities to organize an effective response to fight against the spread of dangerous misinformation that often fuels discrimination, xenophobia and racism, especially in the contexts of COVID-19 and elections [2]. For another, AI can be applied for social good, especially on social

© Springer Nature Switzerland AG 2021
M. Singh et al. (Eds.): IHCI 2020, LNCS 12615, pp. 499–510, 2021.
https://doi.org/10.1007/978-3-030-68449-5_48

media, to address challenges to equality, inclusion and self-efficacy [3]. Social media matters for digital cooperation [2] and common goods [3].

Indeed, mostly generated by individual users, data has enabled social media platforms to automate and recommend interactions regarding what to read, what to buy, what to share, whom to like, etc. Such data includes users' curated content, interactions, and usage data, and then smart and intelligent applications can be developed. It is both the amount and diversity of such data that enable social media platforms for designing and implementing intelligent interaction designs. Indeed, algorithms have been developed for targeted interaction design in social media campaigns [4], and the implications of Big Data and AI in shaping our wider socio-technical systems have been discussed [5]. Thus, social media, as arguably one of the most important user technology interface systems, should be a fruitful site for Human-Computer Interaction (HCI) research.

Because HCI professionals have contributed to the study and design of social media, it is important to understand the ways in which social media users share information, ideas and personal data can be improved with Big Data, AI and cognitive technologies. The importance of social media also reflects on the fact that user research has conducted on social media platforms. Thus, intelligent interaction [6] or data-driven design can learn from the literature at the intersections of social media and AI. There is, however, no systematic review of the theoretical and empirical work. A systematic review of social media and AI literature is expected to provide an overall picture regarding the ways in which AI has been used and misused on social media platforms and data, revealing relevant social media knowledge for better HCI.

2   Research Method and Data

With the aim to identify the main issues and research directions, the research has collected and analyzed literature data at the intersection of AI and social media from the Web of Science (WoS) Collection, covering main journal and book sources such as SCI-EXPANDED, SSCI, A&HCI, BKCI-S, BKCI-SSH, ESCI. The following advanced search query was executed in October 2020:

TS = (("social media" OR "social network*" OR "social web" OR "facebook" OR "twitter" OR "weibo" OR "wechat") AND ("artificial intelligence" OR "cognitive computing" OR "cognitive technolog*" OR "machine learning")).

Overall, 2,443 articles were collected, along with more than 18,000 citations. Python data visualization packages and VOSviewer were used. Starting from the main and emerging disciplines and the related keywords, the research design employs co-citation and co-word analyses to generate an overview and then highlight the significant research issues and directions based on close reading of selected work. A thesaurus and simple taxonomy were developed through the iterative examination of keywords. Special attention is given to the discussions on the future of humans and human conditions on social media, with issues such as ownership and stakes of social media data and algorithms, AI for good (equality, fairness, progress, etc.) [7], and open artificial intelligence models and standards for digital public goods [2]. Such literature review efforts are then organized in such a way to provide a basis for discussions on future research directions on intelligent HCI.

3 Findings

Several facets of the findings are presented based on their disciplines (or research areas), author keywords, and the combined analysis of both disciplines and keywords.

3.1 Main Disciplines and Research Areas

Generally, research fronts can be identified based on bibliographic coupling at various levels such as sources, authors, countries, etc. [8]. With the aim to show how research fronts are structured in relation to disciplines, the article presents findings based on the two WoS categories of each data point: research areas [9] (code: SC) and Web of Science Categories [10] (code: WC). Both disciplinary categorical data points are based on the WoS categories of journals in which these articles have been published. The findings should indicate how disciplines relate to one another in the literature reviewed.

Figure 1 and Fig. 2 show the respective bibliographic coupling visualization outcomes of (a) research areas and (b) Web of Science categories. Figure 1 shows the central cluster of computer and information science, along with other clusters such as the communication and government & law cluster, environmental and public administration cluster, and the health care and medical informatics cluster. Similarly, Fig. 2 shows not only the central cluster of computer science, but also the engineering cluster, the business and management cluster, the communication and political science cluster, environmental studies cluster, the psychology cluster, and the health care cluster.

It should be noted that both Web of Science Categories [10] and research areas [9] categorization schemes are used only by all Web of Science product databases, and thus such categorized outcomes may not be unquestionable, they nonetheless provide consistent disciplinary information from the datasets.

3.2 Main Clusters of Keywords

The content of research fronts can be identified based on keyword co-occurrence analysis [8], for example, showing how often a pair of keywords have been linked by research. Such keywords can be further categorized as concepts, research objects, research concerns, etc., and the network visualization of such keyword co-occurrence can provide insights on how these are connected on another, revealing key ideas and issues.

Revealing the main themes of current research at the intersection of AI and social media, Fig. 3 shows the keyword co-occurrence network map. Note that the top two keywords, "machine learning" and "social media" are removed from this mapping to provide a more readable map. The main cluster (in red color) consists of main keywords such as the main research concept (social networks), research objects (Twitter and Facebook), research purposes (opinion mining, text mining), research methods (sentiment analysis, sentiment classification, natural language processing, classification, data analytics, feature analysis, etc.), and technical methods (naïve bayes, random forest, supervised learning, SVM).

The second-largest cluster (in green color) includes keywords such as main research concept (management, decision making), main research objects (fake news, computer vision), and technical methods (deep learning, neural networks, etc.).

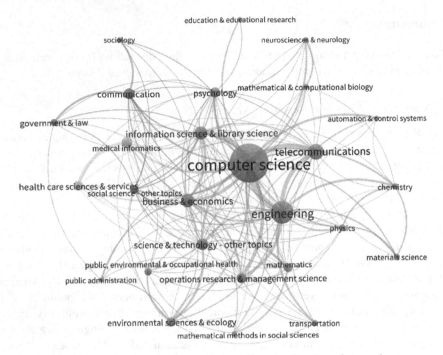

Fig. 1. Top disciplines: bibliographic coupling visualization of research areas

The cluster in blue color includes main research concerns of personality, privacy, and security, surrounding the outcomes of prediction and predictive models. The cluster in yellow color indicates the prominence of technical methods of link prediction and clustering for social network analysis.

The cluster in purple color reveals the emergence of research concerns of COVID-19 and public health, using technical methods of topic modelling. The cluster at the upper-left reveals the emergence of research topics of mental health and depression.

No specific research on intelligent HCI is found in the literature data at the intersection of AI and social media, pointing to a research gap for future research. Nevertheless, research work exists on intelligent systems for and surrounding social media, such as identifying rumor spreaders [11], early risk detection (e.g. depression, rumor or sexual predators) [12], etc., all of which can be seen or further developed as part of the intelligent HCI systems. With the aim to explore the possibilities of social media for intelligent HCI research, the following sections presents more detailed work in the overall picture identified in Fig. 3.

3.3 Main and Emerging Work

Twitter is shown to be the most studied social media site and sentiment analysis the most prominent research method.

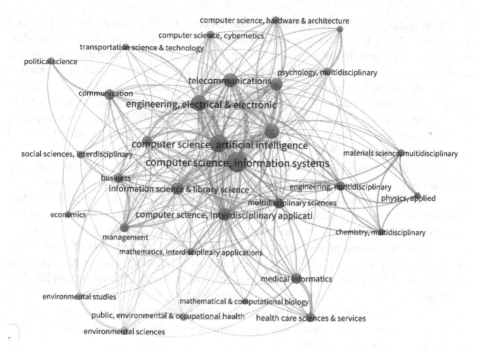

Fig. 2. Top disciplines: bibliographic coupling visualization of Web of Science Categories

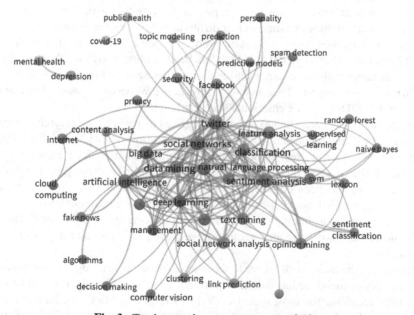

Fig. 3. Top keywords: co-occurrence analysis

Main sentiment analysis work has conducted on platforms such as Twitter [13] and Facebook [14, 15]. For instance, sentiment analysis can be implemented in a Facebook e-learning application with interactive interface, with the aim to consider user's emotional state for adaptive and personalized learning. In addition, a survey on survey on opinion mining and sentiment analysis based on more than one hundred articles during 2002–2015 summarizes the sub-tasks, techniques and applications, as part of the efforts to understand the concept of electronic Word of Mouth (eWOM) [16]. Such development on sentiment analysis leads to the notion of "sentic computing" to better understand online social data on the Web [17].

The sentiment analysis requires personal and social data, with implications on privacy and data governance. Private traits and attributes are found to be predictable on digital records of behavior, including expressing sentiment such as "like" buttons [18], leading to issues such as privacy surrounding self-curation on social media. The possibilities of automatic summarizing data streams result in the notion of social curation, using social media as sensors to put human in the loop of often automatic methods of information processing and filtering [19]. Related issues such as social manipulation [20], political manipulation [21], improvisational manipulations of meaning [22] can be amplified by intelligent agents [20] or detected by machine learning forensics techniques [23]. In short, automation of such curation process may have implications on how interactions, along with attention, are structured and distributed, including the design and policies to ensure a fair and inclusive settings.

Also related to the notion that self-curated content on social media can be seen as forms of self-expressions, main and emerging work on depressions and mental health also attracts research attention. For instance, topics and psycholinguistic clues are extracted from online communities to indicate clear differences between depression communities and other communities, suggesting such predictive power can be used to understand mental health on social media [24], which has been confirmed by one Twitter research [25], one Instagram photo research [26] and one Chinese social media Weibo research [27]. In addition, an empirical research on 17,865 active Web users shows a clear negative impacts of COVID-19 on the emotional state [28].

Social media has been studied or been integrated as part of the research process in relation to the psychological impacts [28, 29] and (mis)information [30, 31]. For instance, a study on psychological stress of medical staffs during the outbreak has been conducted using main Chinese social media platforms to conduct online surveys [29]. For another, an online survey has been conducted to examine the trust of social media and non-reviewed preprints [30]. Social media platforms in China have been implemented as part of health information technologies for epidemic monitoring and control, suggesting that health informatics knowledge and practices have played an important role in COVID-19 response [31].

Altogether, the main work relates to the social, spatial, psychological, and organizational features of social media, and the importance of sentiment analysis based on social media data, including the most recent COVID-19 response. Social media can provide spatial and mobile big data to track and understand the pandemic [32, 33], predict case counts [34], spread misinformation [30, 35], assess psychological conditions of people [28, 29], and coordinate responses [31].

3.4 Emerging Work Related to Interaction Design

Such main and emerging work points to the need for better interaction design at the intersection of social media and AI. Though little work has been conducted in this area, there is still some work on the interaction design of AI on or about social media.

Putting user interface design in a wider context of the socio-technical systems, social media can be seen as one of the developing technologies, along with artificial intelligence, pervasive systems and information integration, that will reshape organizations with multiple user roles, automated work processes and workflows, etc., and thus interaction designers must consider the trends such as more mobile and flexible working patterns, the use of social media in the workplace, and the practice of the virtual organization [5]. Understanding and doing interaction design, from this wider perspective that takes social media, AI, pervasive systems, and information integration into accounts, suggests the need for new skillsets for workers, and possibly higher level of design knowledge such as service design.

Interaction design at the level of the organization thus can be made more intelligent. For instance, a simulation study suggests that organizations can apply a data-oriented method for tailored interactions on social media as part of the social outreach campaigns [4]. Interaction design researchers and practitioners should also pay attention to a critique proposed by Rainer Mühlhoff, that AI technologies such as deep learning have "captured" human cognitive abilities, resulting in AI apparatuses that are likely to involve digital labor exploitation, social control and subjectivation [36]. Overall, social media can be seen as part of the socio-technical systems that require conscious reflections on how interactions are shaped by new technologies.

3.5 Issues of Inclusivity

As social media becomes important to people's daily life, the issues of inclusiveness have been studied and discussed, ranging form content pages to interaction design. Inclusiveness often means equal access to the opportunities, ranging from education to interactions, for all segments of population, and in design, it is related to the guidelines of "inclusive design" or "universal design" [37]. A study has examined Israeli parliament members' Facebook pages and found that coalition members received much more engagement than those of opposition members [38], raising the question of how mainstream media content bias towards those in power also shapes interactions on social media. By analyzing more than 30,000 reports on FixMyStreet (a civic participation platform) in Brussels, the civic participation outcomes on the platform were found to be not inclusive, i.e., marginalizing low-income and ethnically diverse communities [39], providing evidence that informs the more inclusive design of civic participation platforms. Information sharing can also be non-inclusive, as reported by a review on volunteered geographic information platforms, increased flexibility, individual empowerment, and careful management of data quality are important to make platforms more reliable and socially inclusive [40]. Thus, the issues of inclusiveness thus present challenges for interaction designers and researchers, in multiple areas such as page content, interaction arrangement, user participation, information biases, and data priorities.

Using Jörn Bühring's summarized definition of design inclusivity, extending the focus of innovation beyond targeted users so as to address the needs of a wider group of users [37], a more systematic approach is thus needed. Moreover, as innovations require emotional and social intelligence, including self and social awareness, empathy and social skills, it is important to foster emotional and social intelligence-driven universal design, to complement data and AI-driven design for interactions.

3.6 Key Dimensions of Social Media for Intelligent HCI

With the aim to synthesize extant research on the literature reviewed above for intelligent HCI, Table 1 lists the four main dimensions of social media identified by the authors, along with their main issues and implications for intelligent HCI deemed significant by the authors.

Table 1. Social Media Dimensions for Intelligent HCI

Dimensions	Main identified issues	Our suggestions for intelligent HCI
Social interactions and bots e.g. [41, 42]	Unfair distribution of attention and attention data (including user data and user behavior data), and uneven access to the collective intelligence gathered from social interactions	Fair distribution of attention and attention data (including user data and user behavior data-driven intelligent models) and collective intelligence
Data, governance, and wider socio-technical sys e.g. [43]	Data ownership and usage, within and beyond social networking platforms as part of the wider socio-technical systems	Explainable and accountable data processes and governance policies regarding interaction design both within and beyond social networking platforms
Content consumption e.g. [44, 45]	Content recommendation systems that have led to consequences that are unjust and unfair	Explainable and accountable intelligent interaction design between users and content
Curation of self and social content e.g. [19, 38]	Design of meaningful and yet inclusive and fair self-curation and social-curation systems that pay attention to mental health issues	AI- and data-driven mechanisms to improve user experience of self-curation, with additional consideration to the well-being of the users

The first dimension of social media, often overlooked by more technical research, involves the ethical considerations of social interactions and bots. It is important to acknowledge the manipulation can be amplified by Big Data and AI technologies to shape interactions, and countermeasures can be taken to detect such manipulation. The issue of manipulation leads to the questions surrounding fair and just distribution of attention and attention data (including user data and user behavior data-driven intelligent models), and the ensuing generation and use of collective intelligence.

The second dimension of social media requires researchers, designers, and policy makers to see social media platforms as part of the wider socio-technical systems such as workplace, education, and political environments, etc., especially in terms of data ownership and usage. Intelligent HCI should thus better address the issues of explainable and accountable data governance policies regarding interaction design both within and beyond social networking platforms.

The third dimension of social media relates to content consumption, as content recommendation systems on social networking sites have been subject to manipulation, leading to consequences that are unjust and unfair. Intelligent HCI can develop explainable and accountable for intelligent interaction design between users and content.

The fourth dimension of social media relates to the meaningful and fulfilling expressions regarding individual's self-profiles, social relationships, and social lives, with policy and design issues regarding how inclusive and fair self-curation and social-curation systems can be designed, with special attention to mental health issues. Intelligent HCI can explore AI- and data-driven mechanisms to improve the user experience of self-curation, with additional consideration to the well-being of the users beyond traffic and profitability.

4   Conclusion

Given the importance of social media in shaping social interactions, several issues have emerged when Big Data and AI applications have amplified the existing issues of social media, including mental health, potential manipulation, unfair and non-inclusive characteristics, and data governance issues. Based on the scientometric findings, while sentiment analysis appears to be the most prominent research practices and the main platforms for research recently include Chinese social media platforms beyond Twitter and Facebook, questions remain how the issues of privacy, misinformation, depression, mental health, etc. Although the scope of this study is limited to the literature available in the Web of Science database, it nonetheless provides a systematic overview to show what has been done and what needs to be done soon. It appears that overall data- and AI-driven practices of social media have made social media less fair and inclusive. It is then essential for any development of smart interaction design, or intelligent HCI research, to understand and address these issues, with the research agenda to make social media smart *and* inclusive. One possible future research direction is to develop a viable notion of design inclusivity for HCI, so as to allow us to systematically and creatively incorporate human emotional and social intelligence into the existing data and AI-driven design for interactions [37].

Based on the identified dimensions of social media, the article summarizes the main issues and future research directions for intelligent HCI. Overall, intelligent HCI can benefit from a wider socio-technical systems perspective to develop explainable and accountable interaction design, data processes, and data governance that hopefully lead to a fair distribution of attention and attention data and fulfilling and meaningful expressions of individuals and communities. Especially when sentiment analysis has dominated the current research at the intersection of social media and AI, it is important to recognize the need to address mental health issues for a more holistic consideration of sense of well-being, thereby making social media not only smart, but also fair and inclusive. In conclusion, as social media platforms have enormous power to understand and shape their users' interactions, the important dimensions of users, content, data, and intelligent agents need more knowledge for design and policy interventions, especially for a fair and inclusive future.

Acknowledgment. The research is funded by a project of Smart App Design Innovation Research in the Age of New Business, Arts and Engineering Disciplines (2019GXJK186), under the 2019 Guangdong Education Grants, China.

References

1. Griffiths, M.D., Kuss, D.J., Demetrovics, Z.: Social networking addiction. In: Behavioral Addictions, pp. 119–141. Elsevier (2014). https://doi.org/10.1016/B978-0-12-407724-9.000 06-9
2. United Nations General Assembly: Road map for digital cooperation: implementation of the recommendations of the High-level Panel on Digital Cooperation (2020)
3. Chui, M., Harrysson, M., Manyika, J., Roberts, R.: Applying AI for Social Good. Mckinsey Global Institute (2018)
4. Garcia, C.: A nearest-neighbor algorithm for targeted interaction design in social outreach campaigns. Kybernetes **45**, 1243–1256 (2016). https://doi.org/10.1108/K-09-2015-0236
5. Maguire, M.: Socio-technical systems and interaction design – 21st century relevance. Appl. Ergon. **45**, 162–170 (2014). https://doi.org/10.1016/j.apergo.2013.05.011
6. Blandford, A.: Intelligent interaction design: the role of human-computer interaction research in the design of intelligent systems. Expert Syst. **18**, 3–18 (2001). https://doi.org/10.1111/1468-0394.00151
7. Anderson, J., Rainie, L., Luchsinger, A.: Artificial Intelligence and the Future of Humans. Pew Research Center (2018)
8. Garfield, E.: Research fronts. Current Comments (1994)
9. Clarivate Analytics: Research Areas (Categories/Classification). https://images.webofknow ledge.com/WOKRS535R100/help/WOS/hp_research_areas_easca.html. Accessed 01 Nov 2020
10. Clarivate Analytics: Web of Science categories. https://images.webofknowledge.com/WOK RS535R100/help/WOS/hp_subject_category_terms_tasca.html. Accessed 01 Nov 2020
11. Rath, B., Gao, W., Ma, J., Srivastava, J.: Utilizing computational trust to identify rumor spreaders on Twitter. Soc. Netw. Anal. Min. **8**(1), 1–16 (2018). https://doi.org/10.1007/s13 278-018-0540-z
12. Burdisso, S.G., Errecalde, M., Montes-y-Gómez, M.: A text classification framework for simple and effective early depression detection over social media streams. Expert Syst. Appl. **133**, 182–197 (2019). https://doi.org/10.1016/j.eswa.2019.05.023
13. Ghiassi, M., Skinner, J., Zimbra, D.: Twitter brand sentiment analysis: a hybrid system using n-gram analysis and dynamic artificial neural network. Expert Syst. Appl. **40**, 6266–6282 (2013). https://doi.org/10.1016/j.eswa.2013.05.057

14. Ortigosa, A., Martín, J.M., Carro, R.M.: Sentiment analysis in Facebook and its application to e-learning. Comput. Hum. Behav. **31**, 527–541 (2014). https://doi.org/10.1016/j.chb.2013. 05.024
15. Poria, S., Cambria, E., Winterstein, G., Huang, G.-B.: Sentic patterns: dependency-based rules for concept-level sentiment analysis. Knowl.-Based Syst. **69**, 45–63 (2014). https://doi. org/10.1016/j.knosys.2014.05.005
16. Ravi, K., Ravi, V.: A survey on opinion mining and sentiment analysis: tasks, approaches and applications. Knowl.-Based Syst. **89**, 14–46 (2015). https://doi.org/10.1016/j.knosys.2015. 06.015
17. Thelwall, M., Buckley, K., Paltoglou, G.: Sentiment strength detection for the social web. J. Am. Soc. Inf. Sci. **63**, 163–173 (2012). https://doi.org/10.1002/asi.21662
18. Kosinski, M., Stillwell, D., Graepel, T.: Private traits and attributes are predictable from digital records of human behavior. Proc. Natl. Acad. Sci. **110**, 5802–5805 (2013). https://doi.org/10. 1073/pnas.1218772110
19. Kimura, A., Duh, K., Hirao, T., Ishiguro, K., Iwata, T., Au Yeung, A.: Creating stories from socially curated microblog messages. IEICE Trans. Inf. Syst. **E97.D**, 1557–1566 (2014). https://doi.org/10.1587/transinf.E97.D.1557
20. Stella, M., Ferrara, E., De Domenico, M.: Bots increase exposure to negative and inflammatory content in online social systems. Proc. Natl. Acad. Sci. USA **115**, 12435–12440 (2018). https:// doi.org/10.1073/pnas.1803470115
21. Lee, S.: Detection of political manipulation in online communities through measures of effort and collaboration. ACM Trans. Web **9**, 1–24 (2015). https://doi.org/10.1145/2767134
22. Liu, H., Maes, P., Davenport, G.: Unraveling the taste fabric of social networks. Int. J. Semant. Web Inf. Syst. **2**, 42–71 (2006). https://doi.org/10.4018/jswis.2006010102
23. Sandoval Orozco, A.L., Quinto Huamán, C., Povedano Álvarez, D., García Villalba, L.J.: A machine learning forensics technique to detect post-processing in digital videos. Future Gener. Comput. Syst. **111**, 199–212 (2020). https://doi.org/10.1016/j.future.2020.04.041
24. Nguyen, T., Phung, D., Dao, B., Venkatesh, S., Berk, M.: Affective and content analysis of online depression communities. IEEE Trans. Affect. Comput. **5**, 217–226 (2014). https://doi. org/10.1109/TAFFC.2014.2315623
25. Prieto, V.M., Matos, S., Álvarez, M., Cacheda, F., Oliveira, J.L.: Twitter: a good place to detect health conditions. PLoS ONE **9**, e86191 (2014). https://doi.org/10.1371/journal.pone. 0086191
26. Reece, A.G., Danforth, C.M.: Instagram photos reveal predictive markers of depression. EPJ Data Sci. **6**, 15 (2017). https://doi.org/10.1140/epjds/s13688-017-0110-z
27. Cheng, Q., Li, T.M., Kwok, C.-L., Zhu, T., Yip, P.S.: Assessing suicide risk and emotional distress in chinese social media: a text mining and machine learning study. J. Med. Internet Res. **19**, e243 (2017). https://doi.org/10.2196/jmir.7276
28. Li, S., Wang, Y., Xue, J., Zhao, N., Zhu, T.: The impact of COVID-19 epidemic declaration on psychological consequences: a study on active weibo users. IJERPH **17**, 2032 (2020). https:// doi.org/10.3390/ijerph17062032
29. Wu, W., et al.: Psychological stress of medical staffs during outbreak of COVID-19 and adjustment strategy. J. Med. Virol. **92**, 1962–1970 (2020). https://doi.org/10.1002/jmv.25914
30. Gupta, L., Gasparyan, A.Y., Misra, D.P., Agarwal, V., Zimba, O., Yessirkepov, M.: Information and misinformation on COVID-19: a cross-sectional survey study. J. Korean Med. Sci. **35**, e256 (2020). https://doi.org/10.3346/jkms.2020.35.e256
31. Ye, Q., Zhou, J., Wu, H.: Using information technology to manage the COVID-19 pandemic: development of a technical framework based on practical experience in China. JMIR Med. Inform. **8**, e19515 (2020). https://doi.org/10.2196/19515

32. Poom, A., Järv, O., Zook, M., Toivonen, T.: COVID-19 is spatial: ensuring that mobile Big Data is used for social good. Big Data Soc. **7**, 205395172095208 (2020). https://doi.org/10.1177/2053951720952088

33. Zhenghong, P., Wang, R., Liu, L., Wu, H.: Exploring urban spatial features of COVID-19 transmission in wuhan based on social media data. ISPRS Int. J. Geo-Inf. **9**, 402 (2020). https://doi.org/10.3390/ijgi9060402

34. Shen, C., Chen, A., Luo, C., Zhang, J., Feng, B., Liao, W.: Using reports of symptoms and diagnoses on social media to predict COVID-19 case counts in mainland china: observational infoveillance study. J. Med. Internet Res. **22**, e19421 (2020). https://doi.org/10.2196/19421

35. Hua, J., Shaw, R.: Corona Virus (COVID-19) "Infodemic" and emerging issues through a data lens: the case of China. Int. J. Environ. Res. Public Health **17**, 2309 (2020). https://doi.org/10.3390/ijerph17072309

36. Mühlhoff, R.: Human-aided artificial intelligence: or, how to run large computations in human brains? Toward a media sociology of machine learning. New Media Soc. **22**, 1868–1884 (2020). https://doi.org/10.1177/1461444819885334

37. Bühring, J., Patricia, A.M., Torkkeli, M., de Engenharia, F.: Emotional and social intelligence as 'Magic Key' in innovation: a designer's call toward inclusivity for all. J. Innov. Manag **6** (2018)

38. Steinfeld, N., Lev-On, A.: Top-down, non-inclusive and non-egalitarian: characterizing the communication of members of parliament with the public on their Facebook pages. Presented at the June 18 (2019). https://doi.org/10.1145/3325112.3325249

39. Pak, B., Chua, A., Vande Moere, A.: FixMyStreet brussels: socio-demographic inequality in crowdsourced civic participation. J. Urban Technol. **24.0**, 65 (2017). https://doi.org/10.1080/10630732.2016.1270047

40. Haworth, B., Bruce, E., Whittaker, J., Read, R.: The good, the bad, and the uncertain: contributions of volunteered geographic information to community disaster resilience. Front. Earth Sci. **6**, 183 (2018). https://doi.org/10.3389/feart.2018.00183

41. Varol, O., Ferrara, E., Menczer, F., Flammini, A.: Early detection of promoted campaigns on social media. EPJ Data Sci. **6**(1), 1–19 (2017). https://doi.org/10.1140/epjds/s13688-017-0111-y

42. He, F., Pan, Y., Lin, Q., Miao, X., Chen, Z.: Collective intelligence: a taxonomy and survey. IEEE Access **7**, 170213–170225 (2019). https://doi.org/10.1109/ACCESS.2019.2955677

43. Fisher, E., Pearce, W., Molfino, E.: Politics of Science and Technology (2016). http://www.oxfordbibliographies.com/display/id/obo-9780199756223-0192. https://doi.org/10.1093/obo/9780199756223-0192

44. Pentzold, C., Fischer, C.: Framing big data: the discursive construction of a radio cell query in Germany. Big Data Soc. **4.0** (2017). https://doi.org/10.1177/2053951717745897

45. Wu, X., Liao, H.-T.: collective intelligence. In: 2018 IEEE Internet of People, pp. 2005–2010 (2018). https://doi.org/10.1109/SmartWorld.2018.00335

Author Index

Abdirozikov, O. Sh. I-380
Abdurashidova, K. T. I-95
Acuña, Borja Bornail II-172
Ahmed, Anamika I-186
Al-Absi, Mohammed Abdulhakim I-250,
    II-267, II-370
Alimardanov, Shokhzod II-303
Alonso, Miguel Angel Aldudo II-172
Alsaih, Khaled I-132
Amin, Md. Hasibul I-163
Ananthachari, Preethi I-369
Anarova, Sh. A. I-380, I-390
Anis, Sabah Shahnoor II-135
Anzaku, Esla Timothy II-254
Atadjanova, N. S. I-95

Babu, K. Ramesh I-481
Badruddin, Nasreen I-120
Bae, Kyu Hyun II-145
Banothu, Rambabu I-287, I-481
Barnes, Laura I-48
Basha, S. Sadiq I-287
Bente, Britt II-199
Bess, Maurice II-145
Bhuiyan, Abul Bashar I-186
Borade, Jyoti G. I-238
Boyak, Johnathan II-145
Brown, Suzana II-145

Cardenas, Irvin Steve II-406
Cha, Jaekwang II-380
Chaithanya, Velugumetla Siddhi II-73
Chatterjee, Indranath I-403
Chaudhury, Santanu I-142
Chaurasiya, Vijay Kumar II-288
Chin, Eunsuh I-195
Cho, Migyung I-403
Cho, Sangwoo II-357
Choi, Yeonsoo II-319
Chong, Uipil II-303
Chung, Wan Young I-229
Chung, Wan-Young I-104, II-53, II-312
Collazos-Morales, Carlos II-391

Comas-Gonzalez, Zhoe II-391
Cuong, Nguyen Huu II-38

da Costa, Cristiano I-307
De Neve, Wesley II-254
Diván, Mario José I-445
Djumanov, J. X. I-95
Doryab, Afsaneh I-48

Edoh, Thierry II-188
Esfar-E-Alam, A. M. I-274

Farhan Razy, Md. II-135
Fazli, Mehrdad I-48

Gandhi, Tapan I-142
Gangashetty, Suryakanth V. I-287
Gautam, Veerendra Kumar I-287
Gharavi, Erfaneh I-48
Goel, Sharu I-177
Gwan, Jeon I-403

Han, Na Yeon I-208, II-326
Hanafi, Hafizul Fahri II-63
Hasnat, Md. Waliul II-108
Hill-Pastor, Laura II-391
Hoang Long, Nguyen Mai II-312
Huang, Jieqi I-469
Huda, Miftachul II-63
Hussna, Asma Ul I-274
Hwang, Hyeonsang II-326
Hwang, Injang II-208
Hwang, Seok-min II-278

Ibrohimova, Z. E. I-390
Islam, Md. Monirul II-108
Ismael, Muhannad I-342
Ismoilov, SH. M. I-380
Isoird, Carlos Fernández II-172

Jacob, Billy I-262
Jang, Woohyuk I-223, II-248
Jeong, Do-Un II-154, II-160
Jeong, Heeyoon II-167

Jhara, Anita Mahmud   II-84
Jin, Kyung Won   I-223
Jin, Kyungwon   II-27
Jo, Geumbi   I-110
Joe, Hyunwoo   II-172
Joo, Hyunjong   II-319
Joung, Jinoo   II-326
Jung, Je Hyung   II-172
Jung, Sang-Joong   II-154, II-160
Jung, Taewoo   II-254
Juraev, J. U.   I-83
Juturi, Venkateswara   II-73
Juturi, Venkateswara Rao   I-481

Kang, Dae-Ki   I-361
Karim, Mohammad Safkat   I-163
Kashem, Mohammod Abul   II-108
Kaushik, Abhishek   I-262
Khan, Evea Zerin   II-96
Khan, Md. Rezwan Hassan   I-186
Khan, Zia   I-120, I-132
Kim, Chang Bae   II-248
Kim, Chan-il   I-154
Kim, Donggyu   I-195
Kim, Gerard   I-218, II-167
Kim, HyunSuk   II-172
Kim, Jihyun   I-195
Kim, Jong-Hoon   II-406
Kim, Jong-Jin   II-312
Kim, Jong-Myon   I-325
Kim, Mingeon   II-319
Kim, Na Hye   I-208
Kim, Shiho   II-380
Kim, Shin-Gyun   II-363
Kim, So Eui   I-208
kim, So-Eui   II-229
Kim, Taehyung   II-239
Kim, Woojin   II-172
Kim, Yong Jin   I-403
Kim, Young Jin   II-335
Kim, Youngwon   II-27
Kiwelekar, Arvind W.   I-3
Koenig, Christopher   I-342
Koni, Yusuph J.   I-250, II-370
Kothari, Preethi   I-481
Kowsari, Kamran   I-48
Kumari, Reetu   I-12
Kunst, Rafael   I-307

Laddha, Manjushree D.   I-34
Law, Check-Yee   I-431
Laz, Azmiri Newaz Khan   I-274
Le, Giang Truong   II-3
Lee, Boon Giin   I-229
Lee, Byung-Gook   II-267
Lee, Chiwon   I-195, II-319
Lee, Eui Chul   I-110, I-208, I-223, II-27,
    II-219, II-229, II-239, II-248, II-326
Lee, Hooman   II-172
Lee, Hoon Jae   I-250, II-370
Lee, Jee Hang   II-27
Lee, Jeeghang   II-326
Lee, Ji-Su   II-154, II-160
Lee, Jong-ha   I-154, II-278
Lee, Jong-Ha   II-178, II-357, II-363
Lee, Kunyoung   II-27
Lee, Mi Kyung   I-223
Lee, Min Seok   I-403
Lee, Seunggeon   I-110
Lee, Seung-Jun   II-172
Lee, Sinae   II-423
Lenhoff, Caitlyn   II-406
Liao, Han-Teng   I-469, I-499
Liew, Tze-Hui   I-431
Lim, Hyotaek   II-267
Lim, Hyun-il   I-42, I-336
Lim, Wei-Ann   I-431
Lin, Xiangxu   II-406
Lokare, Varsha T.   I-24

Ma, Maodess   II-119
Madan, Shipra   I-142
Magez, Stefan   II-254
Mai, Ngoc-Dau   I-104
Makhmudjanov, Sarvar   II-346
Makhtumov, Nodirbek   I-369
Malviya, Shrikant   I-12
Mariñelarena, Iker   II-172
Masnan, Abdul Halim   II-63
McCall, Roderick   I-342
Meena, Leetesh   II-288
Meriaudeau, Fabrice   I-132
Mishra, Ashutosh   II-380
Mishra, Rohit   I-12
Mok, Ji Won   I-208, II-239
Molakatala, Nagamani   I-287
Monteiro, Emiliano   I-307

Nabiyev, Inomjon I-353
Nagamani, M. I-70, I-481, II-73
Narzulloyev, O. M. I-390
Navandar, Swanand I-3
Nazirova, Elmira I-353
Nematov, Abdug'ani I-353
Netak, Laxman D. I-24, I-34, I-238
Nguyen, Cong Dai I-325
Nguyen, Trong Hai II-21
Nguyen, Trung Trong II-21, II-38
Nguyen, Trung-Hau I-104
Noh, Yun-Hong II-154, II-160

Özbulak, Utku II-254
Ozhelvaci, Alican II-119

Paladugula, Pradeep Kumar II-406
Pan, Chung-Lien I-469
Pandey, Sandeep Kumar I-177
Park, Colin K. II-208
Park, Eun-Bin II-178
Park, Hee-Jun II-278
Park, Jaehee II-47
Park, Jangwoon II-423
Park, Michelle II-406
Park, Tae Sung II-172
Parvez, Mohammad Zavid I-163
Paul, Akash Chandra II-84, II-96
Pawar, Pravin II-208
Petousis, Markos II-145
Prasad, C. Satyanarayana II-73
Primkulov, Shokhrukhbek I-299
Prosvirin, Alexander I-325
Purohit, Neetesh II-288

Qayumova, G. A. I-390

Rabbi, Fazly II-108
Rafsan, Abdullah Al I-163
Rahaman, Ananya II-84, II-96
Rahman, Md Habibur II-53
Rajabov, F. F. I-95
Rajabov, Farkhad II-346
Raju, Shalam I-70
Rama Krishna, M. I-70
Raton, Javier Fínez II-172
Righi, Rodrigo I-307
Ryu, Jihye II-326
Ryu, Jiwon I-218

Sadikov, Rustamjon I-353
Sagar, A. B. I-481
Sanchez, Carlos A. II-391
Saparmammedovich, Seyitmammet
    Alchekov I-250, II-370
Sejan, Mohammad Abrar Shakil II-53
Sek, Yong-Wee I-431
Selamat, Abu Zarrin II-63
Seo, Hyejin II-47
Seo, Ji-Yun II-154, II-160
Seong, Si Won II-248, II-326
Shekhawat, Hanumant Singh I-177
Shin, Myung Jun II-172
Sikder, Md. Shammyo I-274
Singh, Dhananjay I-83, I-307, II-288, II-346
Singh, Kuldeep I-458
Singh, Madhusudan I-299, I-445, II-188,
    II-208
Sitaram, B. II-73
Slijkhuis, Peter Jan Hendrik II-199
Soto, Orlando Rodelo II-391
Suh, Kun Ha II-219, II-229
Sultana, Samiha II-84, II-96
Surovi, Tahmina Rahman I-163
Sutanto, Arief Rachman I-361

Tadjibaeva, D. A. I-95
Talukdar, Aparajit I-59
Tay, Choo Chuan I 431
Thang, Tran Viet II-38
Tinmaz, Hasan I-274
Tiwary, Uma Shanker I-12
Tolendiyev, Gabit II-267
Touhiduzzaman Touhid, Md. II-135
Tran, Nhat Minh II-3
Tran, Thang Viet II-3, II-21
Tripathi, Gaurav I-458
Tulon, Sadid Rafsun I-186

Uddin, Jia I-186, I-274, II-84, II-96, II-108,
    II-135
Um, Dugan II-423
Urolov, Jamshidbek I-299

Vairis, Achilles II-145
van 't Klooster, Jan Willem Jaap Roderick
    II-199
van Gemert-Pijnen, Lisette II-199
van Gend, Joris II-199
Van Messem, Arnout II-254
Velavan, Pankaj I-262

Vimal Babu, Undru   I-70
Vishwakarma, Dinesh Kumar   I-458

Wahab, Mohd Helmy Abd   II-63
Wankhede, Hansaraj S.   I-34
Whitaker, Mark D.   I-414

Xu, Tina Yuqiao   II-406
Xuan, Tan Yu   I-120

Yadav, Dharmendra K.   I-3
Yahya, Norashikin   I-120, I-132

Yoo, Hoon Sik   II-335
Yoon, Daesub   II-172
Yu, Su Gyeong   I-208, II-229
Yusoff, Mohd Zuki   I-120
Yusupov, I.   I-83

Zamora-Musa, Ronald   II-391
Zaynidinov, H. N.   I-83
Zaynidinov, Hakimjon   II-346
Zhou, Yujin   I-488, I-499
Zhou, Zixian   I-488, I-499

Printed in the United States
By Bookmasters